Solid Mechanics and Its Applications

Volume 236

Aims and Scope of the Series

The fundamental questions arising in mechanics are: *Why*?, *How*?, and *How much*? The aim of this series is to provide lucid accounts written by authoritative researchers giving vision and insight in answering these questions on the subject of mechanics as it relates to solids.

The scope of the series covers the entire spectrum of solid mechanics. Thus it includes the foundation of mechanics; variational formulations; computational mechanics; statics, kinematics and dynamics of rigid and elastic bodies: vibrations of solids and structures; dynamical systems and chaos; the theories of elasticity, plasticity and viscoelasticity; composite materials; rods, beams, shells and membranes; structural control and stability; soils, rocks and geomechanics; fracture; tribology; experimental mechanics; biomechanics and machine design.

The median level of presentation is to the first year graduate student. Some texts are monographs defining the current state of the field; others are accessible to final year undergraduates; but essentially the emphasis is on readability and clarity.

More information about this series at http://www.springer.com/series/6557

René Alderliesten

Fatigue and Fracture of Fibre Metal Laminates

Springer

René Alderliesten
Faculty of Aerospace Engineering
Delft University of Technology
Delft
The Netherlands

ISSN 0925-0042 ISSN 2214-7764 (electronic)
Solid Mechanics and Its Applications
ISBN 978-3-319-56226-1 ISBN 978-3-319-56227-8 (eBook)
DOI 10.1007/978-3-319-56227-8

Library of Congress Control Number: 2017934875

Printed on acid-free paper

This Springer imprint is published by Springer Nature
The registered company is Springer International Publishing AG
The registered company address is: Gewerbestrasse 11, 6330 Cham, Switzerland

Preface

After more than one and a half decades since the first status of the development of the Fibre Metal Laminate (FML) GLARE was documented by the FML community in a monograph, this new book presents the state of the art in understanding the fundamental principles of fatigue and fracture of FMLs. By selecting a narrower scope, the subjects of fatigue and fracture are discussed in substantially more depth than the first book, revealing the vast breath of detailed knowledge that has been developed over a time span of three decades.

The timing of publication of this book is aligned with the development of new applications that exploit the benefits of the FML technology for aircraft wing and fuselage structures. After ten years of successful operation of the Airbus A380, demonstrating the maturity and superior performance of the hybrid FML technology, these new applications will push limits, requiring thorough strength and damage tolerance justifications. Hence, a thorough assessment of the current status is necessary.

This book provides the industry with a sound basis to develop new hybrid applications with the corresponding strength, damage tolerance and durability assessment strategies, necessary for certification. In addition, this book provides academia with a benchmark in scientific understanding of this FML technology to enable definition of the necessary steps forward in future research. It is my ambition that this book will aid the FML community in their endeavours to advance research and development (far) beyond the current status.

The reader should be aware that similar to the first book, which was edited by Ad Vlot and Jan-Willem Gunnink, this book, despite having a single author on the cover, is to the greatest extent based on the research efforts of quite a number of MSc and Ph.D. students of the faculty of Aerospace Engineering at TU Delft. Many of them have been inspired personally by Prof. Boud Vogelesang, the 'founding father' of the FML technology, and the driving force behind the development towards successful application. In the end, the primary reason for publishing this monograph is to honour those often tenacious and persevering students for their work and their contribution to the FML community.

In this place, I express my gratitude to two people in particular: Dr. Thomas Beumler and Dr.ir. John-Alan Pascoe. They reviewed the draft version of this book, and their valuable and often detailed comments and criticism were used to improve its quality.

Delft, The Netherlands Dr.ir. René Alderliesten
2017

Contents

Chapter 1
Introduction

Abstract Historically, Fibre Metal Laminates were introduced as a laminated material concept to improve the fatigue and damage tolerance properties of metallic structures in aeronautics. However, the concept can be viewed from different perspectives. This chapter discusses that the FML concept can be seen either as reinforcement of metallic structures or as reinforcement of fibre-reinforced polymer composite structures. Cases are given to illustrate how the concept of damage tolerance can be exploited with FMLs, in particular if the concept is viewed as structural concept rather than a material concept.

1.1 Introduction

The concept of Fibre Metal Laminates was first introduced with the development of ARALL, a laminate comprising aramid fibres embedded in a thermoset adhesive system interspersed between aluminium sheets. The concept originates in the metal bonding technology introduced by Schliekelmann at Fokker and the observation of Schijve that laminated thin sheets have superior fatigue characteristics compared to monolithic panels. Attempts to add fibres to the bondline of 1 mm thick laminated sheets did not provide sufficient benefits. However, further research by Vogelesang and Schijve has led to an optimized FML concept. The history of this development process is well documented by Vlot [1].

The fatigue process appeared to be a balance between the crack growth mechanism in the metal layers, failure of the fibre layers and debonding at the interfaces between these layers. The load transfer from the cracked metal layers through the adhesive to the fibre layers caused significant reduction in crack growth in the metal layers. This transfer is often called 'fibre bridging'.

To tune the fibre bridging and subsequently the overall damage growth under fatigue loading, the resistance against delamination of the adhesive, the stiffness and strength of the fibres and the crack growth resistance of the metal layers are the prime parameters. This has been quantified by the work of Marissen [2] who developed an analytical prediction model for crack growth in the FML ARALL.

R. Alderliesten, *Fatigue and Fracture of Fibre Metal Laminates*, Solid Mechanics and Its Applications 236, DOI 10.1007/978-3-319-56227-8_1

Initially, the development of the FMLs focussed in collaboration with Fokker on wing applications. However, interest from Messerschmitt-Bölkow-Blohm in Hamburg directed the research to fuselage applications. Because fibre failure was observed in ARALL under fuselage load spectra, FMLs containing glass fibres (GLARE) were then developed [3].

1.2 Development Perspectives

Indifferent of the historical development process, the FML concept can be regarded from various perspectives. Traditionally, the concept is treated as additional fibre reinforcement to laminated metallic sheets. Considering the historical development process [1], this perspective seems obvious. However, with the enormous research effort nowadays in fibre reinforced polymer composites, it may be more appropriate to consider the FML concept as the metallic reinforcement to composite materials. That perspective in itself is not new, because various patents have been filed in that context, as will be discussed in Chap. 3. However, the consequence regarding the FML concept from this still rather new perspective may be opening significant opportunities for composite applications.

1.2.1 Increased Damage Growth Resistance of Metal Laminates

Originating from the poor man's solution to build up aluminium structures using bonding technology, instead of using expensive milling equipment, the major driver in the development of FMLs has been to develop a material concept that has inherently higher resistance to (fatigue) cracking then state-of-the-art monolithic aluminium alloys. High fatigue crack growth resistance results in slow crack growth and thus longer inspection intervals, while a high fracture toughness increases the critical crack length, further increasing the damage tolerance of structures.

While developing GLARE, additional beneficial properties were identified that supported its application to primary fuselage structures: high impact resistance and tolerance, high burn through resistance, which could potentially increase evacuation time (safety aspect), and improved corrosion resistance (durability aspect) induced by its layered structure.

Despite all the effort within the GLARE development programme [4] to achieve technology readiness of GLARE with respect to all relevant aspects, the prime focus remained on the fatigue and damage tolerance characteristics.

The literature on this programme illustrates the process from generic understanding of fatigue initiation and crack propagation, towards more detailed understanding of fatigue resistance of mechanically fastened joints, impacted laminates, external and inter-laminar doubler run-outs, edge notches and rows of holes. The latter topic is considered important to address the resistance against widespread fatigue damage.

1.2.2 Utilization in Context of Damage Tolerance

Although clearly explained in the literature, damage tolerance as a development driver has not always been well understood. Academia not related to institutes involved in the development of GLARE has provided useful knowledge and insights. However, its work has never resulted in the development of actual FML applications.

For example, developing structural health monitoring concepts for FMLs without accounting for the inherent high resistance against impacts and (fatigue) cracking [5] makes the developed concepts obsolete or inapplicable for several reasons, especially economic reasons. The need for FML structures originates from the desire to develop damage tolerant structures that are care free (i.e. low inspection and maintenance burden). The need to rely on a health monitoring system to ensure damage tolerance seems to be contradictory to the application of such care (and monitoring) free FML structures. On the other hand, one should realize that such monitoring techniques applied in laboratory experiments may add to the understanding of certain mechanisms. This is illustrated by Austin et al. [6], who utilized fibre Bragg grating sensors to quantify fibre stresses in fatigue crack growth tests; information that previously could only be derived implicitly from other parameters measured.

Inspired by the development of FML as combination of metallic sheet material and fibre reinforced polymer composites, many combinations have been presented in the literature. None of these alternative combinations ever led to actual applications. For example, combining magnesium and carbon fibre reinforced polymer layers as reported by Cortes and Cantwell [7] is inapplicable to actual structures for reasons not addressed in their work. The low stiffness of the magnesium layers and the poor fatigue and corrosion resistance of these materials in comparison with aluminium do not compensate for the lower density, which seems the prime driver for this study [8]. Alternatively, one may think of combining carbon fibres with titanium sheets [9], but that requires addressing titanium pre-treatment challenges together with the high material costs.

This also holds for FML combinations based on thermoplastic matrix systems for which consolidation temperatures need to be applied that exceed thermal stability levels of widely applied aerospace alloys [10, 11]. As a result, only particular applications may be considered for such FMLs. Furthermore, these FMLs often exhibit poor fatigue performance, due to the severe tensile curing stresses in the metallic layers. Here, application of thermoplastic matrix systems should be investigated together with relevant aerospace alloys that can sustain the high curing temperatures.

Another more recent example could be the application of alternative manufacturing processes such as Vacuum Assisted Resin Transfer Moulding (VARTM) presented by Jensen et al. [12]. The introduction of small holes in the metal layers of the FML to enable resin to flow in laminate thickness direction creates a significant number of fatigue critical areas in the FML panel. While usually only the joining areas and cut-outs require a thorough fatigue assessment, this manufacturing concept implies assessment of every square inch. That, together with the fact that every hole implies access from the environment to the composite plies underneath, conflicts with the concepts of damage tolerance and durability, and makes it inapplicable to primary aircraft structures.

The key towards successful development of FML concepts for primary damage tolerant aircraft structures is the capability to provide multiple load paths within one laminated structure. Although crack bridging indeed increases the crack growth resistance of FMLs, this mechanism should not primarily be treated as a way to increase fatigue resistance.

Crack bridging is evidence of multiple structural elements that are joined by the bonding technology that each fulfils their own function within the structure. Whereas the ductility of the metallic layers increases the energy absorption during operational (impact) damage, the fibres provide the second load path in case the metallic constituents crack.

Depending on the combination of constituents, the overall structural performance can be tailored to the structural needs and functions it should provide. Here, the linear elastic fibres may significantly contribute to the static strength of the structures, increasing the strain hardening of the metallic constituents beyond their yield strength.

1.2.3 Increasing Strength of Composites

The aspect of strain hardening relates to viewing FMLs as the reinforcement of metallic structures by adding composite plies. However, the reversed view may solve many issues currently dealt with in fibre reinforced polymer composites. For example, the addition of isotropic metallic layers easily creates quasi-isotropic composite laminates, without the need to place fibres in multiple directions (0, 90, $\pm$ 45°). The additional benefit of having ductility in the composite may increase the damage tolerance by creating the ability to distribute local peak stresses to larger areas of the structure.

Especially where the bearing strength of composites is limiting the joining technology, metallic inserts will help. The higher bearing strength of metals, due to their ductility and isotropy, significantly increases the efficiency of mechanical joining in composite structures. The development of FMLs may provide the understanding necessary to design the appropriate joint areas in composite applications.

1.3 From Material Towards Structural Application

FML is often regarded as denoting a material family, rather than a class of structural concepts. This misperception has led to the development of FML material derivatives, which have mostly never progressed to even a technology readiness level of 3 or 4 [13]. As will be explained in the next chapter, the infinite number of combinations that can be made between the different metals, alloys and their heat treatments, the fibre types and the thermoset and thermoplastic matrices, has led to research and development without evident focus towards structural applications. Nonetheless, the vast number of studies addressing even FML types irrelevant for structural applications has generated further understanding of hybrid material technology principles.

1.4 Contribution to the FML Knowledge

This book aims to contribute to this field of hybrid material technology, by describing the current understanding concerning the hybrid material concept of laminated metallic and composite sheets for primary aeronautical structural applications. The first section of this book aims to provide a general background of the FML technology, indicating the major FML types developed and studied over the past decades (Chap. 2) in conjunction with an overview of industrial developments based on filed patents (Chap. 3).

The second section of this book discusses the mechanical response to quasi-static loading (Chap. 4), as well as the fracture phenomena during quasi-static (Chaps. 5 and 6) and cyclic loading (Chaps. 7–10). To consider the durability aspects related to strength justification and certification of primary aircraft structures, the third section will discuss thermal aspects related to FMLs and their mechanical response to environmental conditions (Chaps. 11–13).

References

1. Vlot A (2001) GLARE—history of the development of a new aircraft material. Kluwer Academic Publishers, Dordrecht, The Netherlands
2. Marissen R (1988) Fatigue crack growth in ARALL, a hybrid Aluminium-Aramid composite material, crack growth mechanisms and quantitative predictions of the crack growth rate. PhD Dissertation, Delft University of Technology
3. Roebroeks GHJJ (1991) Towards GLARE—the development of a fatigue insensitive and damage tolerant aircraft material. PhD Thesis, Delft University of Technology, Delft
4. Vlot A (2000) Towards technology readiness of fibre metal laminates—GLARE technology development 1997–2000. In: Proceedings of the 22nd international congress of aeronautical sciences, Harrogate, United Kingdom, 1–15

5. Kuang KSC, Kenny R, Whelan MP, Cantwell WJ, Chalker PR (2001) Residual strain measurement and impact response of optical fibre Bragg grating sensors in fibre metal laminates. Smart Mater Struct 10:338–346
6. Austin TSP, Singh MM, Gregson PJ, Powell PM (2008) Characterisation of fatigue crack growth and related damage mechanisms in FRP–metal hybrid laminates. Compos Sci Technol 68:1399–1412
7. Cortes P, Cantwell WJ (2006) The fracture properties of a fibre–metal laminate based on magnesium alloy. Compos B 37:163–170
8. Alderliesten R, Rans C, Benedictus R (2008) The applicability of magnesium based fibre metal laminates in aerospace structures. Compos Sci Technol 68(14):2983–2993
9. Burianek DA, Spearing SM (2002) Fatigue damage in titanium-graphite hybrid laminates. Compos Sci Technol 62:607–617
10. Reyes G, Cantwell WJ (2000) The mechanical properties of fibre-metal laminates based on glass fibre reinforced polypropylene. Compos Sci Technol 60(7):1085–1094
11. Carrillo JG, Cantwell WJ (2009) Mechanical properties of a novel fiber-metal laminate based on a polypropylene composite. Mech Mater 41(7):828–838
12. Jensen BJ, Cano RJ, Hales SJ, Alexa JA, Weiser ES, Loos AC, Johnson WS (2009) Fiber Metal Laminates made by the VARTM process. In: Proceedings of the ICCM-17 conference, Edinburgh, Scotland
13. Department of Defence (2005) Technology readiness assessment (TRA) deskbook. Director, Research Directorate (DRD), Office of the Director, Defense Research and Engineering (DDR&E), USA

Chapter 2
Laminate Concepts & Mechanical Properties

Abstract Inspired by the successful application of ARALL and GLARE on aeronautical structures, many researchers and scientists have pursued the development of FML concepts. The fact that the majority of these studies never reached maturity on structural applications may be explained by the observation that FML was mostly treated as a material concept. As a result, not enough consideration was given to the final structural applications. Nonetheless, many FML variants with their properties presented in the literature constitute valuable information for future developments. Therefore, an overview of all the FMLs and the most characteristic properties are given in this chapter.

2.1 Introduction

The early development process of the FML concept has been described in detail by Vlot [1] and Vogelesang [2] who discuss the development of ARALL and GLARE. With these descriptions in mind, it is interesting to look at the FML concepts presented after the introduction of ARALL and GLARE. It is evident that not only the FML community described in [1] contributed to FML development.

Inspired by this successful concept, many researchers all over the world have pursued research on various FML types to investigate the behaviour of these materials under mechanical-, thermal- and environmental loading. As a consequence, the list of internationally refereed journal papers concerning FMLs has become impressive. However, there seems to be an enormous gap between the work reported in these journals and the work being done by institutes and companies on actual FML applications.

To some extent, this gap may be attributed to the apparent lack of relation between the research performed and real applications. However, it appears that some misperceptions often confused the development process of FMLs, both outside and within the FML community described in [1].

A lot of academic research reported in the literature describes a selection of experiments or (numerical) analyses on a particular combination of metallic and fibre-reinforced polymer constituents. The research topic is approached with a

R. Alderliesten, *Fatigue and Fracture of Fibre Metal Laminates*, Solid Mechanics and Its Applications 236, DOI 10.1007/978-3-319-56227-8_2

specific scientific question or objective in mind. Often this leads to detailed understanding of a particular problem. However, despite claims and conclusions, the investigated combination of materials never leads to actual applications, because of the lack of a thorough development process towards applications that addresses all relevant engineering aspects.

Applied research on the other hand, often approaches the material concept from the perspective of particular design criteria or problem. In an attempt to obtain a specific behaviour, the nature of the hybrid laminates seems to be neglected and the scope of research limited. As a consequence, the possible design options remain limited.

The misperception appears to be related to the question whether the FML is primarily a material, or whether it should be considered a structural concept. In later chapters, this is illustrated when discussing the development of theories and prediction models for various FML properties. The FML is considered either a homogeneous material, or the models treat layers individually. The denomination 'first- and second-generation FMLs' and 'next generation FMLs' [3] suggests that the state-of-the-art FMLs (ARALL and GLARE) will be replaced in time by new and advanced FMLs. As a result, the quest was to develop advanced alternatives by applying different fibre types, different metals or alloys and different adhesive types, where the advancement was expressed as increase in material properties. Therefore, the objectives of these developments often were to create FMLs with better static properties and better fatigue resistance than GLARE [4].

In the end, although people often did have potential applications in mind, most research on FML variants has been performed independent of actual applications and their requirements. As a consequence, FML types have been studied that never led to application and never reached a technology readiness level beyond 3 [5]. A more appropriate objective in the FML research would have been to gain better compliance with design criteria for aeronautical structures, and to aim for weight savings for current structural applications.

Nonetheless, the wide range of FML variants studied and reported form an impressive database of information on the potential ranges of properties that can be achieved. It would therefore be foolish not to present a review of the available data in this book.

2.2 Aluminium with Epoxy-Based Adhesive Systems

The first patented FML concept, ARALL, combined aramid fibres with aluminium sheet materials and was initially developed for a Fokker F27 wing application. This concept has been thoroughly investigated especially with respect to its fatigue characteristics [6]. Developing the concept for fuselage applications, however, revealed poor fatigue resistance due to compressive cycle-induced fibre failure. Subsequent development led to the FML based on glass fibres, named GLARE [7].

Because these two FML types, which are often denoted as the first- and second-generation FML [3], have dominated the research and development of the

FML concept in the past decades, an overview of these specific FML types will be given first.

2.2.1 ARALL and GLARE, Codes and Standardisation

In general, FMLs have been treated as a material family, consisting of thin metal layers with in-between fibres embedded in an adhesive system. From the introduction of the first concept (ARALL) onward, trade names have been filed to describe the laminates developed. However, the introduction of trade names for particular configurations seems to add to confusion.

For example, GLARE has been defined by Roebroeks in [8] as the FML based on unidirectional S2-glass fibres embedded in FM94 adhesive (manufactured by Cytec). This is in principle another material system than originally introduced by him in [7] based on the AF163-2 adhesive (manufactured by 3 m). The AF163-2 system was selected at the time based upon the observed high resistance against delamination obtained in an optimization study. Nevertheless, the FM94 adhesive system has been adopted for the FML application on the Airbus A380 for cost reasons.

Indifferent of the reasoning for selecting FM94, it should be noted here that neither the trade name GLARE, nor the grading (GLARE1, GLARE 2, etc.) changed when the other adhesive system was adopted. Technically, the question rises whether the proposed coding is sufficient to describe the material composition. In fact, the aluminium alloy, the fibre type and the adhesive system are only implicitly considered in the name and grading, but are not explicitly defined. It seems that the freedom of tailoring the FML concept similar to the tailoring of fibre-reinforced polymer composites conflicts with the approach adopted in the metal world to code each material.

To illustrate the issue, one may for instance take a closer look at GLARE. As the name indicates, GLARE is a laminate consisting of aluminium reinforced by glass fibres. In general, this could be any combination of glass fibres, aluminium alloys and adhesive systems, as it has not been defined explicitly. Within the community involved in development of GLARE for the A380 [1], the application of unidirectional S2-gass fibre/FM94 epoxy prepreg has always been considered in combination with aluminium 2024-T3, see Table 2.1.

However, a study has been published on the aging effect with fatigue of GLARE [11], where instead of unidirectional prepreg, woven fabric was applied, see Fig. 2.1. It may still be considered GLARE, but the observed behaviour might not directly relate to experimental observations for GLARE based on unidirectional prepreg. For example, the delamination behaviour at the metal/fibre interfaces for unidirectional prepreg and fabric is known to be different [6, 12].

Another example here is the introduction of so-called High Static Strength (HSS) GLARE [10], which is composed of 7475-T761 alloys with the same S2-glass fibres as in GLARE, but in a different high-temperature curing epoxy system. Although presented as an evolution in [10], it mostly comprises a variation similar to

Table 2.1 Standardized ARALL and GLARE grades [7–10]

Grade	Alloy	Metal thickness	Fibre orientation	Fibre	Epoxy	Curing (°C)	Condition
ARALL-1	7075-T6	0.3	0/0	HM aramid	AF163-2	120	0.4% post-stretched
ARALL-2	2024-T3	0.3	0/0	HM aramid	AF163-2	120	As-cured
ARALL-3	7475-T761	0.3	0/0	HM aramid	AF163-2	120	0.4% post-stretched
ARALL-4	2024-T81	0.3	0/0	HM aramid	AF191	175	As-cured
GLARE1	7075-T6	0.3–0.4	0/0	S2-glass	FM94	120	Post-stretched
GLARE2A	2024-T3	0.2–0.5	0/0	S2-glass	FM94	120	As-cured
GLARE2B	2024-T3	0.2–0.5	90/90	S2-glass	FM94	120	As-cured
GLARE3	2024-T3	0.2–0.5	0/90	S2-glass	FM94	120	As-cured
GLARE4A	2024-T3	0.2–0.5	0/90/0	S2-glass	FM94	120	As-cured
GLARE4B	2024-T3	0.2–0.5	90/0/90	S2-glass	FM94	120	As-cured
GLARE5	2024-T3	0.2–0.5	0/90/90/0	S2-glass	FM94	120	As-cured
GLARE6A	2024-T3	0.2–0.5	+45/−45	S2-glass	FM94	120	As-cured
GLARE6B	2024-T3	0.2–0.5	−45/+45	S2-glass	FM94	120	As-cured
HSS GLARE3	7475-T761	0.2–0.5	0/90	S2-glass	FM906	175	As-cured
HSS GLARE4A	7475-T761	0.2–0.5	0/90/0	S2-glass	FM906	175	As-cured
HSS GLARE4B	7475-T761	0.2–0.5	90/0/90	S2-glass	FM906	175	As-cured

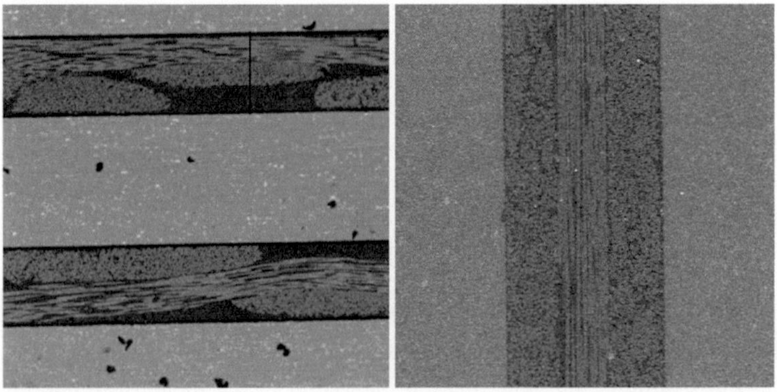

Fig. 2.1 Cross sections of GLARE laminates based on woven fabric [11] (*left*) and based on unidirectional prepreg (*right*)

the variation initially applied to ARALL, where 7475-T761 and 175 °C curing systems were already exploited for specific FML types. In fact, the origin of HSS GLARE is related to the higher static strength compared to standard GLARE, reducing the fatigue crack initiation life provided by the latter one considerably.

Because of the increased laminate variations, Airbus, for example, changed the laminate designations to 2024-FML, 7475-FML and 1441-FML, instead of the traditional names GLARE and HSS GLARE

These examples illustrate that the FML concept, or even the FML family, is not a family of materials, but a structural laminate concept. Depending on the structural applications, a specific combination of alloy, fibre and adhesive may lead to optimal performance of the structure, which does not necessarily translate to an overall increase in mechanical properties at material level. The reader should bear this in mind when reading this chapter on the laminate concepts and their properties.

2.2.2 Aramid Fibres (ARALL)

The FML ARALL, initially developed for lower win skin of the Fokker F27 and F50 [1], was the first laminate concept that achieved a structural application, which was on the C-17 cargo door. This made ARALL the first concept to successfully navigate through the process of material specification and qualification, allowable determination, and development of design and manufacturing principles [10].

Although for the qualified ARALL laminates the constituents are specified, see Table 2.1, different materials have been considered in the early development. Marissen investigated the performance of ARALL laminates containing aramid prepreg and fabric, where in the latter case the laminate was manufactured by placing the dry fabric between structural adhesive films [6]. In his research, the fibre volume fraction of the Twaron HM aramid layer was subject of investigation, while in addition different adhesive systems were applied. Although in the end AF163-2 from 3 m was selected for ARALL-1 to ARALL-3, initially BSL-312-UL from Cyba-Geigy and FM123-5 from American Cyanamid were considered.

In the context of Marissen's research where the shear deformation of the resin-rich layer at the interface between aluminium and fibre layer was deemed important for fatigue crack propagation, Mangkoesoebroto [13] observed that increasing the resin-rich layer thickness increases the delamination resistance of that interface. That, and the observed difference in delamination resistance between unidirectional fibre plies and woven fabrics [6, 12] created different fatigue crack growth characteristics between the ARALL laminates containing prepreg and the ones with woven fabric.

The typical mechanical properties of the commercial ARALL laminates are presented in Table 2.2. The values given in this table relate to the 3/2 lay-up based on 0.3 mm aluminium. A first estimation of the properties for different lay-ups or aluminium layer thicknesses can be made using a rule of mixtures and the appropriate constituent properties, i.e. aluminium and directional prepreg layers.

Table 2.2 Typical mechanical properties of ARALL laminates with 3/2-0.3 lay-up [9, 14, 15]

	Property	Orientation	ARALL-1	ARALL-2	ARALL-3	ARALL-4
Tension	σ_{ult}	L	800	717	828	731
		LT	386	317	373	338
	$\sigma_{0.2}$	L	641	359	587	373
		LT	331	228	317	317
	E_t	L	67.6	64.1	68	64.1
		LT	48.3	49	51	49
	ε_{ult}	L	1.9	2.5	2.2	2.6
		LT	7.9	12.7	8.8	4.6
	ν	L	0.33	0.32		
		LT	0.25	0.26		
Compression	$\sigma_{0.2}$	L	372	262	345	
		LT	393	234	365	
	E_c	L	70	67	66	
		LT	52	52	50	
Shear	$\tau_{0.2}$	L		117		
		LT		114		
	G	L	17	17		
		LT		16		
Blunt notch	σ_{net}	L	497[a]	401[a]		
		LT				
Bearing	$\sigma_{ult,e/D=2.0}$	L	655	531		
		LT	703	545		
	$\sigma_{ult,e/D=3.0}$	L	738	565		
		LT	724	545		
	$\sigma_{yield,e/D=2.0}$	L	586	386		
		LT	607	386		
	$\sigma_{yield,e/D=3.0}$	L	703	455		
		LT	669	441		

[a]Material not produced by ALCOA

From the values for the L and LT orientation of the laminates, the directionality of the laminates is evident. The ARALL laminates, initially developed for lower wing skin panels, have fibres oriented in one direction, providing very high strength and (tangent) stiffness in fibre direction, but significantly less in the transverse direction.

Evident in Table 2.2 and typical for FMLs is the high yield strength for unidirectional laminates. However, a remark must be made here. The yield strength here and in most literature is presented as intersect between the stress–strain curve and the 0.2% strain offset of the elastic curve.

Figure 2.2 illustrates that despite the fact that plasticity in the metal layers of the FML starts at a lower stress compared to monolithic aluminium, the lower stiffness and the higher tangent modulus result in a higher 0.2% strain offset stress for the

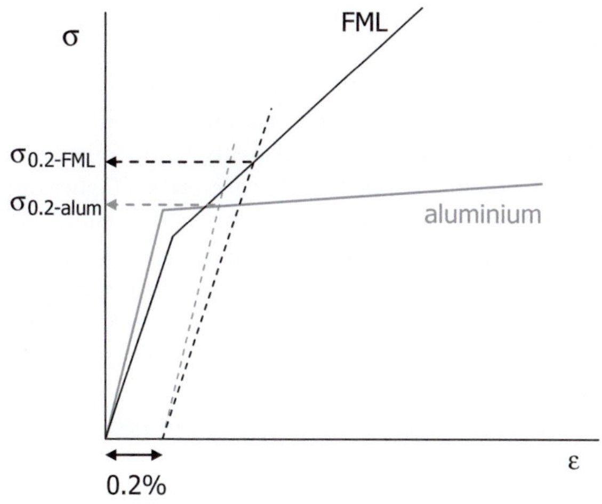

Fig. 2.2 Difference between the 0.2% strain offset yield strength for aluminium and FMLs

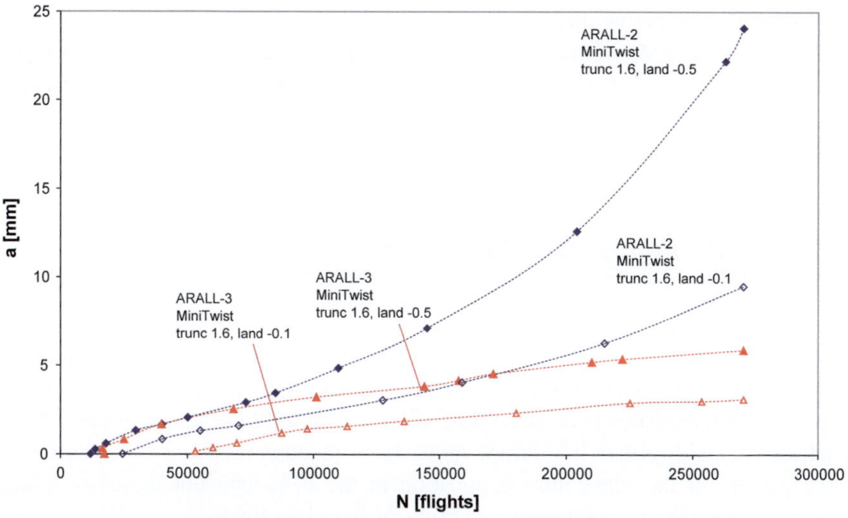

Fig. 2.3 Crack propagation curves for the ARALL-2 and ARALL-3 laminates tested at two different ground stress levels (landing); data from [16]

FML. This means that stress engineers must consider how the yield strength is implemented in their definition of allowables and what that implies for their structural sizing.

As a consequence of the initial objective of FML development, the fatigue resistance was improved with ARALL laminates in particular when post-stretching was adopted. The crack growth resistance for the post-stretched laminates is significantly higher than for monolithic aluminium. Figure 2.3 illustrates the difference in crack growth between as-cured ARALL-2 and post-stretched ARALL-3 subjected to wing load spectra.

The fatigue resistance, or more specifically, the fatigue initiation resistance, of ARALL-2 is less than that of monolithic aluminium, while it is superior for the post-stretched ARALL-3 laminates. The relatively high stiffness of the aluminium layers compared to the overall laminate stiffness creates a higher cyclic strain cycle in the aluminium layers, compared to monolithic aluminium. This means that nucleation of fatigue cracks can occur earlier in as-cured laminates.

On the other hand, the reversal of internal residual stresses induced by the stretching technique creates compressive residual stress in the aluminium layers that effectively reduces the mean stress of the fatigue stress cycle. As a consequence, the nucleation of cracks is delayed.

The durability of ARALL laminates is mainly dictated by the corrosion resistance of the applied aluminium alloys and the moisture absorption characteristics of the aramid fibres. Whereas the 7475-T761 has better exfoliation corrosion characteristics than the 7075-T6, in general the corrosion resistance of ARALL can be increased by applying cladding to the outer aluminium layers. This implies that the laminates are manufactured from bare aluminium inner layers and aluminium with single-sided cladding as outer layers.

Note that the FML properties provided in this section are for laminates tested either in the L or LT direction. The reader should bear in mind that for structural justifications the off-axis properties are of equal importance. The subject of off-axis loading is discussed in later chapters.

2.2.3 Glass Fibres (GLARE, Central)

The GLARE laminates (Table 2.3) were developed to overcome the poor fatigue performance of ARALL laminates in case compressive load cycles occur in the load spectrum. The compressive stability of the ARALL fibres in combination with the adhesion characteristics between fibre and epoxy caused fibre failure under fatigue load spectra with either low stress ratios or with compressive load cycles. The advancement of the glass fibre application in the FML concept therefore relates directly to the higher compressive stability of the glass fibres.

The application of glass fibres instead of aramid fibres in itself would not have been sufficient for filing a new patent [17], because the glass fibres were mentioned in the original ARALL patents. However, the fact that the glass fibres exhibit strain rate effects that significantly increases the impact performance of FMLs opened the door to further develop and tailor the laminated concept.

The following glass fibres have been evaluated for application in FMLs [18]:

- S2 glass (solid and hollow)
- R-glass
- E-glass

Table 2.3 Typical mechanical properties of GLARE laminates with 3/2-0.3 lay-up [14, 19–21]

	Property	Orientation	GLARE 1		GLARE 2[a]		GLARE 3		GLARE 4[a]		GLARE 5		GLARE 6	
			2/1	3/2	2/1	3/2	2/1	3/2	2/1	3/2	2/1	3/2	2/1	3/2
Tension	σ_{ult}	L	1077	1282	992	1101	640	678	831	898	683	916		
		LT	436	352	335	307	627	666	545	562	681			
	$\sigma_{0.2}$	L	525	545	343	343	298	287	304	296	297	276		
		LT	342	333	225	209	267	260	236	225	275			
	E_t	L	66	65	67	64	59.2	56.4	58.5	56	59	51		
		LT	54	50	57	48.5	59	56.2	52.4	48.7	59			
	ε_{ult}	L	4.2	4.2	4.7	4.7	4.7	4.7	4.7	4.7	4.7			
		LT	7.7	7.7	10.8	10.8	4.7	4.7	4.7	4.7	4.7			
	ν	L												
		LT												
Compression	$\sigma_{0.2}$	L	447	424	357	367	295	292	316	317	283			
		LT	427	403	255	237	305	298	278	267	280			
	E_c	L	63	67	67	65.6	61.6	58.9	59.1	57.2	61			
		LT	56	51	57	53.2	62.1	59.6	54.4	51.7	61			
Shear	$\tau_{0.2}$	L			122	112	125	115	111	100				
		LT			122	112	125	115	111	100				
	G	L			16.1	13.6	18.4	16.4	15.6	13.4				
		LT			16.1	13.6	18.4	16.4	15.6	13.4				
Blunt notch	σ_{net}	L			645	694	470	481	562	589				
		LT			278	253	454	467	396	396				

(continued)

Table 2.3 (continued)

	Property	Orientation	GLARE 1		GLARE 2[a]		GLARE 3		GLARE 4[a]		GLARE 5		GLARE 6	
			2/1	3/2	2/1	3/2	2/1	3/2	2/1	3/2	2/1	3/2	2/1	3/2
Bearing	$\sigma_{ult,e/D=2}$	L		832		703		782						
		LT												
	$\sigma_{ult,e/D=3}$	L			1019	980	1050	1018	972	931				
		LT			1013	973	1050	1018	1003	968				
	$\sigma_{yield,e/D=2}$	L		713		530		545						
		LT												
	$\sigma_{yield,e/D=3}$	L			723	729	682	680	651	643				
		LT			557	528	682	680	593	575				

[a]Laminate type A, see Table 2.1

The S-glass and R-glass have similar characteristics, while the E-glass has a lower stiffness and strength. For the FMLs for primary aeronautical structures, the stiffness is important, and often it is desired to have an FML stiffness close to the stiffness of monolithic aluminium. A small stiffness difference between aluminium and FMLs allows the application of FML skin with aluminium backup structure, without having the risk of fatigue damage in frames and stringers.

In an attempt to develop the FML concept further for thick lower wing panels, alternative lay-up concepts have been developed. Building up thick panels from thin metal sheets with composite plies may imply a burden on the manufacturing processes of such large wing panels. Primarily driven by the philosophy that thicker aluminium layers had to be applied in the concept, the CentrAl concept was developed [22].

The CentrAl concept uses GLARE as inner reinforcement in a laminate with aluminium layers of up to 1.6 mm. This concept is basically similar but reversed to an earlier lower wing panel concept, where thick aluminium plates were bonded between FML facings.

Because the application of thicker aluminium layers implied higher load transfer in case of fatigue crack growth, very large delaminations were observed, making fibre bridging ineffective. The required increase in delamination resistance of the aluminium prepreg interface was created by adding more resin to the interface in a so-called resin-rich layer. This concept is based on the earlier mentioned report by Mangkoesoebroto [13] who observed this improvement in behaviour in ARALL. The thicker the resin-rich layer at the interface, the more the shear deformation is unconstrained, reducing the shear stress peak at the delamination tip and thus the delamination growth.

2.2.4 Carbon Fibres (CARE/CARALL)

The apparent benefit of substitution of one fibre type by another in FMLs inspired people to widen the scope of fibre types. Because of the large variation in carbon fibre types available, varying from high strength to high modulus fibres, it appeared to be an excellent step to study the behaviour of FMLs reinforced with carbon fibres [23, 24] (Table 2.4).

Over the years, several studies have been performed [23–33] that all seem to have a recurring theme in the investigation; the issue of galvanic corrosion. The combination of aluminium with carbon fibres is prone to corrosion induced by the difference in potential between the two materials. Preliminary investigations indicated that the issue of galvanic corrosion could be solved if the aluminium remains isolated from the carbon fibres. To provide the isolation, concepts have been proposed in which either aluminium layers are protected with thermoplastic polyetherimide coatings or in which additional glass fibre layers are added between the aluminium and carbon fibre layers [24]. However, these approaches still require additional measures at locations where holes are drilled and fasteners are installed.

Table 2.4 Investigated carbon fibre-reinforced aluminium laminates [24]

Fibre	Adhesive	v_f (%)	t_f (mm)	T_{cure} (°C)	Lay-up	t_{lam} (mm)	σ_{ult} (MPa)	E (MPa)
HM	AF163-2	58	0.22	120	UD –2/1	0.82	800	105
HTA	AF163-2	58	0.27	120	UD –2/1	0.82	984	89.9[a]
T300	DLS1095	60	0.20	120	UD –2/1	0.80	747	85.2
					CP–3/2	1.30	585	71.3
IM600	Fibredux 924C	60	0.105 0.21 0.315	180	UD–2/1	0.705 0.81 0.915	896 1218 1395	87.6[a] 99.3[a] 108.1[a]
T800	Fibredux 924C	60	0.2	180	UD–2/1	0.80	1030	100
					CP–3/2	1.30	728	75.1
FT700	AF163-2	55	0.23	120	UD–2/1	0.83	675	170

[a]Calculated values

A driver behind the investigation of carbon fibre application in FMLs was the excellent crack growth characteristics of such laminates. The stiffness of the carbon fibres provided ply stiffness higher than aluminium, which could not be achieved with glass fibres and aramid fibres.

There was, however, a problem to be encountered; the coefficient of thermal expansion of these fibres differed significantly from the values known for aluminium, and even glass fibres. Adding the high stiffness to the problem, extremely high tensile residual stresses were obtained after curing. In other words, carbon fibres may contribute to the crack growth phase of the fatigue life (damage tolerance), and certainly reduced the initiation life (fatigue).

2.2.5 *Polymer Fibres (HP-PE, Zylon)*

Rather than studying the application of aramid-, glass- or carbon fibres, several studies focused on the performance of polymer fibres in FMLs. For example, in the early days of the ARALL development, Meyers and Roebroeks [34] investigated the application of high-performance polyethylene (HP-PE) fibres in FMLs. They determined the stiffness of the non-impregnated and impregnated fibres to address the performance of the HP-PE fibres in an FML containing aluminium layers. From the tensile tests it appeared that the consolidated fibres had half the stiffness of the non-impregnated fibres, resulting in an FML with similar stiffness to ARALL (Table 2.5).

In addition, because of the low adhesion characteristics between the HP-PE fibres and the epoxy, the fibres failed under low compression loading. This problem was to a lesser extent observed in ARALL when developing the FMLs for wing panel applications. The research was therefore not continued.

Another polymer fibre that has been studied for FML applications more recently was the Zylon fibre [35]. Zylon is a synthetic polybenzoxazole (PBO) fibre that has

Table 2.5 Overview of studied fibre properties

	Fibre properties			Prepreg properties		
Fibre	E (MPa)	σ_{ult} (MPa)	ε_{ult} (%)	ρ (kg/dm^3)	V_f (%)	Adhesive
HM aramid (Twaron)	124	2800	2.5			
Aramid	121	2800	2.0	1.45	50	AF163-2
S2-glass	88	4400	4.7	2.0	60	AF163-2 FM94 FM906 DLS1611
E-glass	66	2350	3.6			
R-glass	88	4400	4.7	1.98	60	
HM	358	2350	0.6	1.79	58	AF163-2
HTA	238	3400	1.4	1.77	58	AF163-2
T300	230	3530	1.5	1.77	60	DLS1095
T800	294	5590	1.9	1.81	60	Fibredux 924C
FT700	700	3300	0.5	2.16	55	AF163-2
IM600	295	5400	1.7	1.79	60	Fibredux 924C
T-650/35	255	4280	1.7			
HP-PE	76–110	2980	2.7		45–92	AF163-2
ZYLON AS	180	5800	3.5			
M5	290	4000	1.5			Epoxy

a strength and modulus almost double that of Kevlar® (p-Aramid fibre) with a density equivalent to aramid fibres [36].

The decision to study the application of Zylon fibres in FMLs was primarily driven by the high stiffness and strain to failure of the fibre. The higher stiffness was required to obtain FML stiffness equivalent to or higher than monolithic aluminium. The strain to failure is generally considered the limiting parameter for residual strength. The slowly propagating cracks, for which FMLs are known, come along with a slowly reducing residual strength. One advantage of GLARE over monolithic aluminium is that the strength is very high, while the reduction in strength induced by cracks or blunt notches such as holes is gradual and relatively small. This is attributed to the strain to failure of the FML in combination with the high strain hardening and strength. The energy absorption characteristics directly influence the residual strength capability. Reduction in the strain to failure by application of high modulus fibres would increase the strength, but significantly reduce the residual strength.

However, research revealed that for an FML containing Zylon the residual stresses after curing were extremely high due to large difference between the coefficients of thermal expansion of the Zylon prepreg and aluminium. This led to high tensile stresses in the aluminium, corresponding to high compressive stresses in the Zylon layers. However, the compressive modulus of elasticity of Zylon appeared to be about half the tensile modulus. As a consequence, the linear elastic tensile behaviour of the as-cured FML was dictated by the compressive modulus of Zylon rather than the tensile modulus, with an FML stiffness equivalent to GLARE.

In production, the high toughness of the fibres induced manufacturing challenges, because milling and drilling FML panels containing Zylon fibres induces significant tool wear, but requires sharp tooling to avoid fibres being pulled out of the laminate.

2.2.6 *M5 Fibres*

To solve the aspect of low compressive strength and even low composite modulus of elasticity related to the polymer fibres, AKZO developed a rigid rod polymer fibre called M5 [37]. The fibre has an excellent combination of strength and stiffness, providing sufficient strain to failure to benefit from this fibre type in FMLs (Fig. 2.4).

The related studies performed at TU Delft, however, showed that the tensile strength of an M5-FML is about 34% lower compared to standard GLARE, which is primarily related to the strain to failure. The stiffness on the other hand was observed to be 13% higher [38, 39, 40]. Interestingly enough, the blunt notch strength of M5-FML is similar to that of standard GLARE, while the residual strength appeared to be somewhat lower.

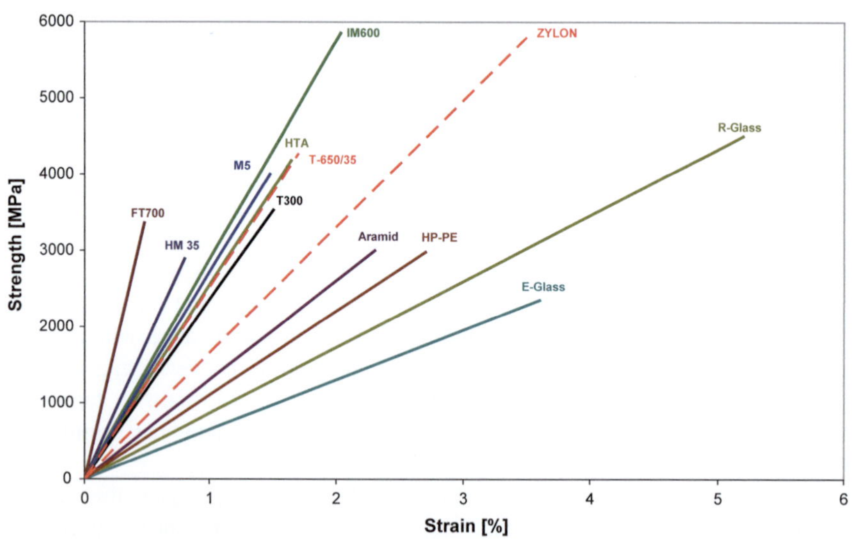

Fig. 2.4 Stress–strain curves for the fibres listed in Table 2.5, tested in previous studies

The major benefit, however, was observed in fatigue crack growth; the crack growth rates were significantly lower than those of GLARE, which were already quite low.

2.3 Other Metal Constituents

Although most work has been performed on FMLs with aluminium as the prime metal constituent, quite a number of studies report the application of other metals and alloys in FMLs. The motivation for the application of other metals may vary. Some studies are driven by seeking better stiffness compatibility with carbon fibres, or by avoiding galvanic corrosion issues when combining with carbon fibres. Other studies aim to develop FMLs for high-temperature applications, like for the high-speed civil transport aircraft or for military applications.

2.3.1 Titanium-Based FMLs

The application of titanium sheet material in FMLs has been studied a number of times. One of the first studies reporting the application of titanium is by Medenblik [27], who combined the commercial alloy Ti-6Al-4 V with a thermoset phenolic triazine resin or with a thermoplastic PEEK resin in a unidirectional FML. In evaluating the mechanical and adhesive properties, Medenblik concluded that the latter properties were greatly influenced by the surface pre-treatment of the titanium alloys. Anodization did not result in a durable oxide layer, leading to adhesive failures rather than cohesive failures.

Concerning the mechanical properties, Medenblik reported that except for the ultimate strength most properties in the longitudinal direction were comparable with the monolithic titanium properties, while in the long transverse direction they were lower. In particular, the blunt notch strength indicated high notch sensitivity, although the residual strength tests seem to show the opposite. In the end, only the fatigue performance was superior compared with the fatigue performance of the monolithic Ti-6Al-4 V.

The latter observation is in agreement with the evaluation of fatigue crack propagation and delamination growth of titanium-based FMLs as reported both by Burianek [41, 42] and by Rans [43, 44]. The fatigue performance, even at elevated temperatures, demonstrated the applicability if these laminates for high-temperature applications. Nonetheless, some of the non-reported work has also illustrated the point raised by Medenblik [27] on the surface pre-treatment of titanium.

2.3.2 Stainless Steel-Based FMLs

Stainless steel could serve as an alternative to titanium in FMLs in combination with carbon fibres. The combination eliminates galvanic corrosion as an issue, while the higher stiffness of stainless steel over titanium creates better compliance with in particular high modulus carbon fibre layers [46]. The downside of stainless steel is the higher density, which puts constraints on the maximum thicknesses of individual layers. Most studies therefore report the application of 0.1 mm AISI 316L stainless steel in FML configurations [45–47], although some studies apply thicker layers, like 0.6 AISI 304L [48].

The tensile static and fatigue properties of these stainless steel-based FMLs are generally high in comparison to aluminium-based FMLs. The high stiffness and failure strength seem to imply a significant benefit over the application of aluminium in an FML, but here one has to be careful. The thin layers, in particular the 0.1 mm thin AISI 316L layers, impose problems in compression, because the stability of these thin layers is insufficient to avoid local sheet buckling in the laminate.

Hence either thicker layers must be adopted, like in [48], which comes at the cost of significant weight, or applications must be considered which are primarily loaded in tension.

2.3.3 Magnesium-Based FMLs

Few studies report the application of magnesium in FMLs. The obvious benefit of magnesium over any of the other metal constituents is its lower density [49]. However, one should be aware that the specific properties may not substantially differ from those of many aluminium alloys. Evaluating the static and fatigue properties of FMLs based on magnesium alloys has illustrated that both the mechanical properties and the fatigue properties are similar to standard GLARE, when presented in the form of specific properties, i.e. normalizing by the densities [50].

Opposite to the application of thin stainless steel sheets in FMLs, where applications primarily loaded in tension should be considered, it seems that magnesium-based FMLs may excel mostly in compression dominated structures. The lower density of the magnesium allows to apply layers which are thicker than traditionally considered for aluminium-based FMLs, which can improve the compression stability of laminates substantially. In particular if a hybrid concept is adopted where the internal metal layers are made of magnesium, with only the outer layers made of aluminium [51].

2.4 Thermoplastic Adhesive Systems

Most of the FML research concerns the utilization of thermoset matrix systems for the composite plies. The application of thermoset-based composites is widely accepted in the aeronautical industry and only recently have structural concepts been developed based on thermoplastic systems.

However, some studies evaluate the application of thermoplastic matrix systems in FMLs. The primary problem to cope with is the required high curing temperatures to consolidate the composite. For most high-strength aluminium alloys, these high curing temperatures imply heat treatments to the aluminium with detrimental consequences for its mechanical properties. For example, a preliminary study by Van Velze [52], who investigating the mechanical properties of ARALL laminates with PEI, PPS and acetal copolymer as matrix system, revealed that the higher curing stresses negatively impact the mechanical and fatigue properties. In post-stretched condition, the thermoplastic ARALL laminates performed similar to post-stretched ARALL based on AF163-2 epoxy matrix. The only benefits that could be demonstrated at the time were absence of fibre failure in fatigue testing as-cured laminates, and the improved fatigue performance at elevated temperatures up to 190 °C.

Because high curing temperatures limit the application of various aluminium alloys, most studies seem to focus on the application of carbon fibre embedded in thermoplastic matrix system combined with titanium in an FML. Titanium alloys can be operated at higher temperatures and allow curing of the thermoplastic composite within the FML at higher temperatures.

Although some initial manufacturing trials and analyses were performed to determine the mechanical performance of FMLs based on titanium and carbon fibres embedded in thermoplastic matrix, [53], most research has focussed on impact performance [54, 55].

2.5 Innovative Hybridization Concepts

The concept of FMLs, combining the world of two distinctly different material classes, was innovative and advanced at the time of introduction. Nowadays, the concept is state-of-the-art and fully mature with the application on the Airbus A380.

The literature illustrates that along with the development of this hybrid concept, research is performed to increase the fundamental understanding of all mechanical and physical aspects involved with the material technology. Often prediction models, initially highly empirical of nature, are improved in parallel with the development of structural applications.

The advantage of that process is that the research receives focus from the structural application, requiring answers to more specific questions regarding the predictive capabilities of the prediction models. In the end, stress engineers need these models to substantiate the development of design principles and stress allowables. However, the drawback is that the academic research changes its horizon from medium- and long term towards the short term. To avoid tunnel vision in research, effort should be put in continuous questioning of the research performed and models developed, for its applicability and validity towards the long-term future.

To that aim, Rensma [45] investigated a selection of highly hybridized material concepts that by no means were chosen for their potential in aeronautical applications, but mainly to identify the blind spots in the understanding and modelling of FMLs at that time. The concepts investigated were selected as theoretical extrapolations on current research trends.

The review in this chapter illustrates that FML concepts were studied in which constituents were varied, but often the thickness of individual constituents was kept the same. Alternatively, studies were performed in which the constituents were kept the same, but thicknesses varied. For example, the CentrAl concept was composed of standard GLARE constituents, but the distribution of ply thicknesses throughout the laminate was changed.

For that reason, Rensma selected hybrid laminate concepts in which both constituents and thicknesses were varied in order to identify the limitations in understanding. An overview of investigated concepts is illustrated in Fig. 2.5. Combinations with aluminium, stainless steel, S2-glass fibre layers and carbon fibre

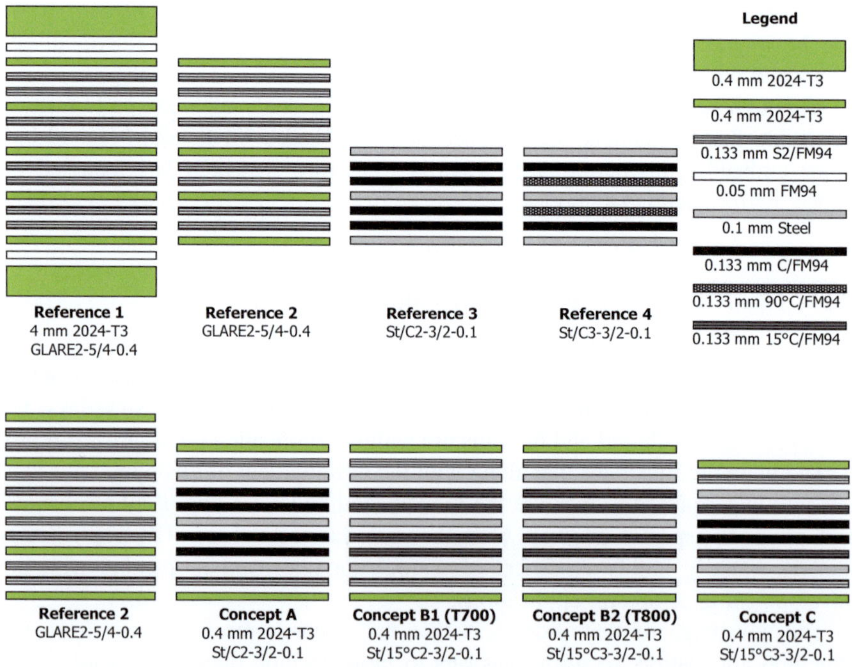

Fig. 2.5 Lay-up of hybrid concepts and the reference laminates (prepreg and adhesive thicknesses are nominal thicknesses after curing), '15°' indicates ±15° orientations

layers were considered. The overall conclusions were that trends could be predicted fairly well, but that the improvements of mechanical and damage tolerance properties highly depended on the constituents used. Some of the specific findings of this research are presented in later chapters.

References

1. Vlot A (2001) GLARE—history of the development of a new aircraft material. Kluwer Academic Publishers, Dordrecht, The Netherlands
2. Vogelesang LB (2003) Fibre metal laminates—the development f a new family of hybrid materials. In: Editor, G, (ed) Proceedings of the 22rd fatigue of aeronautical structures as an engineering challenge, ICAF symposium, vol I, EMAS publishing, UK, p 1–41
3. De Boer T (2001) Next generation fibre metal laminates. In: Vlot A, Gunnink JW (eds) Fibre Metal Laminates—an introduction. Kluwer Academic Publishers, Dordrecht, The Netherlands
4. DIALFAST Development of innovative and advanced laminates for future aircraft structure. Project funded by the European Commission within the 6th framework programme, 2004–2007
5. Department of Defence (2005) Technology readiness assessment (TRA). Deskbook, USA
6. Marissen R (1988) Fatigue Crack Growth in ARALL, a hybrid Aluminium-Aramid composite material, crack growth mechanisms and quantitative predictions of the crack growth rate. PhD Dissertation, Delft University of Technology
7. Roebroeks GHJJ (1991) Towards GLARE—The Development of a fatigue insensitive and damage tolerant aircraft material. PhD Thesis, Delft University of Technology, Delft
8. Vlot A, Gunnink JW (eds) (2001) Fibre Metal Laminates—an introduction. Kluwer Academic Publishers, Dordrecht, The Netherlands
9. Bucci RJ, Mueller LN, Vogelesang LB, Gunnink, JW (1988) Proceedings of the 33rd international sampe symposium. pp 1237–1248
10. Van Hengel C, Kortbeek P (2009) ARALL and GLARE FML's: three decades of bridging the gap between theory and operational practice. In: M. Bos (ed) ICAF 2009, bridging the gap between theory and operational practice, pp 601–615
11. Da Silva DA, Botelho EC, Rezende MC (2008) Hygrothermal aging effect on fatigue behavior of GLARE. J. Reinf. Plast. & Comp 28(20):2487–2499
12. Khan SU, Alderliesten RC, Benedictus R (2009) Delamination in Fibre Metal Laminates under variable amplitude loading. Composites Science and Technology 69(15–16):2604–2615
13. Mangkoesoebroto RH (1987) The Effect of fibre volume fraction on the mechanical properties and the fatigue behaviour of ARALL laminates. Masters Thesis, Delft University of Technology
14. Structural Laminates Company (1994) QA Reports B0319B-2, B1008B-1, B0904A-3, 510 Constitution Blvd., New Kensington, PA 15064
15. Vermeeren CAJR (1990) The blunt notch behaviour of metal laminates: ARALL and GLARE. Report LR-617, Delft University of Technology
16. Wiltink FJ, Bodegom VJW van (1994) Flight-simulation fatigue tests on notched specimens of Fiber-metal laminates. Test results used in LRV-10, Report LRV-11, Delft University of Technology
17. Alderliesten RC (2009) On the development of hybrid material concepts for aircraft structures. Recent Patents on Engineering 3(1):25–38
18. Boertien MFHC (1992) Evaluation of solid S2 and R glass fibers versus hollow S2 glass fibers for Fibre Metal Laminates. Preliminary thesis, Delft University of Technology

19. Roebroeks G (2000) The metal volume fraction approach. Report TD-R-00-003, Structural Laminates Industries
20. Pellenkoft FJ (2000) Initial sizing study of GLARE-based alternative floor panel concepts for the containerised cargo compartment of the A3XX. Report B2 V-00-44, Delft University of Technology
21. Mattousch AC (1991) Bearing strength tests on GLARE sheets for the application in lugs. Preliminary thesis, Delft University of Technology
22. Roebroeks GHJJ (2007) The development of Central. DTAS 2007
23. Vermeeren CAJR (1990) Ultra high modulus carbon fibres in ARALL Laminates. Memorandum M-641, Delft University of Technology
24. Vermeeren CAJR (1991) The application of carbon fibres in ARALL laminates. Report L-658, Delft University of Technology
25. Koos MG de (1990) PEEK carbon fibre reinforced titanium laminates. Master Thesis, Delft University of Technology
26. Kleinendorst RGJ (1990) Corrosion properties of carbon ARALL. Master Thesis, Delft University of Technology
27. Medenblik EW (1994) Titanium fibre-metal laminates. Master Thesis, Delft University of Technology
28. Lin CT, Kao PW, Yang FS (1991) Fatigue behaviour of carbon fibre reinforced aluminium laminates. Composites 22(2):135–141
29. Lin CT, Kao PW, Jen M-HR (1994) Thermal residual strains in carbon–fibre aluminium laminates. Composites 25(4):303–307
30. Lin CT, Kao PW (1995) Effect of fiber bridging on the fatigue crack propagation in carbon fiber-reinforced aluminium laminates. Mater Sci Eng A 190:65–73
31. Lin CT, Kao PW (1996) Fatigue delamination growth in carbon fibre reinforced aluminium laminates. Compos Part A 27A:9–15
32. Silva JMA, Ferreira JAM, Devezas TC (2003) Fatigue damage of carbon–epoxy laminates with embedded optical fibres. Mater Sci Technol 19:809–814
33. Bradshaw RD, Gutierrez SE (2007) Characterization of fatigue crack initiation and growth in hybrid aluminium–graphite fibre composite laminates using image analysis. Fatigue Fract Eng Mater Struct 30:766–781
34. Meyers LG, Roebroeks GHJJ (1986) De HP-PE vezel, enkele experimenten. Technische Hogeschool Delft, Luchtvaart en Ruimtevaarttechniek
35. TOYOBO Co. Ltd (2001) PBO Fiber Zylon®, technical information. (revised 2001.9)
36. Huang YK, Frings PH, Hennes E (2002) Mechanical properties of Zylon/epoxy composite. Compos B Eng 33(2):109–115
37. Jagt OC van der (1998) Mechanical behaviour of a new polymer fibre ('M5') in advanced composite structures. In M.A. Erath (ed) Proceedings of the 19th international conference on SAMPE Europe progress through innovation and cost effectiveness, Paris, April 22–24, 1998. ISBN 3-9520477-5-9 pp 227–239
38. Jagt OC van der, Beukers A (1999) The potential of a new rigid-rod polymer fibre ('M5') in advanced composite structures. Polymer 40:1035–1044
39. Sihombing KM (1999) The application of M5-fibre in fibre-metal laminates. Delft University of Technology, Netherlands
40. Asporo YAR (1999) Towards M5 FML, a preliminary research on the new Fiber Metal Laminate with M5 Fiber. MSc thesis, Delft University of Technology
41. Burianek DA (2001) Mechanics of fatigue damage in Titanium-Graphite hybrid Laminates. PhD dissertation, Massachusetts Institute of Technology
42. Burianek DA, Spearing SM (2002) Fatigue damage in titanium-graphite hybrid laminates. Compos Sci Technol 62:607–617
43. Rans CD (2008) Evaluation of the fatigue crack propagation and delamination growth behaviour of Titanium-Carbon laminates. Report B2 V-08-06, Delft University of Technology

44. Rans CD, Alderliesten RC, Benedictus R (2011) Predicting the influence of temperature on fatigue crack propagation inn Fibre Metal Laminates. Eng Fract Mech 78:2193–2201
45. Rensma E (2007) Investigation of innovative concepts for hybrid structures. MSc thesis, Delft University of Technology
46. Shahinian R (2006) The next generation of fibre metal laminates. A preliminary study of the material properties of carbon fibre-reinforced stainless steel laminates. Internship report, Hogeschool van Amsterdam
47. Rooijen RGJ (2006) Bearing strength characteristics of standard and steel reinforced GLARE. PhD dissertation, Delft University of Technology, Delft
48. Pärnänen T, Kanerva M, Sarlin E, Saarela O (2015) Debonding and impact damage in stainless steel fibre metal laminates prior to metal fracture. Compos Struct 119:777–786
49. Cortes P, Cantwell WJ (2006) The fracture properties of a fibre–metal laminate based on magnesium alloy. Compos B 37:163–170
50. Alderliesten R, Rans C, Benedictus R (2008) The applicability of magnesium based Fibre Metal Laminates in aerospace structures. Compos Sci Technol 68(14):2983–2993
51. Coonen M (2010) Applicability of MgAl-FML's in aerospace. MSc thesis, Delft University of Technology
52. Van Velze TM (1988) Thermoplastics in ARALL laminates—Development, production and testing of ARALL laminates with a thermoplastic matrix. MSc thesis, Delft University of Technology
53. van der Hoeven W (2007) Preliminary evaluation of a high temperature Fibre Metal Laminate, Report NLR-CR-2007-596. National Aerospace Laboratory, The Netherlands
54. Langdon GS, Cantwell WJ, Nurick GN (2005) The blast response of novel thermoplastic-based fibre-metal laminates—some preliminary results and observations. Compos Sci Technol 65:861–872
55. Langdon GS, Cantwell WJ, Nurick GN (2007) Localised blast loading of fibre–metal laminates with a polyamide matrix. Compos B 38:902–913

Chapter 3
Patents and Intellectual Property

Abstract Along with the developments of hybrid laminated configurations comprising metallic and fibre-reinforced polymer constituents, patents have been filed to protect the intellectual properties associated with these developments. This chapter provides an overview of the patents filed worldwide over the past three decades to illustrate the various structural-, material-, and manufacturing concepts developed. These concepts are subsequently discussed with respect to their novelty and innovation.

3.1 Introduction

Since the introduction of the FML concept, numerous patents have been filed on different variants and related manufacturing principles. Taking a closer look at the patents on these hybrid technologies that have been filed over the last two decades illustrates the development of subsequent concepts, but sometimes also witnesses the lack of new inventions. A first review has been published in [1], which is extended into a more complete overview, including recent patents, in this chapter.

3.2 Material Concept Development

3.2.1 Improving Fatigue and Crack Growth

In the early days of the development of the FML concept, the aim at TU Delft was to improve the fatigue behaviour of aluminium as explained by Vlot [2]. At first, a laminated stack of thin aluminium sheets was developed with higher fatigue resistance than equivalent thick monolithic panels [3, 4]. Then, the adhesive was

This chapter is a modification and extension of the original paper, with permission by Bentham Science Publishers.

R. Alderliesten, *Fatigue and Fracture of Fibre Metal Laminates*, Solid Mechanics and Its Applications 236, DOI 10.1007/978-3-319-56227-8_3

reinforced with fibres. This chronology in the development process can be read in the formulation of the original patents [5, 6] that speak about metal layers bonded together with a metal adhesive system, which is reinforced by high modulus fibres.

Schijve et al. [5–7] filed the original patents on the FML concept and corresponding manufacturing method on behalf of Delft University of Technology. The patent describes the application of aramid/Kevlar fibres, but also mentions carbon and glass fibres. The laminate potentially could comprise 3–25 aluminium layers with a thickness between 0.3 and 0.7 mm, where in between the aluminium layers one or more plies of fibre-reinforced polymer could be applied in one or more directions. The aluminium sheet thickness range of 0.3–0.7 mm was added to avoid conflict with prior but not well-documented work on composite-reinforced metals by Fokker, British Aerospace and Boeing [2].

According to the patent, the fibre-reinforced polymer layers consist of parallel filaments of continuous fibres positioned straight within the ply preferably as pre-impregnated fabric. The number of plies and directions specified could indicate for example a cross-ply laminate.

As result of the manufacturing procedure in which fibres were paced between two adhesive films, the fibres in the original concept were often concentrated in the centre of the composite layer, leaving resin-rich layers at both interfaces with the metal layers [8]. The laminate concept, together with this through-thickness fibre distribution, is illustrated in Fig. 3.1a.

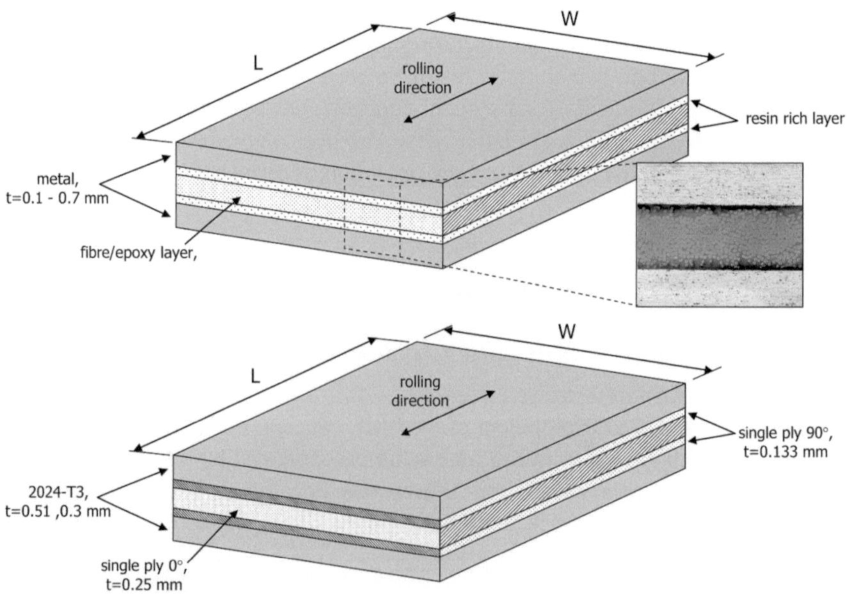

Fig. 3.1 **a** Original FML concept based upon aramid or Kevlar fibres [5–7]. **b** FML concept based on a glass/epoxy reinforcing layer with a FVF of 46% as proposed in [15]; image from [1]

The FML concepts that were developed up until 1990 consisted of laminated aluminium layers bonded by a metal adhesive reinforced with aramid fibres. The FML concept was further developed replacing the aramid fibres by glass fibres by Roebroeks [9]. The replacement related among others to the observation of aramid fibre buckling and failure in ARALL when loading under fatigue at low stress ratios. However, when Vogelesang and Roebroeks [10] introduced the application of glass fibres in the metal adhesive in 1991, they gave as reason to increase the blunt notch behaviour of the FMLs. The blunt notch strength is the strength of a material containing a blunt open hole, also often denoted as open hole strength, see Chap. 5. It was observed that for FMLs this parameter is an important parameter in design.

The basic FML concept comprising metallic layers with unidirectional fibre-reinforced polymer layers oriented in at least two directions had been introduced with the above-discussed patents. The next step in the development was taken by Roebroeks [11]. He introduced an FML containing cross-ply fibre layers between the aluminium layers, which consisted of two different fibre types. As a consequence of the loads on a fuselage skin, the stresses in longitudinal direction are different from the stresses in circumferential direction. To tailor the laminate towards the occurring loads and required fatigue lives, the concept of different fibre types within one laminate was developed.

Another FML concept was proposed by Moreton and Peel [12], who aimed to develop an FML with better fatigue initiation and crack growth properties compared to the original FMLs [5, 6] as well as better compression stability. The increase in fatigue initiation properties could be achieved by pre-straining, or post-stretching the FML, as will be discussed later, but at the expense of higher manufacturing costs. Moreton and Peel improved the FML by application of metal matrix composites containing ceramic particles or whiskers as reinforcement, instead of standard aerospace aluminium alloys.

The higher stiffness of the aluminium sheets increases the buckling stability of such FML, but in order to benefit in fatigue, it required the application of composite layers with stiffnesses at least equivalent to the aluminium. Because the authors observed that both particle and whisker reinforcement provided equal performance, they proposed to reinforce the aluminium with 1–3% lithium in proportion by weight. In practice, they worked with the 8090 aluminium lithium alloy which has a higher stiffness at a lower density compared to the standard aerospace alloys.

In reference to the original FML patent [5, 6], Hilders [13] also developed an FML concept to further increase the fatigue performance. The concept comprises four fibres layers in between the aluminium layers, which are oriented at angles of 45° with one another. The fibres are subsequently stitched together at the point where fibre layers cross each other, to form a cohesive fibre stack bridging the aluminium layers.

An observed benefit of the concept is that the resulting FML almost has an isotropic character, attributed to the quasi-isotropic lay-up of the fibres and the cohesive structure created by the stitching. In application, such a laminate requires a

3/2 lay-up in order to be symmetric with respect to its neutral axis, to avoid bending and warping after manufacturing.

The fact that the FML concept did not remain limited to aerospace applications was illustrated recently by Suarez Bermejo et al. [14] who filed a patent on a basic FML concept where steel layers are reinforced by glass fibre composite plies. They considered their FML concept a good candidate material for shipbuilding.

Because the panels can only be made in certain dimensions, the concept they proposed included connection of panels using some sort of splice configuration. Rather than proposing a splice configuration in which only metallic layers are spliced and fibre layers remain continuous, see Sect. 3.3, they propose to join FML panels stepwise in thickness interrupting both metal and composite layers.

3.2.2 Improving Impact Resistance and Tolerance

Roebroeks and Mattousch [15] introduced an FML concept as impact-resistant material where the fibres are oriented in two perpendicular directions in a symmetric lay-up, Fig. 3.1b. The configuration seems to be very similar to and to great extent covered by the original patent, but they specified that the fibres are either S2- or R-glass fibres, because of the strain rate effect. At the high strain rates during impact, the strain to failure of glass fibres increases, which is not the case for most other fibres. In their concept, the fibre layers are not woven, but continuous parallel filaments pre-impregnated as unidirectional prepregs. Cross-ply laminates are manufactured by placing two unidirectional layers perpendicular to each other. The polymeric material in [15] could be either thermoset or thermoplastic, although epoxy is preferred.

In the attempt to further increase the impact resistance of the FML concept [15], Roebroeks [16] developed a concept where various FMLs are clamped to each other at the edges, but are not connected over the main surface area. The argumentation is that during the impact sequence, the individual FMLs can move freely with respect to one another, enabling a more global deformation of the FMLs at the non-impacted side. The connection at the edges is necessary to avoid overall failure of the concept once impact occurs.

Vogelesang et al. [17] developed an armour plate concept containing ceramic layers at the impact side connected to a sublayer at the non-impact side. This sublayer is composed of an FML of which the aluminium layers are bonded together with an impregnated woven fabric. The thickness of the aluminium layers could range between 0.1 and 2 mm. One of the claimed concepts is illustrated in Fig. 3.2.

An alternative FML panel concept for high impact resistance was developed by Musaefendic [18]. The concept comprises the lay-up of metal layers and fibre layers, whereas in this solution the fibre layers are not oriented in a planar fashion, but in a three-dimensional structure with additional energy dissipating elements. The prime function of these dissipating elements is to redirect the impact load,

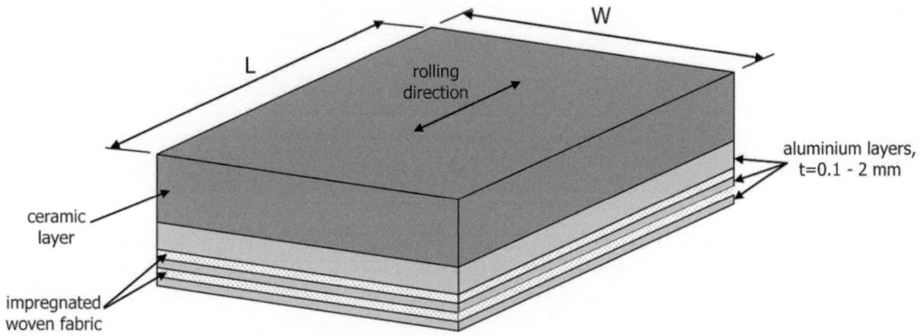

Fig. 3.2 Illustration of an armour plate concept with a ceramic layer at the impact side adhesively connected to a sublayer composed of a FML [17]; image from [1]

which is oriented perpendicularly to the panel, into the plane of the panel, creating tensile and compressive loads in the outer longitudinal layers. The concept was generically described, not limited to FML concepts containing only aluminium and glass fibres.

Gunnink and Evancho [19–21] proposed FML lay-up configurations that have only limited aluminium layers in the FML. The proposed FML lay-ups have a metal volume fraction between 0 and 47%. Both unidirectional prepregs and woven fabrics are considered applicable to the concept. The fibres types considered are primarily aramid, PBO and/or ultra-high-molecular-weight polyethylene fibres, but also glass fibres and carbon fibres are considered applicable.

The applications mentioned for these FMLs are impact-resistant objects like explosive-resistant luggage containers (see Sect. 3.7), and wing leading edges. However, considering the proposed lay-ups also liners could likely be considered.

3.2.3 Thickness Steps

The above-described FML concepts have resulted in the development of FML laminates for commercial applications, especially aircraft fuselage and wing skin applications. Because of the characteristics of these structures and the loads and stresses acting on and in them, thickness variations are needed to limit structural weight. These thickness transitions can be created in various ways.

The earlier-mentioned patent by Roebroeks [22] also describes an external ply drop-off, where the outer aluminium layer and composite ply underneath end at a certain location, see Fig. 3.3a.The concept comprises prepreg layers that terminate while the outer aluminium layer is extended a defined length over that prepreg run-out. The gap is filled with additional adhesive that under the applied curing pressure will squeeze out if not confined. The outer aluminium layer will form itself over the prepreg run-out following a self-forming concept.

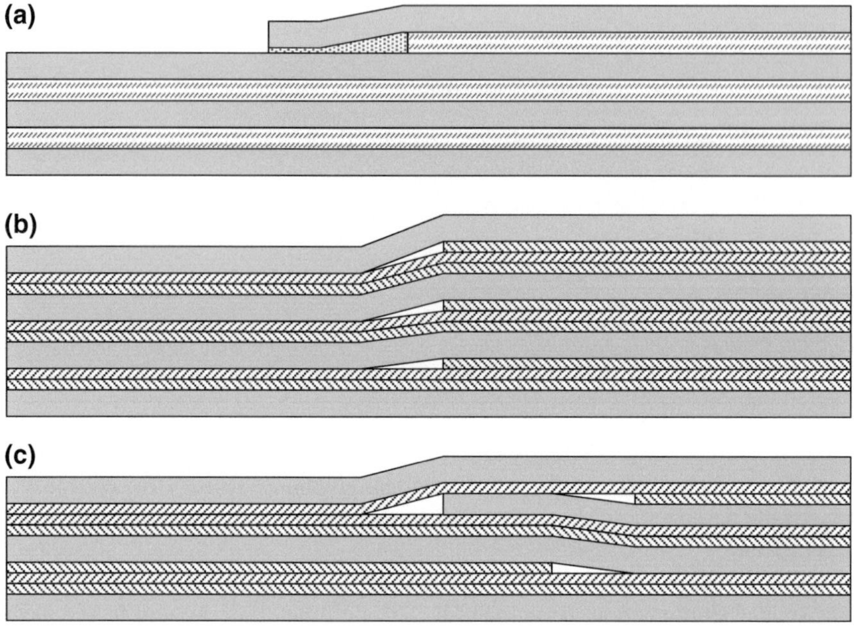

Fig. 3.3 Illustration of an external ply drop-off (**a**) [22], a discontinuity in fibre/epoxy plies only (**b**) and additionally in one aluminium layer (**c**) [23]; image from [1]

An alternative thickness transition was introduced by Rajabali and Ter Steeg [23] based on discontinuous internal layers, often denoted as interlaminar ply drop-offs. The concept is illustrated in Fig. 3.3b, where within the laminate one of more prepreg layers are terminated, while the remaining prepreg and aluminium layers are continued. The concept in Fig. 3.3b gives not only a transition in thickness of the laminate, but also implies changing the grade of the FML. The various FML grades that have been introduced for the aluminium/glass FML GLARE are defined based upon the lay-up. Reducing the number of prepreg plies between two aluminium layers means that the grade changes.

In addition to the concept with discontinuous prepreg layers, Rajabali and Ter Steeg developed a concept where at least one aluminium layer ends within the laminate. An illustration of this concept is given in Fig. 3.3c. In this figure, it can be seen that throughout the laminate prepreg layers are discontinuous in order to keep the overall laminate lay-up at both sides of the transitions equal to the standardized and qualified FML grades.

A concept that combines two different metals within the laminate has been developed by Van Rooijen and Van der Zwaag [24]. The concept comprises an FML with at least two aluminium layers that locally at the edges is reinforced by a strip of metal with a higher elastic modulus, such as stainless steel. The reason for the application of the additional metallic strip is to increase the bearing strength, which might be required for a mechanically fastened joint.

3.2.4 Thick Panel Concepts for Lower Wing Covers

Gunnink [25] introduced a composite laminate with a thick aluminium plate in between FMLs (See Fig. 3.4). This concept was developed for lower wing skin panels, which usually consist of thicker skin panels compared to the fuselage skins. Note that the sublayer in the armour plate concept [17] consisted of an FML with potentially different layers; the thick aluminium layer in the FML in Fig. 3.2 could be up to 2 mm, whereas the other layers could be as thin as 0.1 mm. Although the difference might seem small between the two concepts, the innovation of the concept of Gunnink is in bonding thick tapered aluminium layers to an FML with a fibre-reinforced adhesive (prepreg) instead of with adhesive.

Recently, Roebroeks and Gunnink [26] introduced another similar concept consisting of thick aluminium layers connected to FMLs. They claim that a similar concept can also be applied to fuselage skin panels by reducing the aluminium layer thicknesses. The concepts are illustrated in Fig. 3.5a, b, where the outer aluminium layers are bonded by glass fibre epoxy prepreg to FML straps that are oriented perpendicular to the glass fibre orientation. The straps are made of FML using a Zylon fibre (PBO type) with a higher stiffness than glass fibre-based FMLs. The post-stretching technique that will be discussed later in this chapter could be applied to induce a favourable residual stress system. In case of lower wing skin panels, the thick outer aluminium layers are bonded by an epoxy resin to the FML straps, which for this configuration contain more layers.

Because the FML straps are being constrained within the outer aluminium layers, they envisioned that the discontinuity between the individual straps will not have a negative effect on the overall laminate properties. Although that might be the case, one certainly could expect design constraints, in particular related to positioning mechanical fasteners.

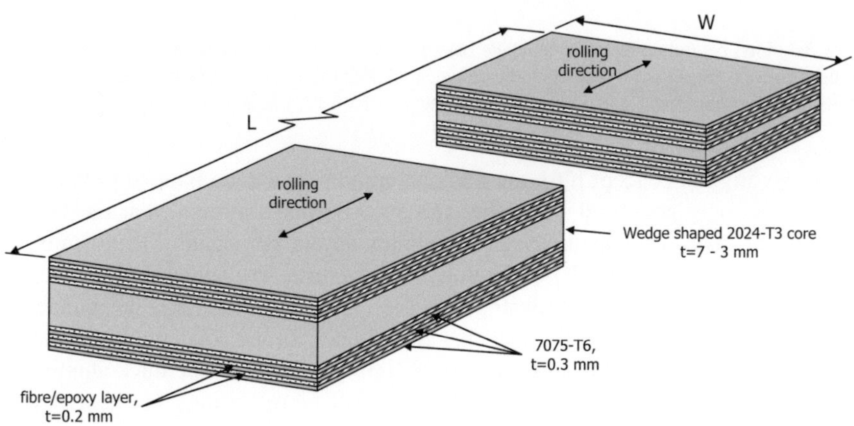

Fig. 3.4 Tapered lower wing skin concept [25]; image from [1]

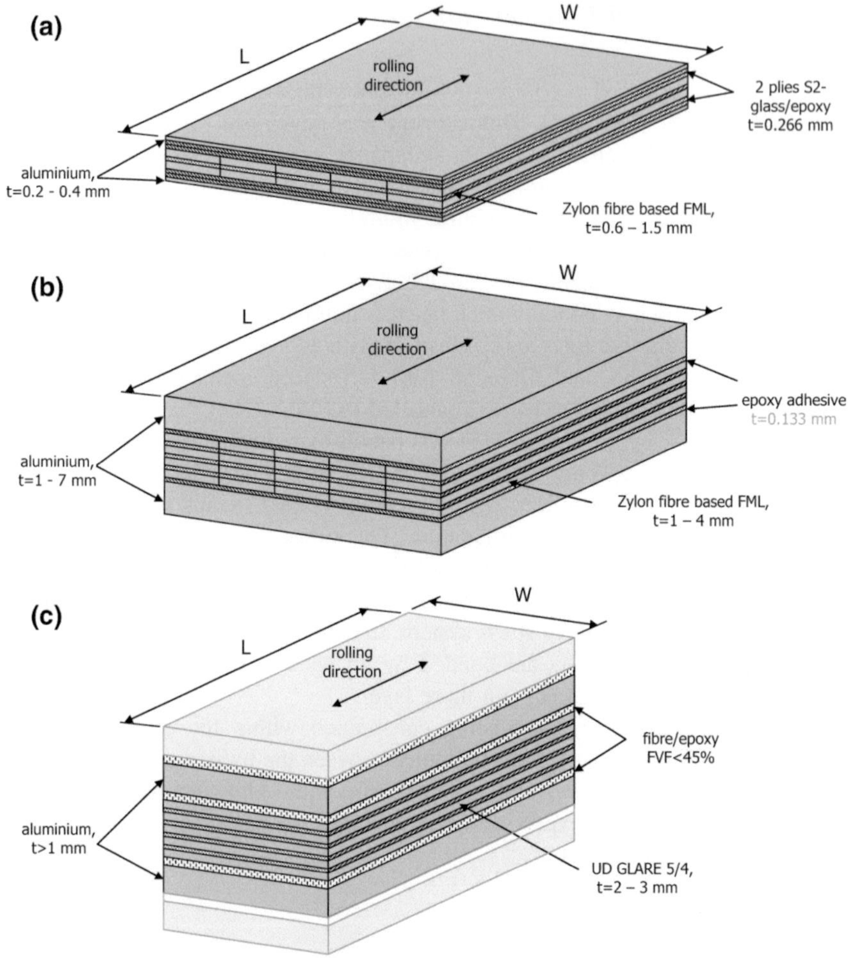

Fig. 3.5 First central concept for fuselage (**a**) and lower wing skin panels (**b**) [26] and modified central concept (**c**); *light grey* layers indicate potential addition of layers at both sides bonded with the same fibre-reinforced epoxy systems with reduced FVF [27]; image from [1]

In the original concept [26], the thick aluminium layers are bonded to the FML straps with only epoxy adhesive. Fatigue experiments illustrated that the fatigue cracks that may occur in the outer aluminium layers propagate directly into the outer layers of the straps. To avoid this, reinforcement was added to the adhesive using high strength fibres, as illustrated in Fig. 3.5a. However, large delaminations at the interface may then still reduce the effectiveness of the concept. As a solution to this problem, Roebroeks and Gunnink [27] propose to bond the thick aluminium layers to the straps using prepreg with a reduced Fibre Volume Fraction (FVF). They explain that the reduced FVF may be obtained by adding a non-reinforced resin layer to a standard prepreg with high FVF.

Although not necessary limited to thick lower wing panel covers, the concept proposed by Alderliesten and Benedictus [28] does apply to the lower wing cover structure. In this concept, a hybrid laminate is proposed that utilized two fibre-reinforced polymer systems of which one may be the state-of-the-art glass fibre-reinforced prepreg, but where the other composite layers are made of high-stiffness fibre reinforcements. Commonly, the lower wing cover FMLs are unidirectional laminates, because of the dominant load case imposed by wing bending. The torsional resistance of the wing box requires tailoring the shear stiffness of the lower wing cover, which is usually low for unidirectional laminate. Hence, the concept proposed in [28] uses high-stiffness fibre layers near the laminate mid-plane but oriented under (positive and negative) angles with respect to the wing span direction. The outer glass fibre layers remain unidirectional in the wing span direction to resist against wing bending. The major advantage of this concept is that the torsional and bending resistance can be tailored individually, by selecting the appropriate angles for the high-stiffness fibres.

3.2.5 Alternative Fuselage Skin Concepts

In general, hybrid FML concepts for fuselage skin applications are rather thin, while for lower wing covers the laminates are often rather thick. An exception to that may be the concept proposed by Alderliesten and Conen [29], illustrated in Fig. 3.6.

The concept comprises a laminate containing magnesium layers near the mid-plane of the laminate with a higher thickness than the aluminium layers in FMLs currently standardized. Because of the lower density of magnesium compared to that of aluminium, thicker magnesium layers can be considered. The range in thicknesses proposed in [29] is 0.2–1.0 mm for the aluminium, with 0.4–2.0 mm for the magnesium layers. With this concept, FMLs can be designed that have an equivalent weight compared to standard FMLs like GLARE, but with higher compression and shear stability properties.

In the concept, the outer metal layers are made of aluminium. With experiments, it was demonstrated that the fatigue crack growth characteristics remain similar to

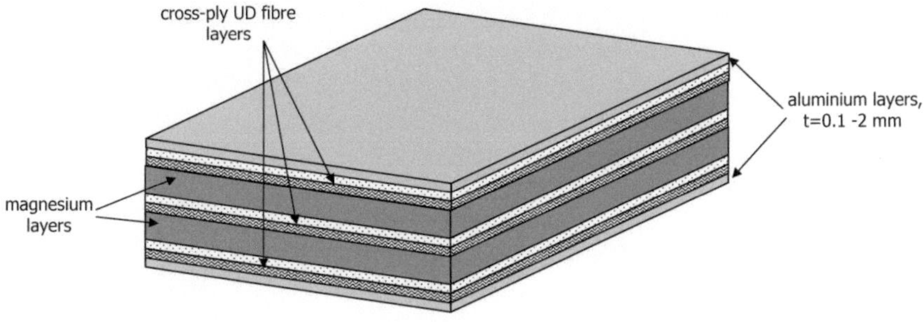

Fig. 3.6 Illustration of an FML concept utilizing inner magnesium layers with outer aluminium layers, representing a concept with excellent compression and shear stability properties [29]

the standard GLARE laminates. The fact that the fatigue properties and crack growth resistance of the magnesium are less compared to the aluminium does not impose a problem in the concept, because magnesium bears less load as result of its lower stiffness compared to aluminium. Hence, fatigue cracks in the outer aluminium layers propagate slightly faster than the cracks in the magnesium layers, which is preferred for reasons of inspectability.

The reason to maintain aluminium as the material for the outer metal layers is related to the fire properties of magnesium. With the concept representing a good candidate for lower fuselage sections, pool fires may impose the risk of igniting the outer FL layer if it is made of magnesium. In the proposed concept, the magnesium layers are protected predominantly by the subsurface fibre epoxy layers, which substantially increase the burn-through resistance of FMLs.

FML-like GLARE and ARALL exhibit in general laminate stiffnesses that are lower compared to the stiffness of the monolithic aluminium. This means that fuselage skin structures made of GLARE will have superior fatigue properties compared to monolithic aluminium, but that they may impose fatigue in longitudinal stringers or circumferential frames if these are made from aluminium. To avoid fatigue in the stringers, the stringers must be made of FMLs as well, as discussed in Sect. 3.6.

An alternative FML concept that allows for monolithic aluminium back-up structures was proposed by Alderliesten et al. [30]. In this hybrid concept, the glass fibre prepreg layers underneath the stringers or frames can be replaced locally by straps of high-stiffness fibre layers (See Fig. 3.7). This will increase the effective

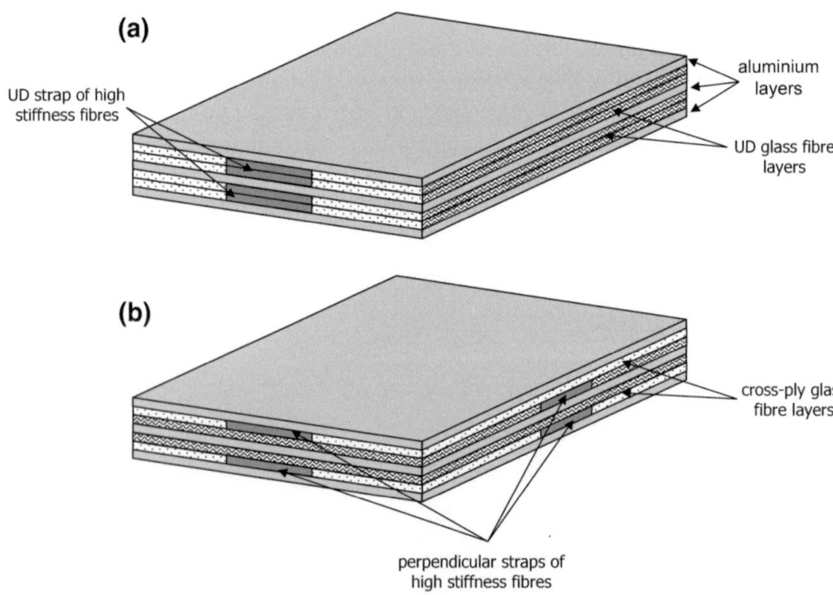

Fig. 3.7 Illustration of local replacement of the UD glass fibre prepreg by a strap of high-stiffness fibres in a unidirectional laminate (**a**) and a cross-ply laminate (**b**) [29]

stiffness of the FML skin, which makes the FML skin more compliant to the monolithic aluminium stiffener. Experiments have demonstrated that the fatigue performance of the monolithic aluminium stringer is substantially improved by the presence of the stiffer straps embedded in the FML underneath the stringer.

3.3 Splicing Concepts

The patents discussed in previous sections were limited to flat sheet material, constrained by the maximum dimensions in which the 0.3–0.5-mm aluminium sheets were available. To obtain larger panels without mechanically fastening small panels, Garesché et al. [31] introduced flat butt-splice as illustrated in Fig. 3.8a.

The gaps between the sheets, about 1 mm, were staggered to limit the reduction in laminate strength. However, further research revealed delamination damage that started at the gap between the outer sheets. To avoid this problem, additional doublers had to be bonded in a second cure cycle over the gaps, to reduce the local stresses. The splice concept containing the additional doubler was proposed by Pettit [32] and is illustrated in Fig. 3.8b.

Another splice concept was introduced by Roebroeks [22] which utilizes the so-called self-forming technique. The high pressures in the autoclave during curing form the thin sheets over internal and external doublers, as illustrated in Fig. 3.8c. This self-forming technique was further explored by optimizing the splice configuration to the so-called overlap splice, also introduced by Roebroeks [33], see Fig. 3.8d. This configuration exhibited various advantages over the older splice configuration, with most important the substantially wider tolerances in the sheet dimensions and the positioning of the sheets.

An alternative solution to the splice solution was proposed by Labordus et al. [34] who introduced the butt-welded splice, see Fig. 3.8e. In their concept, the thin sheets are welded together to larger sheets before being laminated into the panel. This configuration had two advantages over the splice configuration: it provided complete protection of the fibre-reinforced epoxy layers from the environment, and it eliminated the risk of the before-mentioned delaminations.

This welding concept introduces some manufacturing challenges related to pre-treatment of the aluminium layers. If the sheets are welded first, then handling of the very large thin sheets becomes problematic. However, applying the pre-treatment before welding requires removal of the primer layer at the edges to be welded, and as second step the local pre-treatment of the welded area.

Beumler [35] proposed to further modify the optimized splice configuration of Roebroeks [33]. The modification comprises an additional single ply of fibre-reinforced polymer at the outside of the splice to cover the resin-rich area between the two overlapping outer aluminium layers. The resin easily cracks at the end of the terminating aluminium layer, imposing a crack starter into the adhesive joint between the two overlapping layers. Adding the cover ply prevents this adhesive crack formation, significantly extending the durability of the splice joint.

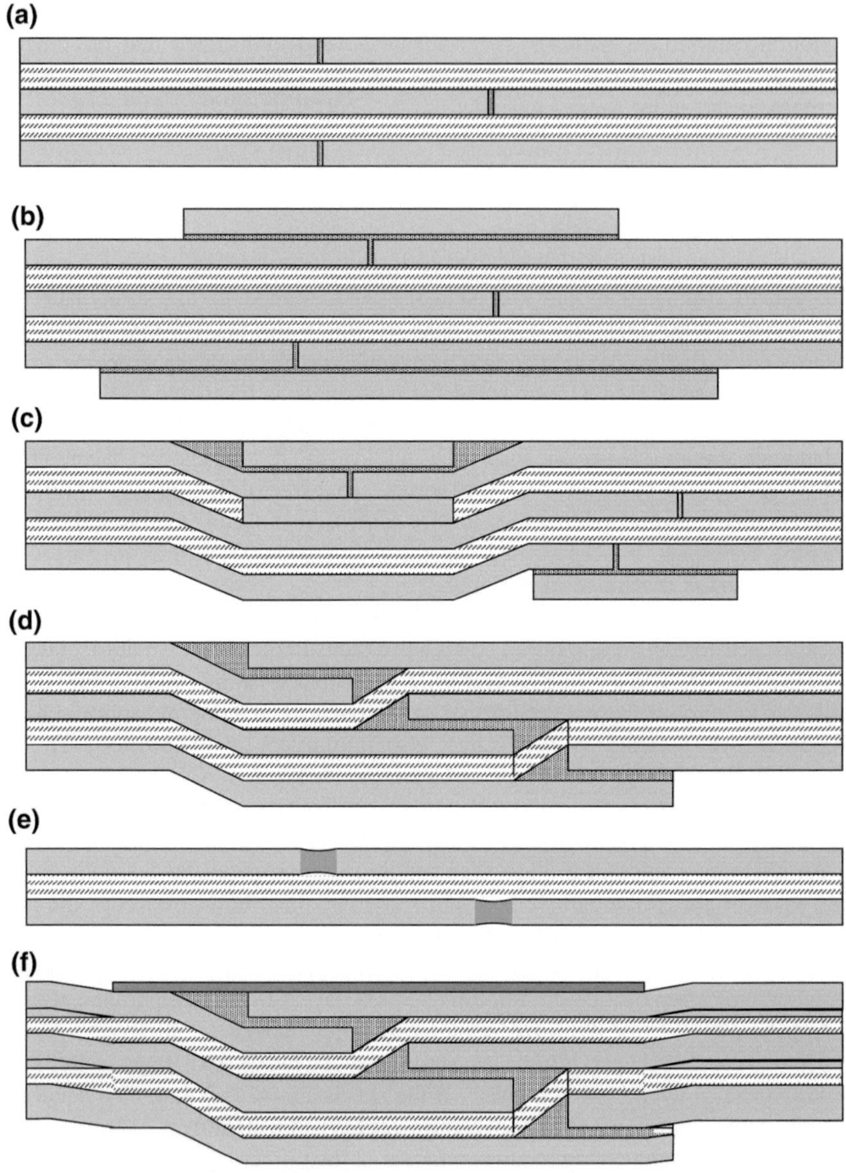

Fig. 3.8 Concepts to build up a FML panel out of smaller metal sheets: butt-joint (**a**) [31], doubler lateral strap-reinforced butt-joint (**b**) [32], splice doubler joint (**c**) [22], overlap splice (**d**) [33], butt-welded joint (**e**) [34] and modified overlap splice joint (**f**) [35]; image adapted from [1]

Although different from the splice configurations illustrated in Fig. 3.8, one may consider the adhesively bonded longitudinal overlap splice between two fuselage shells a variation to the theme. The carbon fibre-reinforced polymer upper shell structure can be joined to an FML lower shell using a longitudinal overlap as

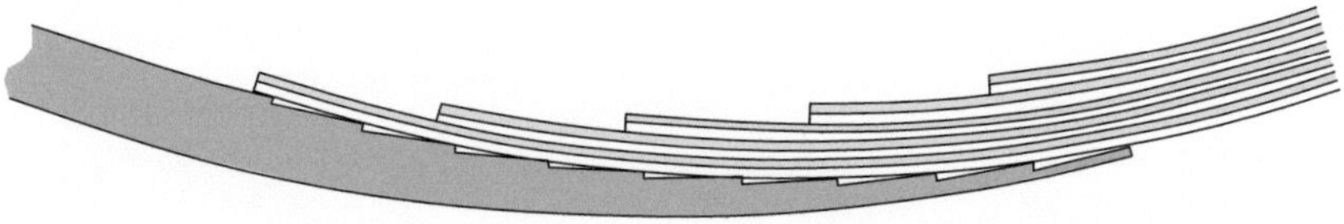

Fig. 3.9 Illustration of an adhesive joint between a carbon fibre composite fuselage shell (*left*) and an FML lower fuselage shell structure (*right*), providing electrical insulation to avoid galvanic corrosion [36]

illustrated in Fig. 3.9. The benefit of this concept, proposed by Beumler [36], is that the adhesive joint between the composite and FML panel provides an electrical insulation that avoids the risk of galvanic corrosion.

3.4 Manufacturing Aspects

3.4.1 Post-stretching Panels After Curing

The curing process of FMLs consists of a thermal cycle in an autoclave in which the FML is heated up to the curing temperature, after which it will cool down to operating- or room temperature. The differences in thermal expansion coefficients shrink the aluminium layers further than the composite layers, inducing tensile residual stresses in the aluminium and compressive stresses in the composite layers. For fatigue, this residual stress system is unfavourable, because the actual stress cycles in the aluminium layers are higher than the applied load cycles, initiating cracks earlier than in monolithic aluminium panels.

This undesired stress system can be changed or even reversed with the post-stretch technique. The principle of stretching the laminate until the aluminium layers plastically deform has been described in the original FML patent [6, 7]. The most common way to stretch the laminate is by clamping both ends and applying a load to the panel until the metal layers yield. With this stretching technique, one should avoid failure at the clamps, for instance by applying tabs to the material, see Fig. 3.10a.

Vogelesang et al. [37] developed a technique to stretch the laminate in a continuous manner as illustrated in Fig. 3.10b. The technique consists of rolling the as-cured laminate through a mill with a slit width smaller than the laminate thickness. This induces a permanent elongation to the FML. The percentage of stretching can be selected by changing the force with which the mills are pressed upon the laminate. Although the main patent describes the application of roll stretching to laminates with a single thickness, it also mentioned roll stretching the tapered laminates as proposed by Gunnink, illustrated in Fig. 3.4. It is interesting that Roebroeks [38] proposed the same technique again with the only difference that this technique is also applied to laminates for which the thickness varies due to discontinuous layers, see Fig. 3.3.

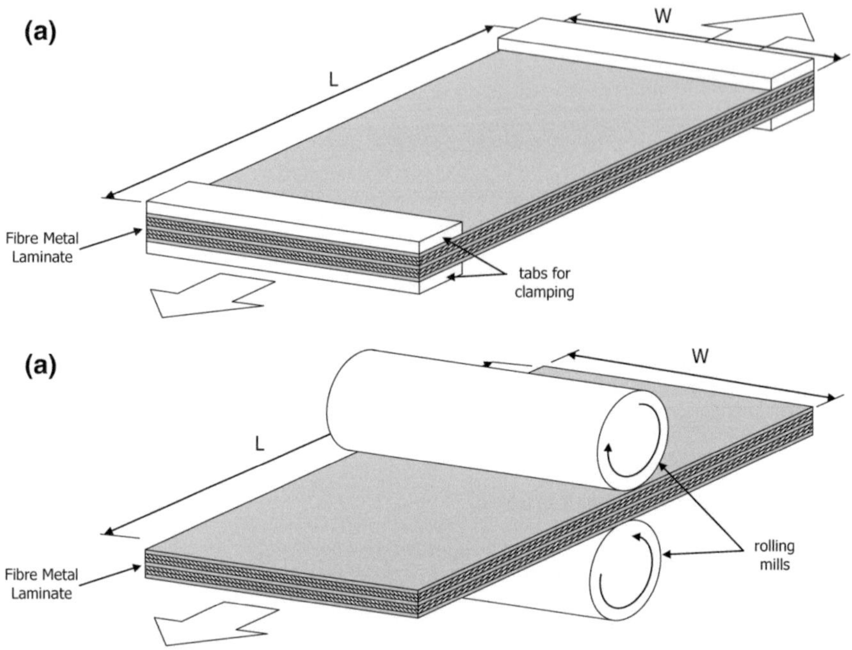

Fig. 3.10 Stretching flat laminated panels by clamping both ends (**a**) and stretching the laminated panel using a roll stretch technique (**b**) [21]; image from [1]

The general disadvantage considered for the post-stretching procedure is the applicability to predominantly unidirectional FMLs. The post-stretch axis is oriented parallel to the fibre direction and maintaining elastic elongation in the fibres, whereas the aluminium layers will plastically deform. However, for biaxially loaded applications, like for example fuselage skin structures, the post-stretch procedure implies the favourable residual stress state can only be created in one direction, even worsening the stress state perpendicular to that direction due to the lateral contraction. Pettit [39] proposed a solution in which the post-stretch axis is oriented at an angle with respect to both primary fibre axes. In this case, the fibres in both directions are stretched, to some extent limited by the induced shear deformation, creating a beneficial stress states in both fibre directions.

3.4.2 Pre-stretching Panels During Curing

The post-stretching techniques discussed in the previous section are applied after curing the laminate. An alternative method was developed by Roebroeks [40], which consists of a reinforced mould that is partially constrained against expanding during the cure cycle.

In the standard cure cycle, all layers expand due to the increase of temperature to the curing temperature. Once the polymer chains start interlocking, the layers become fixed. Hence, the residual stresses develop when the temperature is brought back to room- or operating temperature [41]. The method proposed by Roebroeks [40] prevents the constituents or the moulds from expanding during the increase of temperature in the first phase of the cure cycle. This impedance of strain may induce residual stresses at the end of the cure cycle that are more favourable or even reversed, depending on the actual method of constraining.

3.4.3 Lay-up and Curing Concepts

While curing FMLs at elevated temperatures (often 120 °C or 175 °C), the constituent plies may float a little with respect to each other, because the polymers become fluid. Controlling the position of all individual plies becomes an engineering challenge, in particular for thick laminate stacks and curved moulds. Rajabali [42] proposed a positioning pen to fix the stack in the mould as a solution to this challenge, see Fig. 3.11a.

There are various considerations for choosing the appropriate location of the pen. One prefers not to damage the panel by drilling a hole anywhere in the panel. Therefore, it is proposed to select the pen location such that it is positioned at a location in the panel where in a subsequent manufacturing step larger holes will be drilled for fastening purposes or where a cut-out will be milled. A variation to the theme of using pins in the manufacturing phase to fix the position of individual layers with respect to a forming jig is described in another patent by Rajabali [43].

The concept of mechanically fastening FML skin panels is similar to monolithic aluminium. The performance of the joints can be tailored by creating a stepwise joint to reduce the amount of secondary bending and to better distribute the load transfer over the fastener rows, see Fig. 3.11b. To accommodate such a stepwise joint, FML panels must be manufactured containing a rebate. Rajabali [44] proposed a method for manufacturing such laminates, which uses an auxiliary tool built up from the same constituents and number of plies as the laminate, as illustrated in Fig. 3.11c. Later that year, Rajabali and Ter Steeg [45] filed a second approach to manufacture such laminates using a different auxiliary tool to mutually offset the layers, see Fig. 3.11d.

Discontinuous plies like interlaminar run-outs or splices lead to panel thickness variations. If additional reinforcement layers, such as external doublers, need to be bonded on a basic laminate skin panel, then these doublers must be laminated over a curved contour. Because bonding a flat laminate on top of a curved contour can be problematic, Kroon and Roebroeks [46] proposed a method to solve this problem. In their proposal, the laminate is bonded on top of the curved contour at elevated temperature and pressure. The laminate will deform more easily, forming itself over the curved contour. This concept seems to be a bit arbitrary, because in most cases the subsequent bonding cycles require elevated temperature and pressure.

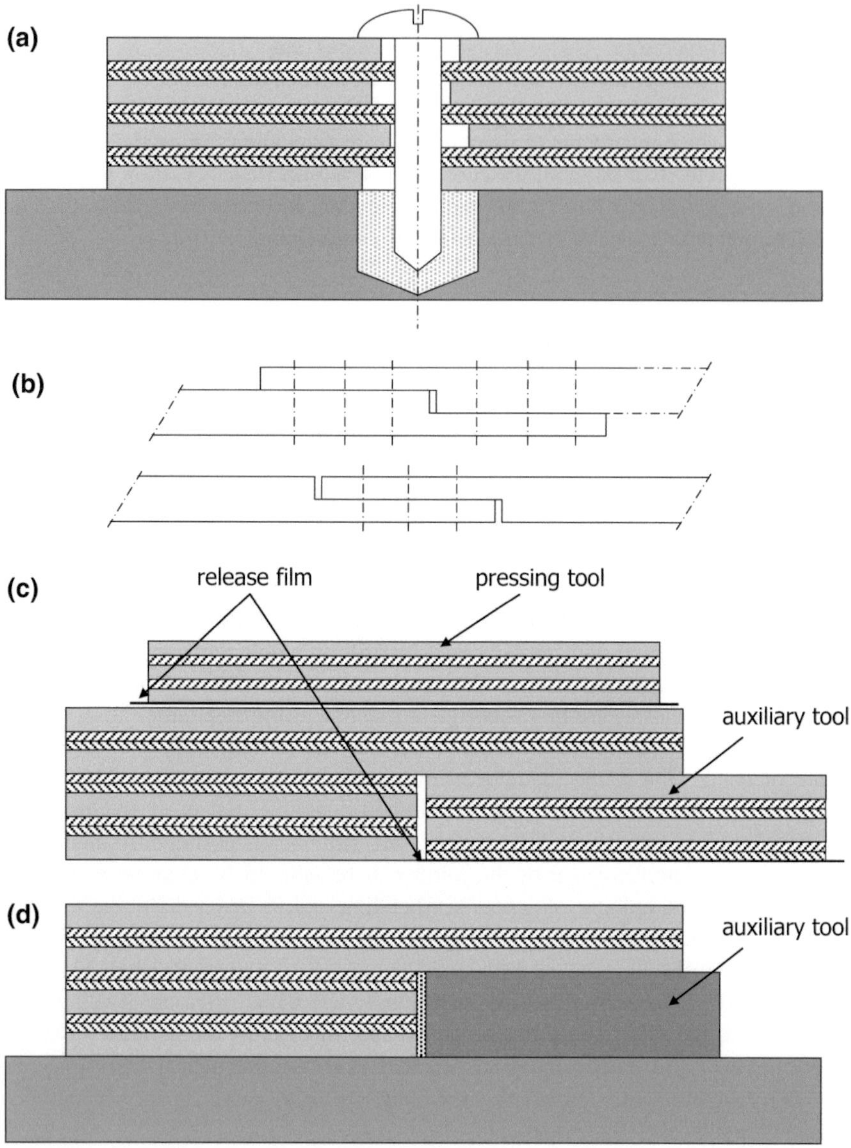

Fig. 3.11 Positioning pen as fixing means in a mould (**a**) [42], stepped mechanically fastened joints (**b**) and two methods to manufacture laminates with such thickness step (**c**) [44] and (**d**) [45]; image from [1]

Most FML panels are laminated in a mould and cured in an autoclave. Despite the availability of very large autoclaves, this means that the available autoclave restricts the maximum panel dimensions. In addition, the production of FML panels remains batch production.

Roebroeks [47] patented a method to make FMLs in a continuous process. Such a process allows for manufacturing FML panels in large volumes for potentially other applications than air- and spacecraft, and reducing the manufacturing costs. Similar to the state-of-the-art production processes, it consists of anodizing the aluminium layers, but instead of the application of adhesive primer, an adhesion layer is added to the metal sheets. After this process, the sheets can either be coiled up again and transported to other production facilities, or they can be fed directly into a continuous process where the sheets are bonded to fibre-reinforced polymer plies. The process of coiling-up after the treatment and subsequent transport to another facility was identified as solution for large scale production, potentially leading to further laminate cost reduction.

To apply the necessary pressure onto the laminate, either the regular process with an autoclave could be used, making the process discontinuous again, or the earlier-mentioned rolling mills could be applied while the laminate is processed at elevated temperatures.

In the standard autoclave cure cycle, a significant amount of time is required to heat up the surrounding air in the autoclave in order to increase the temperature of the laminate to the curing temperature (often 120 °C or 175 °C). Roebroeks and Kroon [48] proposed an alternative method and device for curing composites in an autoclave which does not require heating up the air in the autoclave. The device described in [48] comprises a membrane directly connected to the product that must be heated up. The loss of energy related to heating up the autoclave air first is excluded, and the time needed to heat up the product is significantly reduced. Hence, the processes are cheaper and faster.

More recently, Heinimann et al. [49] proposed a manufacturing procedure for the hybrid concepts in [26, 27], in which the lower thick aluminium layer of the laminate fulfils the function of the mould for the remaining laminate to be cured. Considering the concepts in [26, 27], this seems to be the main innovation of [49] with respect to manufacturing laminates, because the other manufacturing aspects are similar to previous concepts.

To cure FMLs without the need for a large autoclave, Alderliesten and Grashof [50] proposed a process in which FMLs are cured in an oven after vacuum pressure is applied to the panel using two vacuum bags. Although research and development of out-of-autoclave manufacturing processes generally aim to develop polymer systems that are compliant to such a process, initial studies on application of the process to standard GLARE demonstrated that even these FMLs can be made with equivalent quality without the need of an autoclave.

3.4.4 Alternative Impregnation Processes

The majority of FMLs developed and studied were made by either applying pre-impregnated fibre layers between metal sheets using hand lay-up, or by placing two sheets of resin at both sides of dry fibres between the metal sheets.

However, inspired by the development in the world of fibre-reinforced polymer composites, similar processes were proposed and developed for FMLs. An example is given by Labordus and Van Tooren [51], who proposed the application of dry fibres between the metal layers, injecting the resin at a later stage during the production process. The process is proposed for both manufacturing flat panels and for manufacturing curved laminates. The latter is applied for manufacturing stiffener profiles, as is discussed in Sect. 3.6.

Interestingly enough, a similar process was proposed by Cano et al. [52] with some minor amendments to the above-mentioned process. The issue with the resin injection is the required viscosity of the resin to fill up the volume between the metal sheets. In the case of FMLs, the resin can only distribute in two directions. In particular, for larger panels the resin viscosity limits the application of such manufacturing process. The modifications proposed by Cano et al. are to apply perforated metal sheets, and hence sheets contain many small holes. These holes allow transfer of resin not only in planar direction, but also in thickness direction. However, the application of perforated metal sheets obviously contradicts the damage tolerant material philosophy that has driven FML development up until today.

3.5 Design of Fuselage Panels

3.5.1 General Fuselage Panel Concepts

In the current fuselage skin structure of the A380, the splice configuration illustrated in Fig. 3.8d is applied. Especially for wide-body aircraft where the required skin thickness implies the application of multiple metal sheets, this splice configuration consists of a panel region with increased thickness, equivalent to at least a single metal sheet.

However, for single-aisle type of aircraft fuselages, the required skin thickness is fairly small, in the order of FMLs with a 2/1 or 3/2 lay-up. For given thicknesses, the crack-stopping capability required for fatigue crack growth and residual strength justification may locally require additional thickness. To combine both requirements into a panel concept, Popp and Beumler [53] proposed another variant to the splice configuration. They propose to apply the splice overlap for each particular metal sheet in the FML underneath the stiffeners, at which location additional thickness is required for crack-stopping capability, as illustrated in Fig. 3.12b.

The above-mentioned A380 contains fuselage skin panels made of FMLs, in particular made of GLARE. However, not the entire fuselage is made of FMLs, only the upper sections of the front and aft fuselage barrels. The reason is that for the upper fuselage sections the fatigue and damage tolerance requirements are predominant, whereas the lower fuselage sections are often designed for stability,

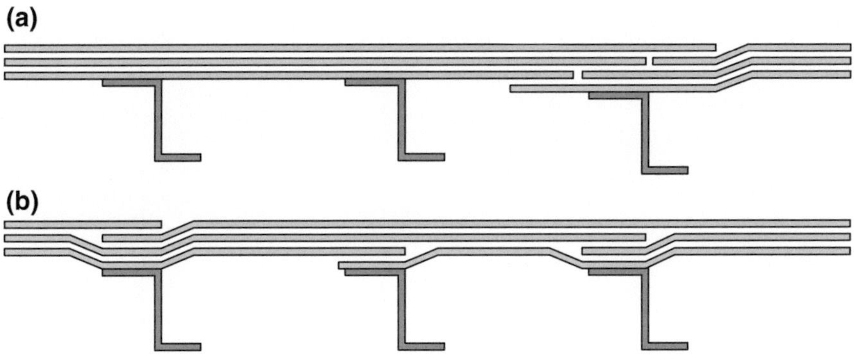

Fig. 3.12 Conventional application of overlap splice configuration in fuselage skin structures (**a**) and an alternative overlap spliced panel configuration as proposed by Popp and Beumler [53] (**b**)

which can easily be achieved with monolithic aluminium panels stiffened with laser beam welded longitudinal stringers.

However, in another filed patent, Haack and Beumler [54] propose to apply GLARE, over the complete circumference of certain barrel sections of the fuselage structure. By doing so, they claim to obtain fire safety zones to which the passengers can escape in the event of fire. The GLARE laminates have proven to be highly fire resistant, offering burn-through times over 4 min, significantly more than the 90 s required for evacuating an entire aircraft.

3.5.2 Interlaminar Reinforcements and Inserts

To design fuselage skin panels with the FML concept required solutions to many detailed issues. For instance, to optimize the panels with respect to the occurring running loads and the weight, thickness variations are necessary. These thickness steps led to the external and interlaminar ply drop-offs and related concepts [22, 23]. However, these concepts often generically describe thickness steps along one direction, while in skin panels thickness variations occur in two directions in the plane of the panel. Hence, these generic concepts have been further developed into detailed design aspects, of which some have been filed as patents.

In this context, Rajabali and Ter Steeg [55] filed a first patent. The overlap splice concept [33] in the longitudinal direction of a fuselage skin panel imposes thickness variations in the circumferential direction. This thickness variation becomes problematic in design, when mechanically fastening such a panel to another panel using a circumferential butt-joint. The concept proposed by Rajabali and Ter Steeg comprises the use of additional fill layers that are placed in the laminate outside the splice area. These fill layers create a single thickness in circumferential direction of

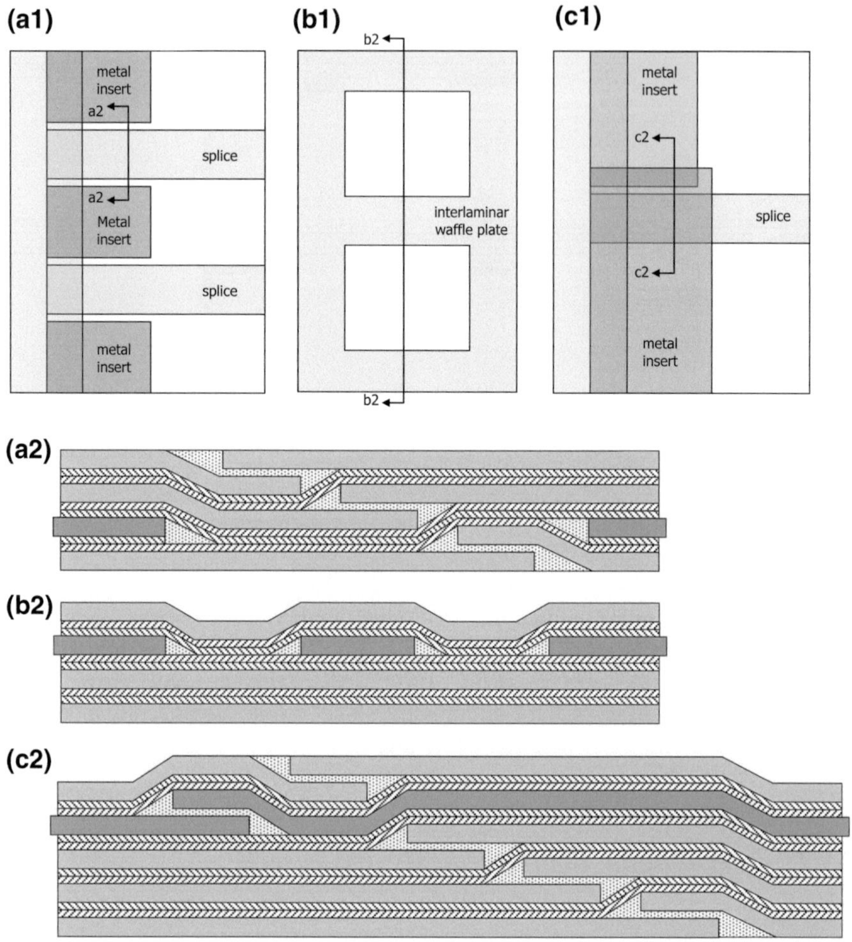

Fig. 3.13 Illustration of additional fill layer to balance the laminate thickness (**a**) [55], local interlaminar reinforcements (**b**) [56] and reinforcement in the joint area (**c**) [57]; image from [1]

the fuselage panel. The concept is illustrated in Fig. 3.13a. Note that in longitudinal direction the added fill layers end as interlaminar ply drop-off within the laminate.

Another concept proposed by Rajabali and Markestein [56] also contains local interlaminar reinforcements, but for different reasons. In case of very thin skin panel thicknesses, additional reinforcement may be required at the locations where for instance reinforcing elements, like stringers, clips or frames, are added. This reinforcement can be added as a so-called waffle-plate, see Fig. 3.13b1. This means that throughout the panel multiple interlaminar reinforcements are added as illustrated in Fig. 3.13b2.

If necessary, the mechanically fastened joint area could be further reinforced by adding additional metallic layers similar to Fig. 3.13a. These layers will then overlap, similar to the other layers within the splice area. Rajabali and Ter Steeg [57] proposed various concepts for adding interlaminar metal and fibre layers to reinforce the mechanically fastened joint area in an FML panel. One of the concepts is illustrated in Fig. 3.13c.

The configurations in Fig. 3.13 illustrate that there are multiple variations that can be developed using the principles shown in these concepts. This can be further illustrated with the concept developed by Martin et al. [58] containing similar structural elements in an FML fuselage skin panel. The principles for the overlap splice configuration as well as the application of external doublers are applied similar to earlier developed concepts.

3.5.3 *Special Design Features*

One of the major benefits concerning fatigue and damage tolerance with FML skin structures is the exploitation of adhesive bonding technology. The increased thickness required for specific panel areas is obtained by bonding additional doublers to the panel, rather than using mechanical fastening. The reduced rivet usage decreases the number of fatigue critical areas in the structure, reducing the maintenance burden and contributing to the damage tolerance design philosophy.

Another example of the exploitation of adhesive bonding technology is provided by Pellenkoft and Gennai [59], who propose adhesively bonding window frames to FML skin structures. The concept of bonded window frames in FML skin structures was developed by the authors in conjunction with the Dutch FML technology partners.

The bonded window frame concept does not only make mechanical fastening obsolete, further increasing the damage tolerance characteristics of given structures, but it also complies with all requirements coming from static load cases, including the (shear) buckling cases. For stability considerations, it is known that bonded structures often increase the strength.

3.6 Design of Panel Stiffening Elements

In the development of FML skin panels, it was observed that monolithic aluminium stringers may lead to insufficient fatigue performance. The higher stiffness of the aluminium stringers would attract more load from the FML skin panels, creating fatigue problems in the stringers rather than in the skin. Stiffening elements with an elastic modulus similar to the skin panel or with significantly increased fatigue resistance were thus required.

This necessity has led to the development of FML stringers, which are proposed in [60] together with the description of a manufacturing procedure. The concept comprises pre-forming laminates with a 2/1 lay-up (single fibre-reinforced polymer between two metal layers) and afterwards bonding them together with either polymer adhesive or fibre-reinforced polymer, Fig. 3.14a. Later, Livi and Puccini [61] filed a different concept to manufacture FML skin panels reinforced with FML stringers. They proposed to first lay-up the complete laminate stack and pre-form it before curing, see Fig. 3.14b.

A couple of years later, two patents were filed parallel to each other describing the application of Zylon fibres in fuselage skin stringers. Beumler [62] and Roebroeks et al. [63] both propose the concept of a skin panel comprised of an FML containing Zylon fibres. Roebroeks et al. additionally describe the manufacturing procedure, which is quite similar to the original procedure [60]; however,

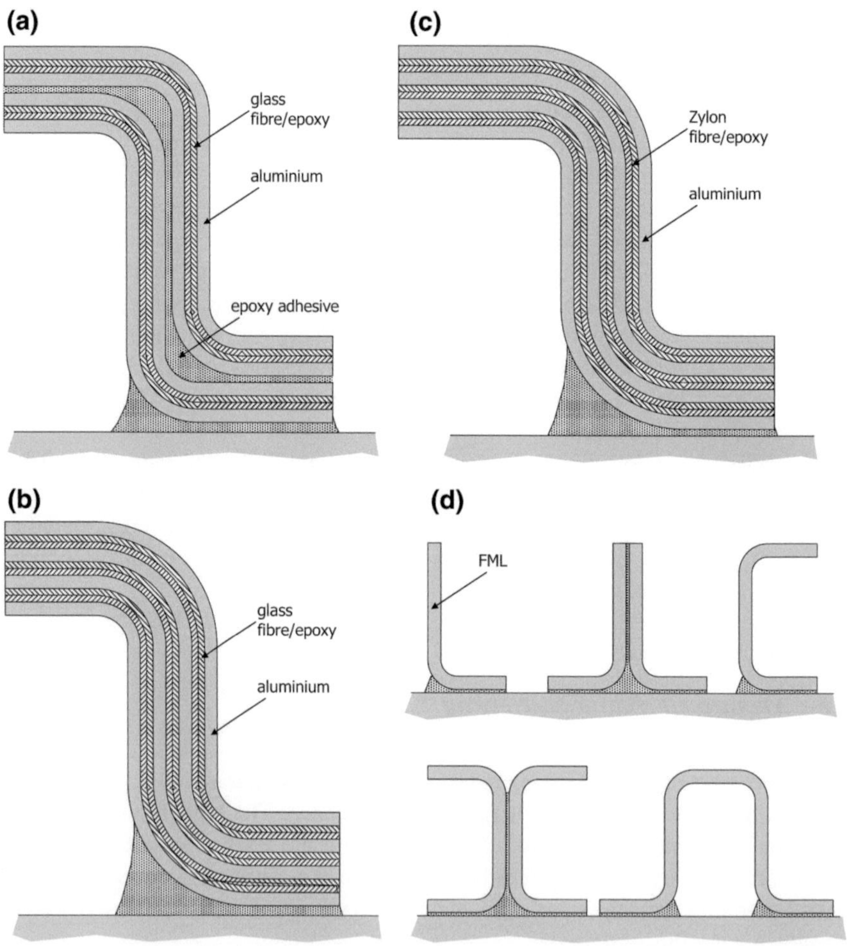

Fig. 3.14 Illustration of the FML stringer configuration described in [60] (**a**), the configuration described in [61] (**b**), the FML stringer based on Zylon fibres (**c**) [62] and a selection of FML stringer profiles described in [63] (**d**); image from [1]

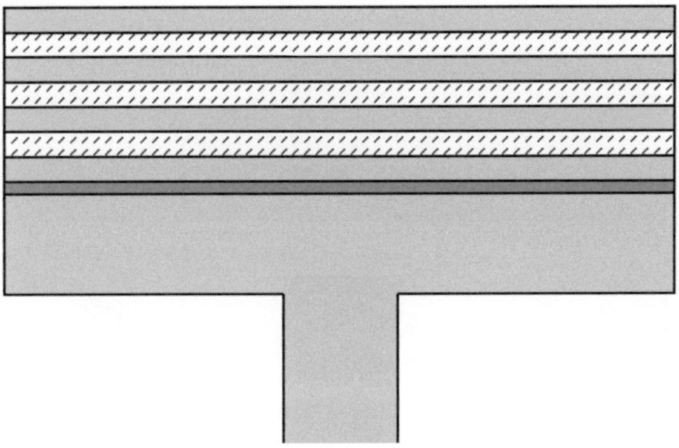

Fig. 3.15 Illustration of the frame flange (*lower*) reinforced with an adhesively bonded FML strap (*upper*) [64]

the stretching procedure from [37] has been added to create the favourable residual stress distribution within the individual layers.

Another patent related to stiffening shell structures is the frame reinforcement proposed by Ohrloff et al. [64]. Their concept relates to the design of the A400M fuselage frames that are made of monolithic aluminium. However, the fatigue design loads in these frames are of such magnitude that fatigue design allowables restrict the design space, leading to heavy frame structures.

The solution proposed is to replace some of the frame thickness by an FML strap that substantially extends the fatigue life of the structure. Initiation of fatigue cracks has to take place in each FML metallic layer individually, because the growth of the crack through the thickness is prohibited by the fibre-reinforced prepreg layers, which is discussed in Chap. 9. The concept is illustrated in Fig. 3.15.

3.7 FML Components

Most of the reviewed patents discussed in this chapter relate to development of material concepts and stiffened skin panels, rather than actual components or products. One exception is the concept of an explosion resistant luggage container for transportation of luggage in aircraft, proposed by Kroon and Gunnink [65]. In the container design, they exploited the excellent impact performance of FMLs, which has been the subject of other patents. The container is made from FML panels that have been formed such that the number of necessary mechanically fastened edges is limited, see Fig. 3.16. The FML comprises aluminium layers with in-between glass fibre epoxy plies, mainly oriented under ±45°. Under these angles, the impact resistance of the laminates was observed to be the highest.

Because explosions induce extremely high strain rates, glass fibres were selected for their strain rate effect. The strain to failure and the failure strength of the glass fibres are significantly higher at these strain rates, increasing the impact resistance.

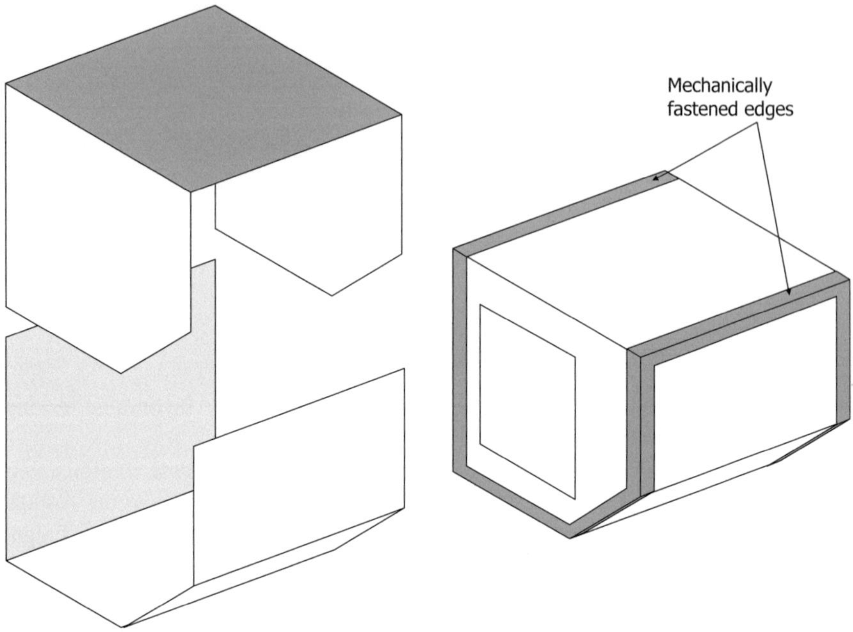

Fig. 3.16 Illustration of the explosion resistant luggage container [65]; image from [1]

3.8 Discussion

3.8.1 Flat Material Concepts

The review of patents filed on flat laminate concepts clearly illustrates the steps that have been made in the development of this technology. Here, the chronology of patents filed particularly illustrates the historical process of the FML development. From the need to increase the fatigue resistance of aerospace aluminium alloys, the technology was developed from laminates of thin sheets bonded with adhesive towards fibre-reinforced metal laminate concepts.

In view of this development history, one may observe that subsequent patents often describe an enhancement of previous concepts, rather than a completely novel or innovative idea. For example, the concept filed by Roebroeks and Mattousch [15] comprises an FML that is basically equivalent to the original patented concept, but the fibre-reinforced layer consists of unidirectional prepregs that are oriented in two or more directions rather than a cross-ply woven fabric. In addition, the proposed fibre type is different.

The small enhancement is particularly evident is the concept filed by Roebroeks [11]. Whereas the basic FML patents [5–7] describe the FML concept utilizing a single fibre type within the laminate, though identifying the use of different fibre types in the concept, Roebroeks proposes to use two of these different fibre types within one laminate. Although the idea is relevant considering the background (principal loads in a structure in both directions are not equal), the technology itself remains the same as described in the previous patents, especially as described in [15].

Similarity in concepts can also be observed when comparing the concept of locally reinforcing the FML for increased bearing strength, proposed by Van Rooijen and Van der Zwaag [24]. Their concept comprises an FML internally reinforced at the edges by a metal strip with higher modulus than the aluminium layers in the FML. This improves the pin/hole behaviour of mechanically fastened joints. A similar concept has been proposed for fibre-reinforced polymer composite material by Labordus et al. [66] where also additional strips are applied for increased bearing strength that could be made of, for example, metal alloy.

This similarity can be further illustrated in detail with the concepts that have been filed recently for lower wing skin panels by Roebroeks and Gunnink [26, 27]. Looking back at previous patented concepts and technologies, some additional observations can be made.

The original patent [26] describes reinforcing FMLs based on a PBO fibre/epoxy system. This PBO fibre is currently manufactured by Toyobo, but was developed by the US Air Force [67, 68] as a heat resistant polymer that is better than Kevlar [69]. The structures of Kevlar and the PBO fibre Zylon are illustrated in Fig. 3.17.

One should keep in mind that the application of Kevlar in FMLs has been described in the original patent by Delft University [5] along with a family of aramid based fibres. The original Dutch patent lists also provide a list of families and types of fibres, such as Aramid, Kevlar, Carbon and Glass [5, 6], whereas the European patent only lists the ranges of fibre moduli [7]. Academically, one can argue that the modulus of Zylon HM with its 270 GPa [70] exceeds the listed range of less than 250 GPa in [5, 7]. However, it is evident that the innovation of applying this fibre type in FMLs remains limited considering the claims in previous patents. Even more so if one considers that the application of this fibre type in FMLs was already proposed for stringer applications [62, 63]. Similarly, one can argue that replacement of the thermoset resin system by a thermoplastic resin system represents limited innovation [71]. Probably, the corresponding manufacturing method would make the actual difference.

The difference between the original lower wing skin concept [26] and the modified concept [27] is mostly related to the connection between thick aluminium layers and the centre reinforcing FMLs. The original patent proposed to bond the

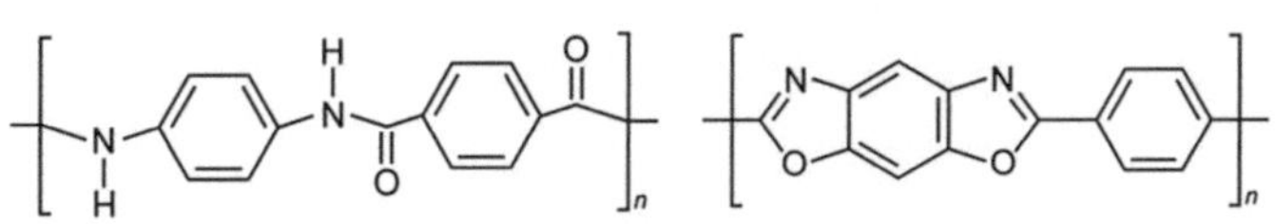

Fig. 3.17 Structure of poly-para-phenylene-terephthalamide [Kevlar] (*left*) and the structure of poly-para-phenylene-2,6-benzobisoxazole [Zylon] (*right*); image from [1]

thick aluminium layers with an epoxy adhesive to the FML, see Fig. 3.5a, b, while the modified patent proposes a fibre-reinforced polymer prepreg.

Although this modification should be considered an advancement in terms of material performance, the step itself is not novel. In fact, the concept of bonding thick aluminium layers with FMLs using a prepreg was patented before by Vogelesang [17] and Gunnink [25] who developed a lower wing skin concept using a thick tapered aluminium layer between two reinforcing FMLs, as illustrated in Fig. 3.3.

Alternatively, the novelty could be attributed to the fact that the modified concept [27] has a fibre-reinforced polymer layer connecting the thick aluminium layers with a reduced FVF less than 45%. It is mentioned that this lower FVF can be achieved by combining a polymer adhesive layer with a fibre-reinforced polymer layer containing a higher FVF.

The so-obtained reduced FVF does not give an even distribution of fibres through the polymer layer, but concentrates the fibres in the centre or at one side, keeping a resin-rich layer at the interface with the thick aluminium layer.

In fact, although the patent formulates the reduced FVF as being the innovative principle, the advantage of higher delamination resistance is achieved by this high resin content at the interface [72].

This aspect, however, is not new or innovative, but has been described in the early patents on FMLs; see Fig. 3.1a. Due to the manufacturing procedure at that time [73], fibre layers were put in between two polymer adhesive layers, creating a resin-rich volume at both interfaces with the aluminium.

In fact, the increasing delamination resistance for lower FVF at interfaces between metal and fibre-reinforced polymer layers was investigated before at Delft University of Technology by Mangoesoebroto [74]. One of the two authors of [26, 27] was involved as supervisor at the time that Mangoesoebroto investigated the effect of the FVF of the prepreg within the ARALL laminates.

The fact that it was already known that the delamination resistance depends on the FVF or on the presence of resin-rich layers can also be illustrated by the reported investigations of these aspects for composite materials in general in the open literature [75–78].

Despite that most of the variations to the hybrid concept of FMLs has been patented or presented in literature, there seems still room to file new patents to flat FML concepts. This is demonstrated with the patents filed by Wilson et al. [79] and Gunnink [80]. In these patents, the knowledge developed over the past decades on the hybrid concept has been utilized to tune and tailor the performance of specific FMLs. A range of FML configurations is claimed for which certain performance parameters meet specific ranges not covered in prior patents. Nevertheless, the FMLs following from these concepts have similar characteristics as FMLs proposed before.

3.8.2 Design Aspects

Concerning the design aspects, some observations can be made on the FML stringer concepts that have been filed. The basic principle of manufacturing thin FMLs that are pre-formed and subsequently stacked has been reported in the open literature [60], but has not been patented. The manufacturing methods proposed by Livi and Puccini [61] are different and can be utilized as a continuous process, but the generic FML stringer concept remains. One should recall, however, that the continuous process for manufacturing FMLs was already disclosed by Roebroeks [47].

The similarity between the original stringer concepts and recent concepts is even more the case for the concepts filed by Beumler [62] and Roebroeks et al. [63]. Interestingly enough, almost at the same time, two patents have been filed in two different countries. This could hardly be a coincidence as the authors of both patents are known to have been working together in the past [2].

Taking a closer look for instance at the claims by Roebroeks et al. [63], the concept describes a stiffened skin panel of which both the skin and stiffening element are made of a laminate comprising metal and fibre-reinforced polymer layers. The fibre-reinforced polymer layers are characterised by describing elastic moduli and ratios of compressive over tensile moduli, while for the metal layers all known aluminium alloys are listed.

The described manufacturing procedure of the laminated stiffener consists of forming into a three-dimensional shape and subsequently stretching to reverse curing stress distributions, or performing these steps in the reverse order. The stiffener is either bonded to the skin with a polymer adhesive containing a carrier, or by a fibre-reinforced polymer with reduced FVF.

The skin panel is related to aircraft or spacecraft fuselage panels, containing cross-ply laminates in the skin, and reinforced with either three-dimensional stiffeners or flat straps, known as crack retarders.

Based on the overview given in this paper and the previously discussed developments, it can be argued that the majority of the claimed topics are either described in previous patents, or have been reported before in the open literature.

3.9 Concluding Remarks

From the overview of patents on the hybrid FML technology, it may be concluded that there is hardly any room left for new innovative patents on plain material concepts. The fibre types and the metallic materials that can be selected to be combined in a hybrid panel are more or less known. However, the fundamental understanding of the hybrid technology is only recently emerging to the level that FMLs can be fully tailored to specific structural applications.

Some of the understanding on quasi-static fracture and fatigue damage evolution will be discussed in the following chapters, to illustrate the state of the art. With this knowledge and with the growing interest in the aviation industry to apply these hybrid technologies, it may be expected that further research and development will lead to new and dedicated structural applications. It therefore also may be expected that ne developments will focus on structural details or components and related manufacturing techniques, rather than hybrid materials in general.

References

1. Alderliesten RC (2009) On the development of hybrid material concepts for aircraft structures. Recent Pat Eng 3(1):25–38
2. Vlot A (2001) History of the development. Kluwer Academic Publishers, Dordrecht, The Netherlands
3. Schijve J, Van Lipzig HTM, Van Gestel GFJA, Hoeymakers AHW (1979) Fatigue properties of adhesive-bonded laminated sheet material of aluminium alloys. Eng Fract Mech 12: 561–579
4. Schijve J (2001) Fatigue of structures and materials. Kluwer Academic Publishers, Dordrecht
5. Schijve J, Vogelesang LB, Marissen R (1981) NL8100087
6. Schijve J, Vogelesang LB, Marissen R (1981) NL8100088
7. Schijve J, Vogelesang LB, Marissen R (1982) EP0056289
8. Marissen R (1988) Fatigue crack growth in ARALL. A hybrid aluminium-aramid composite material. Crack growth mechanisms and quantitative predictions of the crack growth rate. PhD dissertation, Delft University of Technology
9. Roebroeks GHJJ (1991) Towards GLARE—the development of a fatigue insensitive and damage tolerant aircraft material. PhD thesis, Delft University of Technology, Delft
10. Vogelesang LB, Roebroeks GHJJ (1991) US5039571
11. Roebroeks GHJJ (2003) NL1022706
12. Moreton R, Peel CJ (1992) GB2253185A1
13. Hilders HG (2000) WO00/53408 A1
14. Suarez Bermejo JC, Santiago MA, Diez de Ulzurrun Romeo I, Lopez Martin F, Herreros Sierra MA, Illescas Molina J, Soria Bartolome A, Garcia Nunez A (2006) WO06/103309 A2
15. Roebroeks GHJJ, Mattousch AC (1996) US5547735
16. Roebroeks GHJJ (2001) NL1019574
17. Vogelesang LB, Verbruggen LCE, Paalvast CG (1989) US4836084
18. Musaefendic J (2004) AU2004001004 A1
19. Gunnink JW, Evancho JW (2008) EP2085215 A1
20. Gunnink JW, Evancho JW (2009) WO2009/095381 A1
21. Gunnink JW, Evancho JW (2011) US2011/0052910 A1
22. Roebroeks GHJJ (1998) US6736919
23. Rajabali AF, Ter Steeg WJN (2005) US2005/0089704A1
24. Van Rooijen RGJ, Van der Zwaag S (2006) WO2006/123928 A1
25. Gunnink JW (1990) US4935291
26. Roebroeks GHJJ, Gunnink JW (2007) WO2007/061304A1
27. Roebroeks GHJJ, Gunnink JW (2007) WO2007/145512A1
28. Alderliesten RC, Benedictus R (2011) EP2627505B1
29. Alderliesten RC, Conen M (2011) NL2007603(C)
30. Alderliesten RC, Benedictus R, Rans CD, Rodi R (2010) NL2005028(C)
31. Garesché CE, Roebroeks GHJJ, Van Wimmersma-Greidanus B, Van Oost RC, Gunnink JW (1995) US5429326
32. Pettit RG (1996) US5567535
33. Roebroeks GHJJ (2002) WO02/078950A1

34. Labordus M, Verhoeven CG, Van Tooren MJL (2004) US2004/0151921A1
35. Beumler T (2014) EP2907654A1
36. Beumler T (2011) WO2012/049021A1
37. Vogelesang LB, Smulders FE, Chen D (1998) EP0323660
38. Roebroeks GHJJ (2007) WO2007/037695A1
39. Pettit RG (1993) US5227216
40. Roebroeks GHJJ (2007) WO2007/027093A1
41. Laliberté J, Mahendran D, Djokic D, Li C, Kratz J (2007) Effect of process-induced residual stresses on mechanical properties and fatigue crack initiation in fibre metal laminates. In: Proceedings of the 24th ICAF symposium, Naples, Italy, ICAF Doc. No. 2417
42. Rajabali AF (2003) WO03/011594A1
43. Rajabali AF (2003) WO2004/041519A1
44. Rajabali AF (2007) US7223318B2
45. Rajabali AF, Ter Steeg WJN (2007) US7279062B2
46. Kroon EJ, Roebroeks GHJJ (2001) NL1018120
47. Roebroeks GHJJ (2002) NL1022247
48. Roebroeks GHJJ, Kroon EJ (2007) WO2007/035100A3
49. Heinimann MB, Kulak M, Chu EW, Siemon JT (2008) WO2008/054876A2
50. Alderliesten RC, Grashof BA (2010) NL2005667(C)
51. Labordus M, Van Tooren MJL (2004) US2004/0053027 A1
52. Cano RJ, Grimsley BW, Weiser ES, Jensen BJ (2006) US7595112 B1
53. Popp V, Beumler T (2010) US2010/0086804 A1
54. Haack C, Beumler T (2010) US2010/0206987 A1
55. Rajabali AF, Ter Steeg WJN (2006) US2006/0159886A1
56. Rajabali AF, Markestein MA (2006) US2006/0159887A1
57. Rajabali AF, Ter Steeg WJN (2007) US2007/0042214A1
58. Martin A, Tillich A, Pellenkoft F (2008) US2008/0006741A1
59. Pellenkoft F, Gennai A (2006) EP1932757A1
60. De Jong TW, Kroon E, Sinke J (2001) Formability. In: Vlot A, Gunnink JW (eds) Fibre metal laminates—an introduction. Kluwer Academic Publishers, Dordrecht, The Netherlands
61. Livi F, Puccini G (2002) EP1336469A1
62. Beumler T (2008) WO2008/053041A1
63. Roebroeks GHJJ, Gunnink JW, Kroon EJ (2008) WO2008/033017A1
64. Ohrloff N, Beumler T, Daverschot D, Plokker M (2009) WO2010/043516A1
65. Kroon EJ, Gunnink JW (2003) NL1022709
66. Labordus M, Van Tooren MJL, De Graaf TJM (2004) WO2004/085141A1
67. Evers RC (1982) US4359567
68. Dotrong M, Dotrong M, Evers RC (1995) US5426173
69. Kitagawa T, Murase H, Yabuki K (1998) Morphological study on poly-p-phenylene-benzobisoxazole (PBO) fiber. J Polym Sci Part B Polym Phys 36:39–48
70. www.toyobo.co.jp/e/seihin/kc/pbo/Technical_Information_2005.pdf
71. Schulze K, Hausmann J, Schmitz D (2013) DE2011082697A1
72. Heinimann M, Kulak M, Bucci R, James M, Wilson G, Brockenbrough J, Zonker H, Sklyut H (2007) Validation of advanced metallic hybrid concept with improved damage tolerance capabilities for next generation lower wing and fuselage applications. In: Lazzeri L, Salvetti A (eds) Proceedings of the 24th ICAF symposium, Naples, Italy, ICAF Doc. No. 2417
73. Vogelesang LB, Marissen R, Schijve J (1981) A new fatigue resistant material: aramid reinforced aluminum laminate (ARALL). In: De Jonge JB, Van der Linden HH (eds) Proceedings of the 11th ICAF symposium, Noordwijkerhout, The Netherlands, ICAF Doc. No. 1216
74. Mangkoesoebroto RH (1987) The effect of fibre volume fraction on the mechanical properties and the fatigue behaviour of ARALL laminates. MSc thesis, Delft University of Technology (not in public domain)

75. Bradley WL, Cohen RN (1985) Matrix deformation and fracture in graphite-reinforced epoxies. ASTM STP 876:389–410
76. Hunston DL, Moulton RJ, Johnston JJ, Bascom WD (1987) Matrix resin effects in composite delamination: mode I fracture aspects. ASTM STP 937:74–94
77. Khan B, Rao RMVGK, Venkataraman N (1995) Low velocity impact fatigue studies on glass epoxy composite laminates with varied material and test parameters—effect of incident energy and fibre volume fraction. J Reinf Plast Compos 14:1150–1159
78. Li X, Carlsson LA, Davies P (2004) Influence of fiber volume fraction on mode III interlaminar fracture toughness of glass/epoxy composites. Compos Sci Technol 64: 1279–1286
79. Wilson G, Alderliesten R, Gunnink JW (2012) EP2763849B1
80. Gunnink JW (2012) WO2012/074394A1

Chapter 4
Stress and Strain

Abstract The in-plane mechanical properties of FMLs have been described with methods ranging from simple engineering methods to scientific laminated plate theories. This chapter describes the current understanding of stress and strain in FMLs, explaining the laminated plate theories adapted for the different thermal expansion properties of both metal and composites, and the plasticity induced by the metallic constituent. The application and limitations of the various methods are addressed.

4.1 Introduction

One important aspect in the development process of Fibre Metal Laminates (FMLs) has been the evolvement of fundamental understanding about the observed phenomena. This understanding was often established by developing engineering and scientific prediction tools for specific loading types and failure mechanisms.

This chapter describes the state-of-the-art methods developed to describe the relationship between strains and stresses in FMLs. In some cases, the fundamental methods will be related to engineering methods also published in the literature, discussing the applicability and limitations.

4.2 Stress–Strain in Orthotropic Materials Under Plane Stress

The constitutive equations describing the general relation between stresses and strains for orthotropic materials in their principal directions are described by

$$\begin{aligned} \overline{\sigma} &= S\overline{\varepsilon} \\ \overline{\varepsilon} &= C\overline{\sigma} \end{aligned} \tag{4.1}$$

R. Alderliesten, *Fatigue and Fracture of Fibre Metal Laminates*, Solid Mechanics and Its Applications 236, DOI 10.1007/978-3-319-56227-8_4

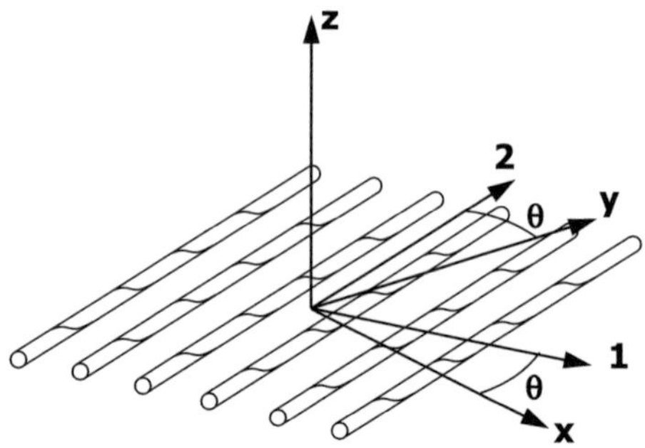

Fig. 4.1 Definition of principal material coordinates and the transition to laminate coordinates

where S and C are, respectively, the stiffness and compliance matrix.[1] For the plain stress condition in orthotropic lamina, the stress and strain components can be reduced to [1]

$$\overline{\sigma} = \begin{bmatrix} \sigma_1 \\ \sigma_2 \\ \tau_{12} \end{bmatrix}; \quad \overline{\varepsilon} = \begin{bmatrix} \varepsilon_1 \\ \varepsilon_2 \\ \gamma_{12} \end{bmatrix} \tag{4.2}$$

Here, the subscripts 1 and 2 indicate the principal material coordinates of the material (Fig. 4.1).

The corresponding stiffness and compliance matrix are given by

$$S = \begin{bmatrix} \frac{E_1}{1-\mu_{12}\mu_{21}} & \frac{\mu_{12}E_2}{1-\mu_{12}\mu_{21}} & 0 \\ \frac{\mu_{12}E_2}{1-\mu_{12}\mu_{21}} & \frac{E_2}{1-\mu_{12}\mu_{21}} & 0 \\ 0 & 0 & G_{12} \end{bmatrix}; \quad C = \begin{bmatrix} \frac{1}{E_1} & \frac{-\mu_{12}}{E_1} & 0 \\ \frac{-\mu_{12}}{E_1} & \frac{1}{E_2} & 0 \\ 0 & 0 & \frac{1}{G_{12}} \end{bmatrix} \tag{4.3}$$

To describe the stresses and strains for orthotropic materials in arbitrary directions (other than the principal directions), a transformation must be applied. The transformation from principal material coordinates towards laminate coordinates over an angle θ, as illustrated in Fig. 4.1, is defined by

$$\begin{aligned} \overline{\sigma}_\theta &= M\overline{\sigma} \\ \overline{\varepsilon} &= M^{-1}\overline{\varepsilon}_\theta \end{aligned} \tag{4.4}$$

M is the transition matrix given by

$$M = \begin{bmatrix} m^2 & n^2 & 2mn \\ n^2 & m^2 & -2mn \\ -mn & mn & -n^2 \end{bmatrix} \tag{4.5}$$

[1]Note that the definition of S and C here is opposite to the definition of Jones [1].

with $m = \cos(\theta)$ and $n = \sin(\theta)$. The stiffness and compliance matrix for laminates under an angle θ with the principal material coordinates can be obtained by substituting Eq. (4.4) into Eq. (4.1)

$$\begin{aligned} \overline{S}_\theta &= M\,S\,M^T \\ \overline{C}_\theta &= \left[M^{-1}\right]^T C\,M^{-1} \end{aligned} \tag{4.6}$$

which gives

$$\begin{aligned} \overline{\sigma}_\theta &= \overline{S}_\theta \overline{\varepsilon}_\theta \\ \overline{\varepsilon}_\theta &= \overline{C}_\theta \overline{\sigma}_\theta \end{aligned} \tag{4.7}$$

4.3 Classical Laminated Plate Theory

For laminated plates such as FMLs, the theory of deformation in relation to the applied loads and moments has been generalized into

$$\begin{bmatrix} N_i \\ M_i \end{bmatrix} = \begin{bmatrix} A_{ij} & B_{ij} \\ B_{ij} & D_{ij} \end{bmatrix} \begin{bmatrix} \varepsilon_j \\ \kappa_j \end{bmatrix} \tag{4.8}$$

where N_i and M_i are the vectors describing, respectively, the in-plane loads and the moments on the plate. Together, the in-plane strain vector ε_j and the curvature vector κ_j define the strains of each particular layer within the plate. The extensional stiffness A_{ij}, the bending–extension coupling stiffness B_{ij} and the bending stiffness D_{ij} are defined by

$$\begin{aligned} A_{ij} &= \sum_{k=1}^{N} \left(\overline{S}_{ij}\right)_k (z_k - z_{k-1}) \\ B_{ij} &= \frac{1}{2}\sum_{k=1}^{N} \left(\overline{S}_{ij}\right)_k \left(z_k^2 - z_{k-1}^2\right) \\ D_{ij} &= \frac{1}{3}\sum_{k=1}^{N} \left(\overline{S}_{ij}\right)_k \left(z_k^3 - z_{k-1}^3\right) \end{aligned} \tag{4.9}$$

where z represents the vertical position in the ply measured from the mid-plane.

4.4 Residual Stresses

The classical laminated plate theory (CLT) in principle does not account for differences between curing and operating temperatures. The relations between the stresses and strains of each individual layer are geometrically related to the deformation of the complete laminate. For fibre-reinforced polymer composites that often is sufficient, but the magnitude of residual stresses as result of curing FMLs is often greater compared to fibre-reinforced polymer composites. This is due to the significant difference in coefficient of thermal expansion between the metallic and the fibre/matrix constituents. To add the effect of thermal residual stresses within each individual layer of the FML, Honselaar [2] and Homan [3] introduced additional components to Eq. (4.2)

$$\overline{\sigma} = \begin{bmatrix} \sigma_1 \\ \sigma_2 \\ \tau_{12} \end{bmatrix}; \quad \overline{\varepsilon} = \begin{bmatrix} \varepsilon_1 - \alpha_1 \Delta T \\ \varepsilon_2 - \alpha_2 \Delta T \\ \gamma_{12} \end{bmatrix} \tag{4.10}$$

where α is the coefficient of thermal expansion for indicated principal material direction, while ΔT indicates the temperature difference between curing temperature T_{cur} and the ambient (operational) temperature T. As a consequence of equilibrium in overall laminate strain, these additional components will induce residual stresses in the individual layers. Depending on the magnitude of α, these stresses are either tensile or compressive stresses. For FMLs in general, α of metallic constituents is greater than that of the fibre-reinforced polymer constituents. This creates tensile stresses in the metal layers and compressive stresses in the fibre layers.

When looking at the fatigue characteristics of FMLs, the residual stresses induced by the cure cycle remain constant and add to the stresses in each layer by shifting the stress cycles. However, one has to keep in mind that the fatigue evaluation for real aircraft load spectra requires consideration of operating environmental temperatures that range from about −55 °C at cruising altitude (high ΔT) and up to about 80 °C on the platform in the desert (low ΔT) [4] (Fig. 4.2).

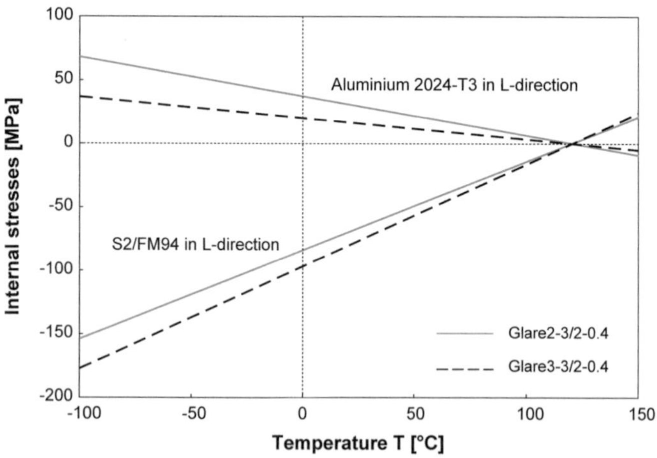

Fig. 4.2 Illustration of internal stresses in the metal and fibre layers of unidirectional and cross-ply FML

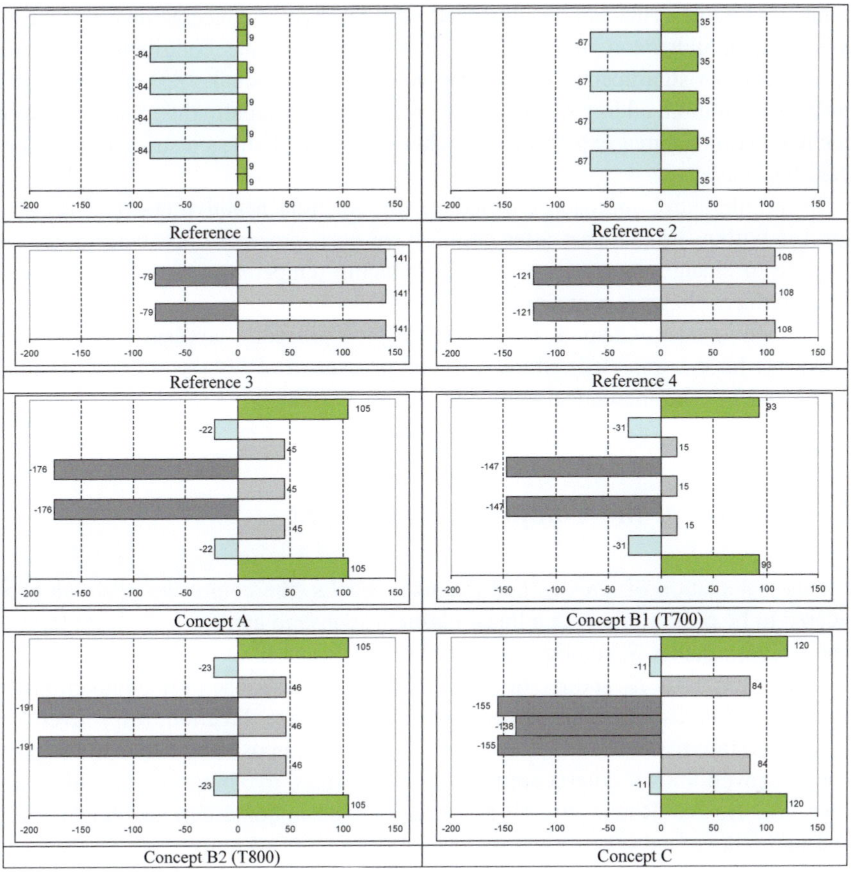

Fig. 4.3 Calculated thermally induced residual stresses in the concepts investigated by Rensma [6]

The above addition to the stress–strain equation holds for in-plane strain equilibrium only. Whenever a curved FML panel is being cured, release of pressure will impose an equilibrium state that may cause spring-back behaviour [5]. If so, out-of-plane deformations need to be considered in the equilibrium.

The effect of residual stresses may also need to be considered for hybrid structures, i.e. stiffened structures that contain components made of materials with different coefficients of thermal expansion. An example here could be a stiffened skin structure containing aluminium stringers or frames as backup structure to an FML skin. Because of the mismatch in material stiffness, this combination is prone to fatigue issues in the backup structure, but that may be worsened by the present residual stresses.

The exact level of residual stresses within the laminate depends on the constituents' coefficients of thermal expansion, the stiffness and the contribution to the laminate volume. This can be illustrated with the hybrid concepts investigated by Rensma [6], illustrated in Fig. 2.5. Changing the volume contribution of each constituent can significantly change the value of the residual stress obtained after curing. In addition, adding other constituents with substantially different stiffness or coefficients of thermal expansion may alter the residual stress distribution entirely (See Fig. 4.3).

In the literature, approaches to reduce the potential negative effect of residual stresses within the laminate after curing have been presented. Laliberté et al. [7] propose to change the cure cycle by introducing temperature steps, which effectively reduces the cure temperature T_{cur} and thus the temperature difference ΔT.

As mentioned in Chap. 3, post-stretching is a procedure applied after the cure cycle has been completed, which changes or even reverses the internal residual stress system. The theoretical approach will be presented hereafter.

4.5 Failure of the Composite Constituent

For the calculation of stresses and strains in FMLs, the composite plies are considered to be orthotropic with a linear elastic response to applied loads until failure. The stiffness matrix S, see Eq. (4.3), for the composite plies contains the Young's moduli of elasticity E_1 and E_2, describing the elastic response of a unidirectionally reinforced composite in fibre direction and in transverse direction, respectively.

Various failure criteria have been developed for composite materials in the past. The most well-known criteria are the maximum stress, maximum strain, Hill–Tsai, Tsai–Wu, Hashin–Rotem and Hashin criteria. The major difference between the first two criteria and other criteria is in the assumption of no interactions between stresses and strains in case of maximum stress or strain criteria.

Bouwman [8] proposed to use Hashin–Rotem failure criteria for failure of the matrix in the composite plies. The criterion incorporates strain interaction only partially and is given by

$$\left(\frac{\varepsilon_2}{\varepsilon_{2\max}}\right)^2 + \left(\frac{\gamma_{12}}{\tau_{\max}/G_{12}}\right)^2 > 1 \tag{4.11}$$

Once failure is determined, the transverse and shear stiffness of the composite layers are set to zero, maintaining the achieved stress. This implies that the stress–strain behaviour of the matrix is represented by an ideal plasticity type in which the resulting stiffness matrix for the kth layer reduces to

$$S_k = \begin{bmatrix} E_1 & 0 & 0 \\ 0 & 0 & 0 \\ 0 & 0 & 0 \end{bmatrix}_k \tag{4.12}$$

Fibre failure is assumed to occur when the fibres exceed their maximum strain in longitudinal direction, i.e.

$$\frac{\varepsilon_1}{\varepsilon_{1\max}} > 1 \tag{4.13}$$

It has been observed that the application of any of the other failure criteria instead of Hashin–Rotem results in minimal differences in the results. The metallic layers in FMLs redistribute the loads within the laminate differently as compared to composites based on fibre-reinforced polymers only. In particular, matrix failure criteria appear to have negligible influence for certain loading conditions, which sometimes justifies considering fibre failure only.

4.6 Plasticity of the Metal Constituent

The CLT is based on a linear elastic material relationship (Hooke's law), which does not account for deforming a laminate beyond the yield strength of the metal constituents. When the metals yield at certain strain levels, the above-described theory is no longer applicable, as the stress distribution will become different. To account for that, Honselaar [2] added a third (plastic) strain component to Eq. (4.10)

$$\overline{\sigma} = \begin{bmatrix} \sigma_1 \\ \sigma_2 \\ \tau_{12} \end{bmatrix}; \quad \overline{\varepsilon} = \begin{bmatrix} \varepsilon_1 - \varepsilon_{1,\text{pl}} - \alpha_1 \Delta T \\ \varepsilon_2 - \varepsilon_{2,\text{pl}} - \alpha_2 \Delta T \\ \gamma_{12} \end{bmatrix} \tag{4.14}$$

The strains in layer k at a distance z_k (see Fig. 4.4) from the middle surface are given by

$$\begin{bmatrix} \varepsilon_x \\ \varepsilon_y \\ \gamma_{xy} \end{bmatrix}_k = \begin{bmatrix} \varepsilon_x \\ \varepsilon_y \\ \gamma_{xy} \end{bmatrix}_0 + z_k \begin{bmatrix} \kappa_x \\ \kappa_y \\ \kappa_{xy} \end{bmatrix}_0 + \begin{bmatrix} \varepsilon_{x,\text{pl}} \\ \varepsilon_{y,\text{pl}} \\ 0 \end{bmatrix}_0 \tag{4.15}$$

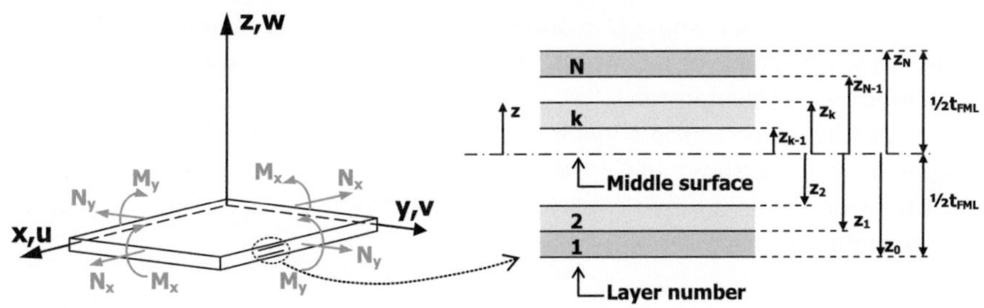

Fig. 4.4 Definition of the geometry of an n-layered laminate

where the subscripts 0 indicate the (plastic) strain ε and curvature κ of the middle surface of the laminate.

In unidirectional FMLs (i.e. ARALL, GLARE1 and GLARE2), the principal material coordinates 1, 2 correspond to the laminate coordinates x, $y(\theta = 0)$. In the other standardized FMLs (GLARE3, GLARE4), some fibre layers correspond to the laminate coordinates with $\theta = 90$. Combining Eq. (4.1) with (4.14) and substituting Eq. (4.15) yield

$$\begin{bmatrix} N_1 \\ N_2 \\ N_{12} \end{bmatrix} = \begin{bmatrix} A_{11} & A_{12} & A_{16} \\ A_{12} & A_{22} & A_{26} \\ A_{16} & A_{26} & A_{66} \end{bmatrix} \begin{bmatrix} \varepsilon_1 + \varepsilon_{1,\text{pl}} \\ \varepsilon_2 + \varepsilon_{2,\text{pl}} \\ \gamma_{12} \end{bmatrix}_0 + \begin{bmatrix} B_{11} & B_{12} & B_{16} \\ B_{12} & B_{22} & B_{26} \\ B_{16} & B_{26} & B_{66} \end{bmatrix} \begin{bmatrix} \kappa_1 \\ \kappa_2 \\ \kappa_{12} \end{bmatrix}_0 - \begin{bmatrix} N_1^T \\ N_2^T \\ N_{12}^T \end{bmatrix} \quad (4.16)$$

where N_i is the external force and N_i^T the force as result of the thermal extension

$$\begin{bmatrix} N_1^T \\ N_2^T \\ N_{12}^T \end{bmatrix} = \Delta T \left(\frac{t_f}{1 - \mu_{12,f}\mu_{21,f}} \begin{bmatrix} E_{11,f}\alpha_{1,f} + \mu_{12,f}E_{22,f}\alpha_{2,f} \\ \mu_{12,f}E_{11,f}\alpha_{1,f} + E_{22,f}\alpha_{2,f} \\ 0 \end{bmatrix} + \frac{E_{11,m}t_m^2}{1 - \mu_{12,m}} \begin{bmatrix} 1 \\ 1 \\ 0 \end{bmatrix} \right) \quad (4.17)$$

Similarly, the relation for the bending moments is given by

$$\begin{bmatrix} M_1 \\ M_2 \\ M_{12} \end{bmatrix} = \begin{bmatrix} B_{11} & B_{12} & B_{16} \\ B_{12} & B_{22} & B_{26} \\ B_{16} & B_{26} & B_{66} \end{bmatrix} \begin{bmatrix} \varepsilon_1 + \varepsilon_{1,\text{pl}} \\ \varepsilon_2 + \varepsilon_{2,\text{pl}} \\ \gamma_{12} \end{bmatrix}_0 + \begin{bmatrix} D_{11} & D_{12} & D_{16} \\ D_{12} & D_{22} & D_{26} \\ D_{16} & D_{26} & D_{66} \end{bmatrix} \begin{bmatrix} \kappa_1 \\ \kappa_2 \\ \kappa_{12} \end{bmatrix}_0 - \begin{bmatrix} M_1^T \\ M_2^T \\ M_{12}^T \end{bmatrix} \quad (4.18)$$

The extensional stiffness A_{ij}, the bending–extension coupling stiffness B_{ij} and the bending stiffness D_{ij} are defined by Eq. (4.9).

4.7 Generalized Theories of Plasticity

An alternative approach to the previously presented method for plasticity in the metallic layers was proposed by Bouwman [8]. Initially, the metallic layers respond linear elastically in an isotropic manner up to yielding, but beyond yielding, the elastic–plastic response of the metallic layers is described with anisotropic yielding and anisotropic hardening rules.

To describe the anisotropic yielding and hardening, so-called yield surfaces are composed from measured stress states in which the linear elastic response of the metal transitions into an elastic–plastic response. For this purpose, the von Mises yield function for plane stress is utilized

$$f = \sigma_e - \sigma_0 = \sqrt{(\varepsilon_x - \alpha_x)^2 + (\varepsilon_y - \alpha_y)^2 - (\varepsilon_x - \alpha_x)(\varepsilon_y - \alpha_y) + 3(\tau_{xy} - \alpha_{xy})^2} - \sigma_0 \tag{4.19}$$

in which σ_e is the equivalent stress, σ_0 the yield stress and α the kinematic shift of the yield surface centre. Because the yield surface shifts at the point where first yielding occurs in the FML, the proposed approach is incremental of nature. After each load increment, the transition of the yield surface must be calculated, to determine the input for the next load increment.

Bouwman developed a tool to predict stress–strain curves for arbitrary GLARE laminates based on this theory. The generic applicability for FML panes loaded in arbitrary directions is in theory covered by the method, but has never been validated to that extent.

4.8 Post-stretching

If the residual stresses induced by the curing cycle are large, there is an alternative process in some cases to compensate for that, as discussed in Sect. 3.4.1. This process is unknown for fibre-reinforced polymers, due to the absence of metallic constituents that at some point plastically yield, remaining deformed even after unloading.

The post-stretching procedure, sometimes denoted as pre-straining, utilizes the elastic–plastic stress–strain relationship of the metallic constituents to change or even reverse the internal residual stress distribution obtained after curing. The effect on the stress–strain relationship is illustrated in Fig. 4.5. The fibre/epoxy layers remain elastic during the stretch procedure, while the metallic constituents yield. The tangent modulus implies a different internal stress distribution within the plastic region of the curve. After unloading the metallic layers are permanently deformed, which causes a different internal stress equilibrium. The amount of stretching, i.e. the deformation into the plastic region, determines how much the residual stress after curing is altered and whether it is reversed.

As mentioned in Sect. 3.4.1, this process is applicable predominantly to unidirectional FMLs, for example GLARE2. What further limits the applicability of the process in practice is that it only applies to FMLs without any design details or complexity. For example, panels containing interlaminar doublers or bonded splices cannot be post-stretched. The cross section of the laminate lay-up has to be constant and continuous.

Post-stretching is a tensile procedure without any bending or torsion of the laminate. This means for Eq. (4.16) that the tension bending coupling stiffness matrix B_{ij} should be equal to zero. The main requirement for this matrix to be zero is that the laminate itself is balanced and symmetric. In other words, the distribution of the layers at both sides of the neutral line should be perfectly in equilibrium.

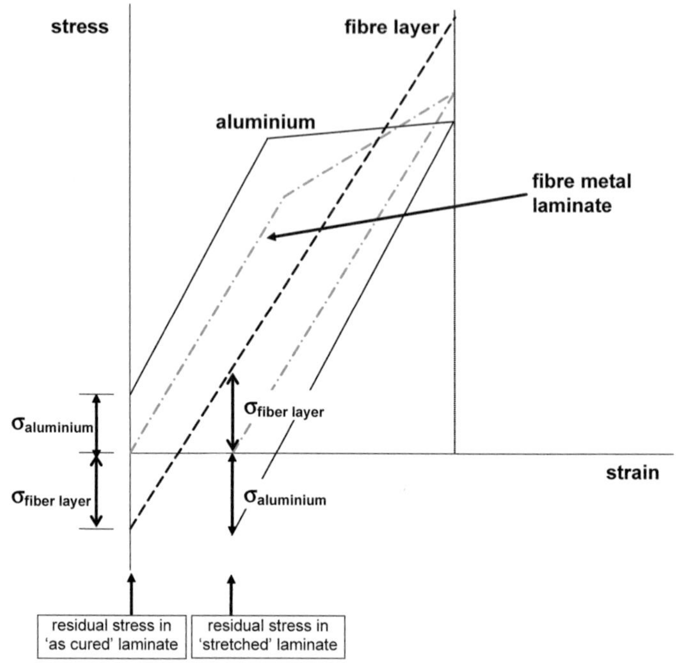

Fig. 4.5 Illustration of the post-stretching procedure on the uniaxial stress–strain relationship of the FML and its constituents

Most common laminates (ARALL and GLARE) contain an (almost) symmetric lay-up for practical reasons; an asymmetric laminate will tend to warp after the curing process due to the mismatch of the coefficients of thermal expansion.

Thus, the common unidirectional laminates (ARALL, GLARE1 and GLARE2) are balanced and symmetric. This enables the simplification of Eq. (4.16) to obtain the stresses in the metal and fibre/epoxy layers [9]

$$\begin{bmatrix}\sigma_1\\ \sigma_2\\ \tau_{12}\end{bmatrix}_m = S\left(\begin{bmatrix}\varepsilon_{1,\mathrm{pl}}\\ \varepsilon_{2,\mathrm{pl}}\\ 0\end{bmatrix}_{\mathrm{lam}} + A^{-1}\begin{bmatrix}N_1+N_1^T\\ N_2+N_2^T\\ N_{12}+N_{12}^T\end{bmatrix}_{\mathrm{lam}} - \alpha_m\Delta T\begin{bmatrix}1\\1\\0\end{bmatrix} - \begin{bmatrix}\varepsilon_{1,\mathrm{pl}}\\ \varepsilon_{2,\mathrm{pl}}\\ 0\end{bmatrix}_m\right) \tag{4.20}$$

$$\begin{bmatrix}\sigma_1\\ \sigma_2\\ \tau_{12}\end{bmatrix}_f = S\left(\begin{bmatrix}\varepsilon_{1,\mathrm{pl}}\\ \varepsilon_{2,\mathrm{pl}}\\ 0\end{bmatrix}_{\mathrm{lam}} + A^{-1}\begin{bmatrix}N_1+N_1^T\\ N_2+N_2^T\\ N_{12}+N_{12}^T\end{bmatrix}_{\mathrm{lam}} - \alpha_{1,f}\Delta T\begin{bmatrix}1\\1\\0\end{bmatrix}\right) \tag{4.21}$$

The deformation in the individual layers and the laminate induced by the post-stretching procedure must be coupled with an extra equilibrium equation. If the external forces are equal to zero, this equation states

$$\begin{bmatrix} \sigma_1 \\ \sigma_2 \\ \tau_{12} \end{bmatrix}_m t_m + \begin{bmatrix} \sigma_1 \\ \sigma_2 \\ \tau_{12} \end{bmatrix}_f t_f = 0 \tag{4.22}$$

For isotropic yielding in the metal layers, yielding may be calculated with the von Mises criterion

$$\sqrt{\frac{1}{2}\left[(\sigma_1-\sigma_2)^2+(\sigma_2-\sigma_3)^2+(\sigma_3-\sigma_1)^2\right]} = \sigma_{0.2} \tag{4.23}$$

At yielding, the plastic strains $\varepsilon_{i,\mathrm{pl,lam}}$ and $\varepsilon_{\mathrm{i,pl},m}$ may still be considered to be equal to zero, which simplifies Eq. (4.20) to

$$\begin{bmatrix} \sigma_1 \\ \sigma_2 \end{bmatrix}_m = \begin{bmatrix} S_{11} & S_{12} \\ S_{12} & S_{22} \end{bmatrix}_m \left(\begin{bmatrix} A_{11} & A_{12} \\ A_{12} & A_{22} \end{bmatrix}^{-1} \begin{bmatrix} N_1 + N_1^T \\ N_2 + N_2^T \end{bmatrix}_{\mathrm{lam}} - \alpha_m \Delta T \begin{bmatrix} 1 \\ 1 \end{bmatrix} \right) \tag{4.24}$$

Here, the forces N_1^T and N_2^T are calculated with the temperature range ΔT, i.e. the difference between the temperature during post-stretching and the curing temperature. With these stresses, the deviator stresses [10] can be calculated with

$$\begin{aligned} S_{T1} &= \frac{2}{3}\sigma_1 - \frac{1}{3}(\sigma_2 - \sigma_3) \\ S_{T2} &= \frac{2}{3}\sigma_2 - \frac{1}{3}(\sigma_3 - \sigma_1) \\ S_{T3} &= \frac{2}{3}\sigma_3 - \frac{1}{3}(\sigma_1 - \sigma_2) \end{aligned} \tag{4.25}$$

FMLs contain thin layers each facing a plain stress condition, which implies that $\sigma_3 = 0$. The plastic strain distribution can be described with the deviator stresses

$$\begin{bmatrix} \varepsilon_{1,\mathrm{pl}} \\ \varepsilon_{2,\mathrm{pl}} \\ \varepsilon_{3,\mathrm{pl}} \end{bmatrix}_m = \Delta\lambda \begin{bmatrix} S_{T1} \\ S_{T2} \\ S_{T3} \end{bmatrix} \tag{4.26}$$

One should keep in mind that this one-step approach including the von Mises criterion is only valid for relatively small strains in metallic sheets containing small anisotropy in the yield stresses. For metals with high plastic anisotropy, for example some aluminium–lithium alloys, a more complicated iterative calculation must be considered to determine the development of stress and strain during the post-stretch procedure.

In the unloaded (and un-stretched) situation, the internal stresses in the metal and fibre/epoxy layers satisfy Eqs. (4.20) and (4.21) with $N_i = 0$. When stretching the laminate only in direction 1, the strain $\varepsilon_{1,\mathrm{pl,lam}}$ is known. During yielding, the force $N_2 = 0$, which means that N_1 can be solved with Eqs. (4.23) and (4.24). Using Eqs. (4.22) and (4.24)–(4.26), one may obtain

$$\Delta\lambda = \frac{\varepsilon_{1,\mathrm{pl,lam}}}{t_m\left[\left(A_{11}S_{11,m} + A_{12}S_{12,m}\right)S_{T1} + \left(A_{11}S_{12,m} + A_{12}S_{22,m}\right)S_{T2}\right]} \tag{4.27}$$

while the plastic strain $\varepsilon_{2,\mathrm{pl,lam}}$ can be calculated with

$$\varepsilon_{2,\mathrm{pl,lam}} = \varepsilon_{1,\mathrm{pl,lam}} \frac{\left(A_{12}S_{11,m} + A_{22}S_{12,m}\right)S_{T1} + \left(A_{12}S_{12,m} + A_{22}S_{22,m}\right)S_{T2}}{\left(A_{11}S_{11,m} + A_{12}S_{12,m}\right)S_{T1} + \left(A_{11}S_{12,m} + A_{12}S_{22,m}\right)S_{T2}} \tag{4.28}$$

When the FML is stretched in direction 1 and direction 2 simultaneously, both plastic strains $\varepsilon_{1,\mathrm{pl,lam}}$ and $\varepsilon_{2,\mathrm{pl,lam}}$ are known. The correlated plastic strains in the metal layers can directly be calculated from Eq. (4.22)

$$\begin{bmatrix} \varepsilon_{1,\mathrm{pl}} \\ \varepsilon_{2,\mathrm{pl}} \end{bmatrix}_m = \frac{1}{t_m} \begin{bmatrix} S_{11} & S_{12} \\ S_{12} & S_{22} \end{bmatrix}_m^{-1} \begin{bmatrix} A_{11} & A_{12} \\ A_{12} & A_{22} \end{bmatrix} \begin{bmatrix} \varepsilon_{1,\mathrm{pl}} \\ \varepsilon_{2,\mathrm{pl}} \end{bmatrix}_{\mathrm{lam}} \tag{4.29}$$

while the residual internal stresses follow from Eqs. (4.20) and (4.21) without the use of the von Mises criterion.

Aside from the theoretical method involved in describing the stress redistribution during the post-stretching procedure, one has to carefully verify whether the stresses actually induced in the laminate correspond with the calculated values. Take, for example, the preliminary study performed by Verbruggen [11] involving post-stretching (or pre-straining) of ARALL laminates. Verbruggen observed quite a significant variation in residual stresses in samples taken from the same post-stretched plate. Such variation suggests that the stress distribution during the post-stretch procedure is non-uniform.

4.9 Shear Stress and Strain

The shear properties of structural materials are of particular interest when the compression and shear stability of shell structures are assessed. The determination of these properties for FMLs has been treated as a separate task compared to the tensile and compressive strength properties [12].

The shear stress–strain properties are not easily determined by experiments, as is illustrated by the large number of shear experiments described in the literature. Because the shear properties are generally only considered for structural applications in combination with tensile and compressive stresses for stability assessments, the behaviour beyond the shear yield strength is often of less interest. As a consequence, often only the elastic shear response is determined and described up to the point that the shear yield strength, defined with the 0.2% offset in strain, can be determined.

With the formulation of the CLT, explained for FMLs in the previous sections, the shear properties are included in the description of the laminates. For the elastic part of the shear stress–strain curve, this follows directly from the formulation of Eq. (4.2).

Important for the appropriate description using CLT is the definition of the axial and shear properties of the laminate in its material axes and the formulation of the properties at any intermediate angle of orientation. For example, the relations for the Young's moduli may be written as [13]

$$\begin{aligned} \frac{1}{E_1^*} &= \frac{1}{E_1}\cos^4\theta + \left(\frac{1}{G_{12}} - 2\frac{\nu_{12}}{E_1}\right)\sin^2\theta\cos^2\theta + \frac{1}{E_2}\sin^4\theta \\ \frac{1}{E_2^*} &= \frac{1}{E_1}\sin^4\theta + \left(\frac{1}{G_{12}} - 2\frac{\nu_{12}}{E_1}\right)\sin^2\theta\cos^2\theta + \frac{1}{E_2}\cos^4\theta \end{aligned} \tag{4.30}$$

With for the shear modulus

$$\frac{1}{G_{12}^*} = 2\left(\frac{2}{E_1} + \frac{2}{E_2} + 4\frac{\nu_{12}}{E_1} - \frac{1}{G_{12}}\right)\sin^2\theta\cos^2\theta + \frac{1}{G_{12}}\left(\sin^4\theta + \cos^4\theta\right) \tag{4.31}$$

and for the Poisson's ratio

$$\nu_{12}^* = E_1^*\left(\frac{\nu_{12}}{E_1}\left(\sin^4\theta + \cos^4\theta\right) - \left(\frac{1}{E_1} + \frac{1}{E_2} - \frac{1}{G_{12}}\right)\sin^2\theta\cos^2\theta\right) \tag{4.32}$$

4.10 Out-of-Plane (Bending and Torsion)

Equations (4.16) and (4.18) can be formulated as a single matrix relation [14] with

$$\begin{bmatrix} N_i \\ M_i \end{bmatrix} = \begin{bmatrix} A_{ij} & B_{ij} \\ B_{ij} & D_{ij} \end{bmatrix} \begin{bmatrix} \varepsilon_j^o \\ \kappa_j \end{bmatrix} - \begin{bmatrix} N_i^T \\ M_i^T \end{bmatrix} \tag{4.33}$$

In case the laminate deforms out-of-plane, the matrix B_{ij} in this equation is no longer equal to zero. The previous sections dealing with plasticity in the metal layers started with the assumption that the laminates to be considered are balanced and symmetric. The advantage of this assumption is that if $B_{ij} = 0$ this simplifies the equation significantly, enabling the derivation of relations including plasticity. With $B_{ij} \neq 0$, the development towards this ABD matrix including plasticity is not applicable [14].

Nevertheless, the CLT can still be applied to practical problems related to elastic out-of-plane deformation such as the above-mentioned spring back after curing [5]. One must take into account, however, that simple geometrical shapes can still be easily implemented in analytically formulated codes, but complex three-dimensional shapes require more robust computer assisted approaches (i.e. finite element analyses).

In addition to the above-mentioned failure criteria for the matrix and fibres in composite plies, and the yielding and hardening rules for metallic layers, failure criteria must be adopted for the interfaces between the individual layers. Whereas in-plane loading and deformations usually imply compatibility between individual layers, out-of-plane deformations, such as bending, create interlaminar shear stresses that may cause delaminations. This failure mode must be considered when describing the incremental increase in deformations.

Aside from laminate forming models, these interlaminar shear failure criteria are also to be considered when calculating the deformations during low-velocity impact sequences. However, in the latter case, because quasi-static deformations may not always be applicable, other mechanisms contribute to the complexity of the problem.

4.11 Simple Methods for Design Purposes

The theories presented in this chapter often require fundamental understanding and programming of matrix type calculations. In an engineering environment, especially those related to certification and material qualification, simple methods are often preferred.

4.11.1 Metal Volume Fraction

Here a well-known and convenient method is the rules of mixtures, in case of FMLs often referred to as the metal volume fraction (MVF) method [15]. This method relates the laminate properties to the properties of the individual constituents equivalent to their contribution in volume or thickness. In fact, this relation has a similar basis as the equilibrium given by Eq. (4.22).

One has to be aware, however, that this approach can only be applied to individual constituent and laminate properties P in the form

$$P_{\text{lam}} = MVF \cdot P_m + (1 - MVF)P_f \tag{4.34}$$

With respect to the stress–strain behaviour of the laminate, only specific points on the curve can be determined with Eq. (4.34). A simple method based on this MVF approach that has been described in [16] considers the stress–strain curve as a bilinear curve. The stress and strain for the yield point and final failure are calculated for the laminate using Eq. (4.34), which provide a bilinear curve for the laminate. For preliminary design purposes or material selection criteria, this approach may be considered very convenient and fast. For detailed analysis, however, this approach must be considered inaccurate requiring too many experiments to determine the required properties.

For example, stability analyses require a more detailed description of the stress–strain curve, especially near the yield point [17]. For these particular design cases, the application of the Ramberg–Osgood equation has proved to be sufficiently accurate.

The MVF approach represents straightforward rules of mixtures approach. Obviously, this approach is most accurate when both the constituent parameters represent a similar state as the parameter to be determined for the FML. For example, the ultimate strength of the FML can be linearly related to the ultimate strength of both constituent materials, but in the laminated composition, one has to be aware that the strength of the FML is dictated by the failure strain of the fibre layers. For standard GLARE, the glass fibres have a failure strain of approximately 4.7% [18], which implies that the ultimate strength of the aluminium is not yet reached at the moment the fibres and FML fail. This can be illustrated by performing a uniaxial stress–strain test in displacement-controlled conditions. After reaching the maximum strength, the force drops significantly (corresponding to fibre failure), allowing the metallic layers to further extend until failure. On most cases, the fibre fracture in such tensile test is clearly visible, because this failure mode is accompanied by significant delamination between these fractured fibres and the intact aluminium layers. In general, however, the tests are terminated after the drop at ultimate strength, because in force-controlled conditions that represents entire laminate failure.

To perform the rule of mixtures properly, Müller [18] therefore proposed to use instead of the ultimate strength for aluminium, the stress at 4.7% strain. The metal volume fraction then becomes

$$S_{\text{ult,FML}} = \text{MVF} \cdot S_{m,(\varepsilon=4.7\%)} + (1 - \text{MVF})S_{f,\text{ult}} \tag{4.35}$$

Although Eq. (4.35) seems more appropriate than using the ultimate strength of the metal, the MVF approach has generally been adopted in a rather pragmatic engineering approach. To calculate the yield strength of an FML, for example, requires an equivalent yield strength for the fibre layers, which generally are linear elastic until failure. Therefore, the parameters that can be established individually for the constituents have been taken, but often adapted while fitting the linear trend through the FML properties. That and the requirement of having conservative values for design led to validated MVF methods in industry based on modified engineering constituent data.

4.11.2 Determination of Shear Properties Using Uniaxial Material Data

It has been explained that the shear properties of FMLs are related to the uniaxial material properties in formulations such as the CLT. A method that is based on such relationships is the method to determine the shear properties of materials up to the yield strength using the transformation equations for plane stress conditions proposed by Timoshenko [19]

$$\tau_{xy}(\theta) = -\frac{\sigma_x - \sigma_y}{2}\sin(2\theta) + \tau_{xy}\cos(2\theta) \quad (4.36)$$

This transformation may only be applied to homogeneous materials, which FMLs evidently are not. However, since the methodology aims to be quick rather than accurate, one may assume the FML responds in a homogeneous manner. In uniaxial tensile tests, the uniaxial stress state implies that the shear stress and the stress perpendicular to the applied load are both equal to zero. Equation (4.36) then simplifies to

$$\tau_{xy}(45^\circ) = \frac{1}{2}\sigma_y \quad (4.37)$$

For FMLs containing fibres oriented only in the major material directions, and thus in loading direction of the test, see Fig. 4.6a, the contribution of the fibres to the shear properties of the FML is fairly low, related to the in-plane shear stiffness of the unidirectional prepreg layer. For tests performed on FMLs with the fibres positioned under an angle of ±45°, see Fig. 4.6b, for example GLARE6 (see Table 2.1), the relation should be taken as

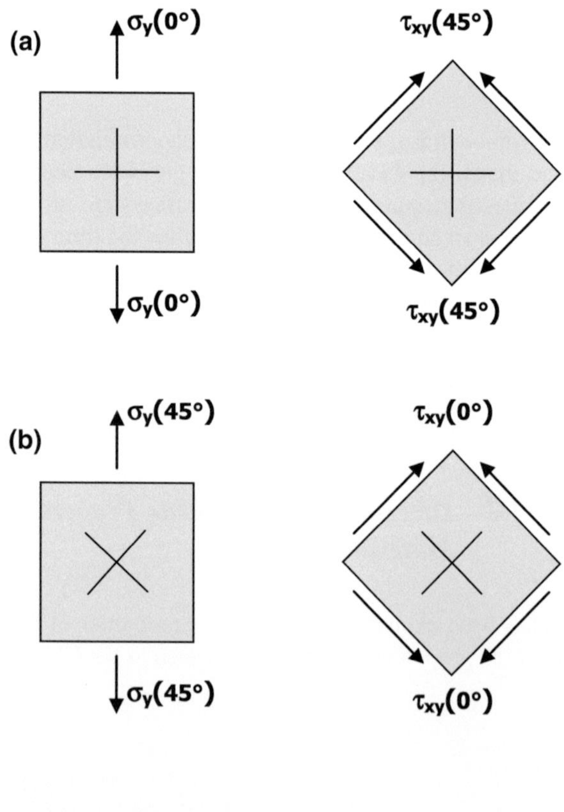

Fig. 4.6 Definition of the shear stress with respect to the material axes and fibre orientation

$$\tau_{xy}(0^\circ/90^\circ) = \frac{1}{2}\sigma_y(45^\circ) \tag{4.38}$$

Note that these relations can only be assumed to be valid for linear elastic material behaviour. This implies that the accuracy of the method is fair to the point of first yielding, but may be less accurate when defining yielding with the 0.2% offset rule; see Chap. 2.

To determine the shear yield strength using the uniaxial yield strength data, Hagenbeek [20] investigated the above-described method using both Tresca and von Mises criteria. He concluded that the Tresca criterion was more accurate than von Mises, but both were inaccurate for predicting laminate yield strength values defined as $\sigma_{0.2}$. For the onset of yield strength, the predictions were better, in particular when they were applied to the metallic constituents only. However, such a formulation would contradict with the objective to have a simple engineering method.

4.12 Limit of Validity of CLT and MVF

The theories presented in this chapter assume plane stress conditions. Considering the FMLs developed with metal layer thicknesses of 0.2–0.5 mm, this assumption is appropriate. However, for certain applications, it could be preferred to work with metal layer thickness beyond that range. Consider, for example, lower wing panels of commercial transport aircraft, in particular near the wing root. With aluminium thicknesses of 0.3–0.5 mm, the total panel thickness could require the placement of many aluminium sheets and composite plies, substantially increasing manufacturing costs.

For this purpose, various studies proposed the application of thicker aluminium sheets [21–24]. To identify whether the CLT or the MVF method, both presented in this chapter, is limited to the aluminium layer thicknesses considered, Rickerd [25] studied the limits of validity of both these methodologies with respect to aluminium thickness. For the in-plane tensile and compression properties, both methods were proved to be even accurate for aluminium thicknesses all the way up to 1 mm.

References

1. Jones RM (1999) Mechanics of composite materials, 2nd edn. Taylor & Francis Inc., Philadelphia
2. Honselaar C (1986) The residual internal stress distribution in ARALL. Master's thesis, Delft University of Technology, Delft
3. Homan JJ (2006) Fatigue initiation in fibre metal laminates. Int J Fatigue 28:366–374
4. Beumler TH (2004) Flying GLARE®, a contribution to aircraft certification issues on strengths properties in non-damaged and fatigue damaged GLARE® structures. PhD dissertation, Delft University of Technology

5. Sinke J (2003) Manufacturing of GLARE parts and structures. Appl Compos Mater 10:293–305
6. Rensma E (2007) Investigation of innovative concepts for hybrid structures. MSc thesis, Delft University of Technology
7. Laliberté J, Mahendran M, Djokic D, Li J, Kratz J, Effect of process-induced residual stresses on mechanical properties and fatigue crack initiation in fibre metal laminates. In: Proceedings of the 24th symposium of the international committee on aeronautical fatigue, 16–18 May 2007, Naples, Italy, 2007
8. Bouwman VP (2002) A calculation tool for structural analysis of general laminates, Report NLR-CR-2002-44. National Aerospace Laboratory NLR, The Netherlands
9. Khan SU, Alderliesten RC, Benedictus R (2009) Post-stretching induced stress redistribution in fibre metal laminates for increased fatigue crack growth resistance. Composites Science and Technology 69(3–4):396–405
10. Mendelson A (1968) Plasticity: theory and application. The Macmillan Company, USA
11. Verbruggen M (1983) Relaxation due to temperature, moisture and external loading—preliminary results, Memorandum M-491. Delft University of Technology, Delft
12. Hagenbeek M (2001) Shear yield strength. In: Vlot A, Gunnink JW (eds) Fibre metal laminates—an introduction. Kluwer Academic Publishers, Dordrecht, The Netherlands
13. Hagenbeek M (2000) Estimation tool for basic material properties, Report B2v-00-29. Delft University of Technology, Delft
14. De Jong TW (2004) Forming of laminates. PhD dissertation, Delft University of Technology
15. Roebroeks GHJJ (2000) The metal volume fraction approach, TD-R-00-003, SLI, Delft
16. van Hengel C (2001) Stress-strain curve. In: Vlot A, Gunnink JW (eds) Fibre metal laminates, an introduction. Kluwer Academic Publishers, Dordrecht, The Netherlands
17. Verolme JL (1995) The development of a design tool for fibre metal laminate compression panels. PhD dissertation, Delft University of Technology, Delft
18. Müller RPG (1995) An experimental and analytical investigation on the fatigue behaviour of fuselage riveted lap joints, the significance of the Rivet Squeeze Force, and a comparison of 2024-T3 and GLARE 3. PhD dissertation, Delft University of Technology, Delft
19. Timoshenko SP, Gere JM (1985) Mechanics of materials. Stanford University, England
20. Hagenbeek M (2000) Derivation of shear yield values from uniaxial tensile tests—shear yield strength of GLARE, Report B2v-99-38. Delft University of Technology, Delft
21. Heinimann M, Kulak M, Bucci R, James M, Wilson G, Brockenbrough J, Zonker H, Sklyut H (2007) Validation of advanced metallic hybrid concept with improved damage tolerance capabilities for next generation lower wing and fuselage applications. In: Lazzeri L, Salvetti A (eds) Proceedings of the 24th ICAF symposium, Naples, Italy, ICAF doc. no. 2417
22. Gunnink JW (2006) Hybrid primary aircraft structures. In: Proceedings of the SAMPE Europe conference, Toulouse, France, September 13th–14th
23. Roebroeks GHJJ, Hooijmeijer PA, Kroon EJ, Heinimann MB (2007) The development of central. In: First international conference on damage tolerance of aircraft structures, Delft, The Netherlands
24. Fredell RS, Gunnink JW, Bucci RJ, Hinrichsen J (2007) Carefree hybrid wing structures for aging USAF transports. In: First international conference on damage tolerance of aircraft structures, Delft, The Netherlands
25. Rickerd G, Alderliesten RC, Rans CD, Benedictus R (2010) Considering fiber metal laminate aircraft wings—what should we know? In: Aircraft structural integrity program conference, San Antonio, Texas, USA

Chapter 5
Blunt Notch Strength

Abstract The blunt notch strength of FMLs is discussed with respect to the deformation and failure phenomena of the individual constituents. The influence of metal plasticity, the splitting and delamination phenomena in the composite layers, and fibre failure are discussed. Various failure criteria and theories are presented to describe the blunt notch strength in the major material axes and under off-axis angles with respect to these axes.

5.1 Introduction

The blunt notch strength, or the open-hole tension strength, has become an important design parameter for structures using composite materials or FMLs. The open-hole tension strength of ductile metallic structures was often not considered a design parameter, for the reason that the reduction in strength due to the introduction of a blunt notch or hole would be limited [1].

Ductile alloys, like most aluminium alloys applied in aeronautical structures, have a low notch sensitivity, which implies that the strength reduction could be less than, for example, the difference between ultimate and yield strength. Since permanent deformations are often not allowed, the yield strength of aluminium alloys would be sufficient for structural sizing, ignoring reduced strength imposed by holes and notches. However, with the introduction of notch-sensitive composites and composite plies in FMLs, the open hole implies a strength reduction that can no longer be ignored.

For this reason, the blunt notch strength has been defined as the net or nominal strength of a structure containing a hole

$$S_{\mathrm{BN}} = \frac{W}{W-D} S_{\max} \tag{5.1}$$

During the development of FMLs, this net blunt notch strength has been introduced specifically to be determined with the standardized specimen geometry illustrated in Fig. 5.1a [2]. The standard test procedure is fairly similar to the

R. Alderliesten, *Fatigue and Fracture of Fibre Metal Laminates*, Solid Mechanics and Its Applications 236, DOI 10.1007/978-3-319-56227-8_5

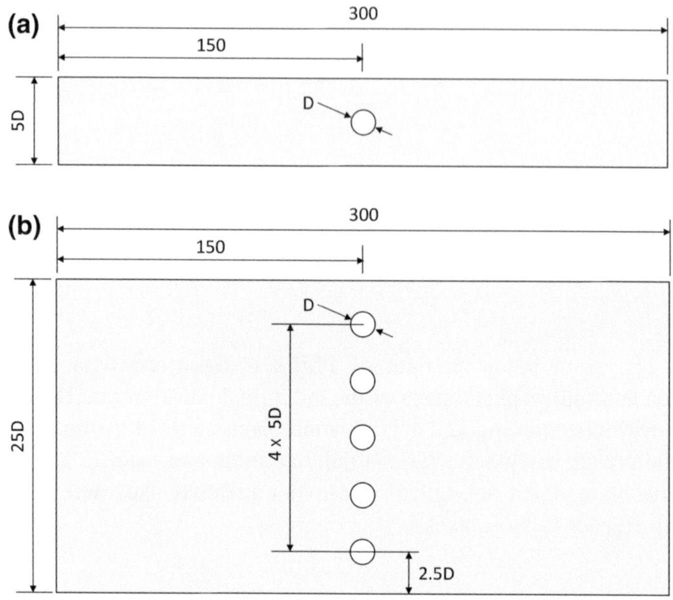

Fig. 5.1 Single open-hole blunt notch specimen configuration as defined in [2] (**a**) and five-hole blunt notch specimen (**b**)

standard test procedure for polymer matrix composites [3], but with slightly different specimen dimensions. Since then an extensive amount of research has been performed on this characteristic material property, from which many prediction methods have been derived and proposed.

Later, also another blunt notch specimen geometry was proposed, illustrated in Fig. 5.1b, which could also be used to test fatigue initiation, crack propagation and residual strength [4]. Using a specimen geometry with multiple holes turns out to provide more representative results than the geometry containing a single hole. For example, Beumler [4] explains that the stresses measured around the holes are not identical for all five holes. Towards the edge, holes seem to experience slightly lower stresses, which could imply that the single-hole geometry illustrated in Fig. 5.1a does not well represent a hole in a row of holes in a large structure or component.

Nonetheless, most research seems to relate to the original single-hole geometry [5–15]. Hence, this chapter presents the most important experimental observations together with an overview of the methodologies [16–22].

5.2 Definitions and Failure Phenomena

To predict the blunt notch strength of materials, first the failure mechanisms must be understood. Several authors have studied the strength and failure of monolithic metals and FMLs in the presence of circular holes, reporting the characteristic observations. This section presents major definitions and qualitatively describes the failure phenomena providing a basis for the analytical description of failure.

5.2.1 Definitions

The blunt notch strength represents a reduced strength property to capture the influence of a blunt notch or hole in thin panels, like fuselage skin panels. The blunt notch strength in its definition is a nominal strength, equivalent to an average stress in the net section. This is illustrated in Fig. 5.2a, for which Eq. (5.1) defines the blunt notch strength.

In these panels, stresses are present in multiple directions, as illustrated in Fig. 5.2b. Thus failure is defined by an interaction of biaxial and shear stresses. Each of the biaxial stress components illustrated in Fig. 5.2b may represent either tension or compression, while shear may also occur in the direction opposite to the illustrated direction.

For FMLs, the two principal material directions are defined by the longitudinal rolling and longitudinal-transverse direction of the metal layers, denoted, respectively, by L and LT. Aligned with these principal material directions, the corresponding stresses are defined in Fig. 5.2b.

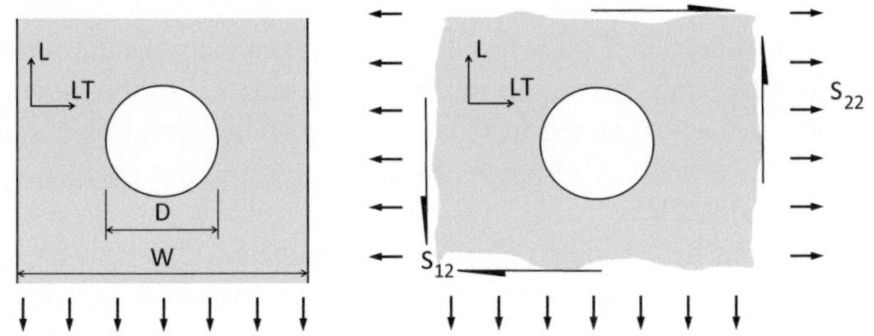

Fig. 5.2 Definition of biaxial blunt notch stresses relative to the FML material directions

5.2.2 Notch Sensitivity and Ductility

In the design of metallic components, the notch sensitivity of materials has been a key parameter in the design against fatigue and failure. The stress concentration factor K_t, discussed in more detail for fatigue initiation in Chap. 7, assumes a linear elastic material response. This assumption implies that failure occurs once the peak stress at the notch equals the failure strength of the material. For a circular hole in a large or infinite panel, this implies that gross or nominal stresses are roughly a factor three lower compared than the failure strength.

However, many low strength and ductile metal alloys appear to be less sensitive to the occurrence of blunt notches in the structure. This has been illustrated by the introduction of a notch factor K_f that is often lower than the stress concentration factor K_t in

$$\sigma_{\text{peak}} = K_f S_{\text{nom}} \tag{5.2}$$

which implies

$$S_{\text{BN}} \sim \frac{S_{\text{ult}}}{K_f} \tag{5.3}$$

The notch sensitivity is then expressed as

$$q = \frac{K_f - 1}{K_t - 1} \tag{5.4}$$

Because both the K_t and K_f are dependent on the notch radius or diameter, the notch sensitivity is also dependent on the notch geometry [23] as illustrated in Fig. 5.3. Key to the notch insensitivity is the low yield strength and the almost perfect plastic behaviour beyond yielding of metals. This forces redistribution of the concentrated stresses near the root of the notch towards a larger surrounding area; see Fig. 5.4.

Evidently, the fibres in FMLs respond linear elastically to the applied load, which increases the strain hardening in the FML compared to the monolithic aluminium constituent, as explained in Chap. 4. The higher strain hardening makes FMLs more sensitive to the occurrence of stress concentrations than the monolithic metal constituent alone.

This is illustrated by the data reported by Vermeeren [25] who compared the strength reduction as result of blunt notches for two common aerospace aluminium alloys and unidirectional FMLs, combining these two alloys with R-glass fibre layers in a 3/2 lay-up.

This comparison is illustrated in Fig. 5.5 where the ratio of nominal blunt notch strength over ultimate strength is plotted against the K_t calculated with the solutions for isotropic material behaviour, i.e.

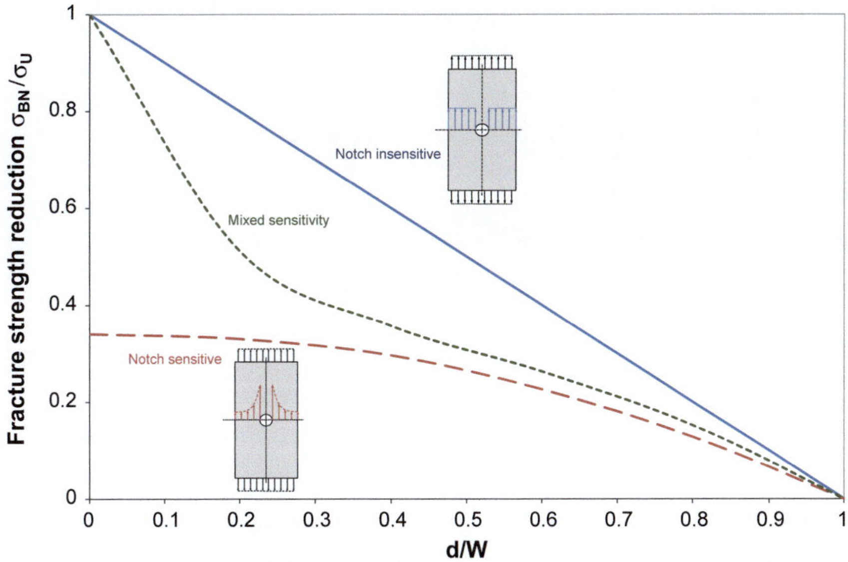

Fig. 5.3 Illustration of notch sensitivity [24]

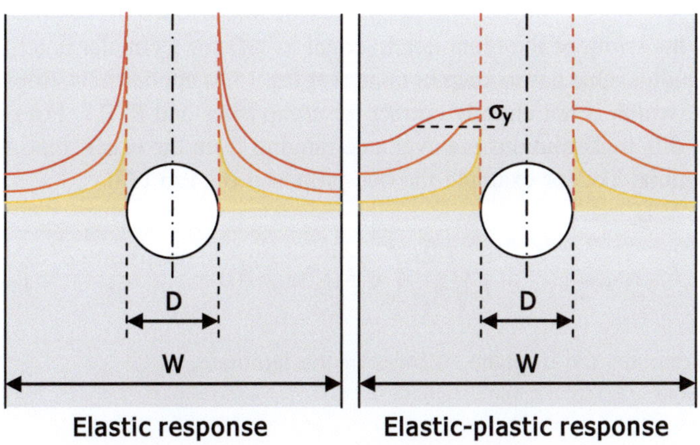

Fig. 5.4 Illustration of elastic and elastic–plastic stress distribution near a notch

$$
\begin{aligned}
K_{t,\text{circle, inf}} &= 3 \\
K_{t,\text{ellipse,inf}} &= 1 + 2\sqrt{\frac{a}{\rho}} \\
K_{t,\text{circle,fin}} &= 2 + \left(1 - \frac{D}{W}\right)^3
\end{aligned}
\tag{5.5}
$$

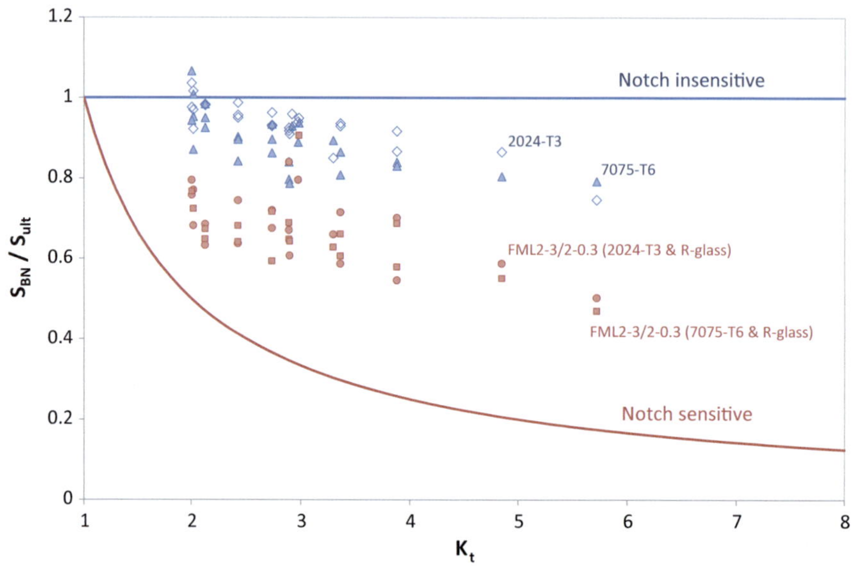

Fig. 5.5 Strength reduction as result of blunt notches for monolithic aluminium alloys and early fibre metal laminates; data from [25], K_t based on Eqs. (5.5)

Here, a is the width of the blunt notch (equal to ½D for a circular notch), and ρ is the notch radius. One has to keep in mind that Eq. (5.5) applies to isotropic material behaviour, which is not entirely correct for composites and FMLs. For orthotropic laminates, different solutions are available ranging from the exact solutions [26] to approximations like for example the one proposed by Tan [27]

$$K_{t,\text{ellipse,inf}} = 1 + \frac{a}{b}\sqrt{\frac{2}{A_{22}}\left(\sqrt{A_{11}A_{22}} - A_{12} + \frac{A_{11}A_{22} - A_{12}^2}{2A_{66}}\right)} \tag{5.6}$$

where A_{ij} denotes the in-plane stiffness of the laminate.

5.2.3 *Biaxial Loading Using Uniaxial Data*

The blunt notch strength values that are determined with the tests standardized for FMLs [28, 29] are uniaxial blunt notch strengths. This means that the contribution of a transverse load is neglected in the determination of the material response.

However, in most aeronautical shell structures for which FMLs are considered, biaxial loading occurs; see Fig. 5.2. Because the stress field surrounding the blunt notch can be described with stress concentration factors, the biaxial cases can be similarly approximated by superposition, i.e. adding the stresses from two uniaxial

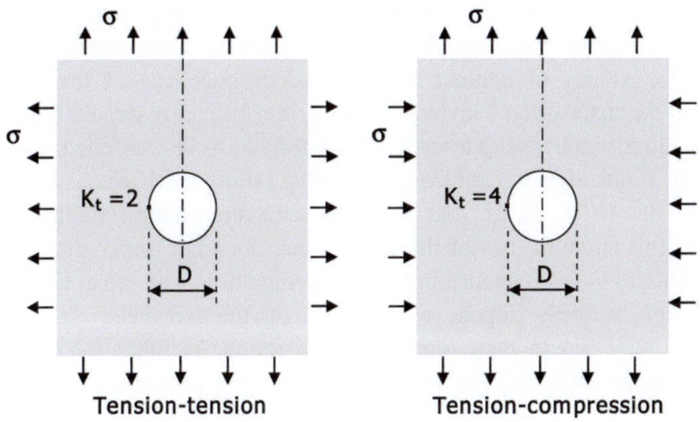

Fig. 5.6 Illustration of the influence of biaxial loading on the stress concentration factor in an infinite isotropic sheet containing a hole

load cases at a particular location in the stress field; see for example Fig. 5.6. Theoretically, the biaxial load case then can be evaluated using uniaxial blunt notch strength. Describing the blunt notch strength with a notch factor K_f is evidently less straightforward, because the value for that factor is generally obtained empirically.

5.2.4 Composite Failure Modes

The major failure modes that have been observed during the blunt notch strength experiments are the formation of plasticity at the notch root, as discussed in Sect. 5.2.1, and characteristic failure modes that are typical for fibre-reinforced polymer composites, i.e. longitudinal splitting of fibre layers, formation of transverse matrix cracking, and delamination between the layers. These composite failure modes are illustrated in Fig. 5.7

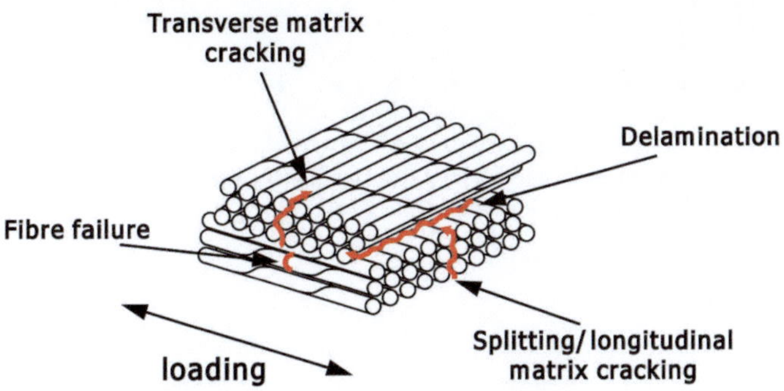

Fig. 5.7 Typical failure mechanisms for fibre-reinforced polymer composites

The fibre-reinforced layers in FMLs have a low stiffness and strength perpendicular to their fibre direction. Therefore, transverse matrix cracking occurs at relatively low values of applied stress. The consequences of this cracking are negligible at the macroscopic level. Therefore, this failure mode, although occurring during the blunt notch strength tests, does not have to be modelled.

However, if one aims to capture all occurring failure mechanisms in a theoretical model, one has to be aware that although transverse matrix failure may occur, resulting in full strength loss of the matrix, this does not imply that the fibres no longer contribute to the response in transverse direction. The intact fibres with their relatively high stiffness impose a constraint on the transverse deformations by restricting contraction in their longitudinal direction. In other words, despite the transverse matrix cracking, the stiffness and strength properties in fibre direction remain.

Longitudinal matrix cracking, or splitting, results from the shear stresses occurring along the fibres in loading direction adjacent to the notch root. As illustrated in Fig. 5.8a, the fibres that are cut by the notch are not loaded, while the intact fibres adjacent to the notch root take up the full peak stress near the notch root. The shear deformation that occurs as a consequence corresponds to the shear stresses at the interface indicated in Fig. 5.8a. Once these shear stresses reach a critical level, splitting may occur.

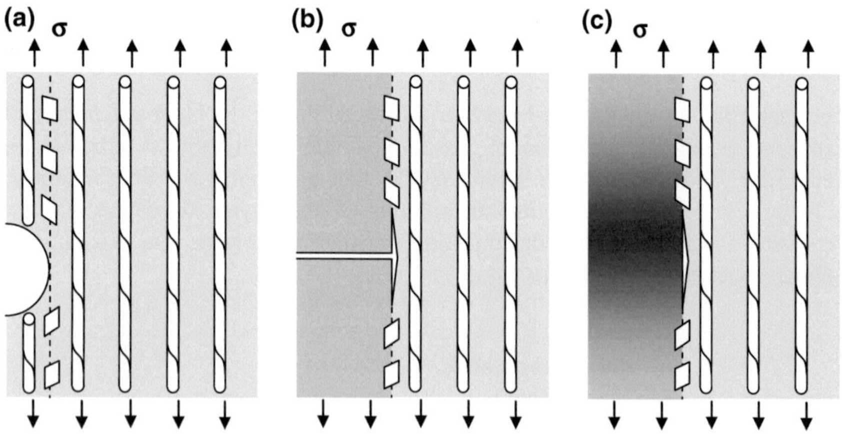

Fig. 5.8 Splitting induced by shear deformation between intact and interrupted fibres (**a**) and analogy with delamination induced by shear deformation between intact and interrupted layers (**b**) or induced by plastic deformation of metallic layers (**c**)

5.2.5 Plasticity-Induced Delamination

Several reports on blunt notch strength experiments report the observation of delamination formation near the notch root of the specimens. The formation of delaminations under quasi-static shear stresses is often described by the critical strain energy release rate. This parameter relates to the critical shear stress and defines the upper limit before delamination initiation or extension may be expected.

The shear stresses between the plies may build up, because the two layers at both sides of the interface are no longer compliant in deformation. This mismatch in deformation field may be induced by the fact that one of the layers is interrupted by, for example, a crack or notch; see Chaps. 8 and 9, but it may also occur because one of the layers is no longer deforming linear elastically. In the latter case, the elastic–plastic deformation of the metallic layers must be compensated for by interlaminar shear to keep the two layers jointed. Both cases are illustrated in Fig. 5.8; the delamination induced by interrupted layers is illustrated in Fig. 5.8b, while the delamination induced by plastic deformation is illustrated in Fig. 5.8c.

Some evidence of the observed delamination shapes in blunt notch specimens prior to failure is given in Table 5.1. Aside from fibre splitting, delaminations form at the root of the notch once the aluminium layers are significantly plastically deforming.

Table 5.1 Observations of the delamination formation during blunt notch failure [32]

Specimen	% σ_{BN}	Failure mechanism	Appearance of etched specimen
1	87	Fibre splitting at notch root	
2	98	Plasticity-induced delamination	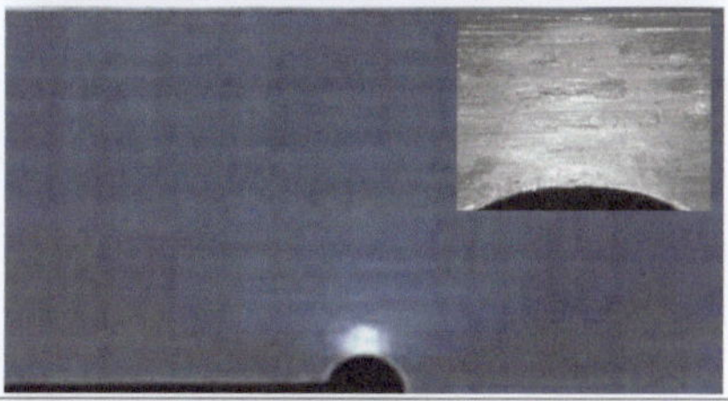
3	100	Dynamic delamination at final failure	

One has to be aware that the formation of this delamination is a key mechanism for the blunt notch strength characteristics of FMLs. Absence of delamination would imply less length over which the fibre layers could distribute their strain, restraining the fibre strain locally at the notch root. As a consequence, the delamination is delaying the failure of the fibres, which has been observed to be the onset of final failure.

In that respect, the delamination shapes observed after final failure (also added to Table 5.1) are not considered to be characteristic for understanding what determines the blunt notch failure strength. They are merely the result of final failure in which all stored energy is dynamically dissipated, mainly in dynamic delamination.

It may therefore be argued that showing the delaminations after blunt notch strength failure has occurred, as several authors have done recently [30, 31], has no particular meaning to the problem. The shapes observed post-mortem are characteristics for dynamic delamination failure, but have no causal relationship to how the load builds up in the specimen prior to failure. It therefore has also no meaning to the recorded blunt notch strength values.

5.2.6 Other Failure Phenomena

Depending on the combination of constituents in the hybrid concept, various failure phenomena may be observed during the blunt notch test in addition to the observations reported in the previous section. The concepts studied by Rensma [33] illustrated in Fig. 2.5, revealed phenomena not observed with standard GLARE. Aside from the obvious ovalization of the open hole, the concept B1 exhibited significant delaminations after failure, but also failure of the inner steel–carbon plies, while the outer aluminium layers remained intact despite significant lateral contraction and yielding. These observations are illustrated in Fig. 5.9.

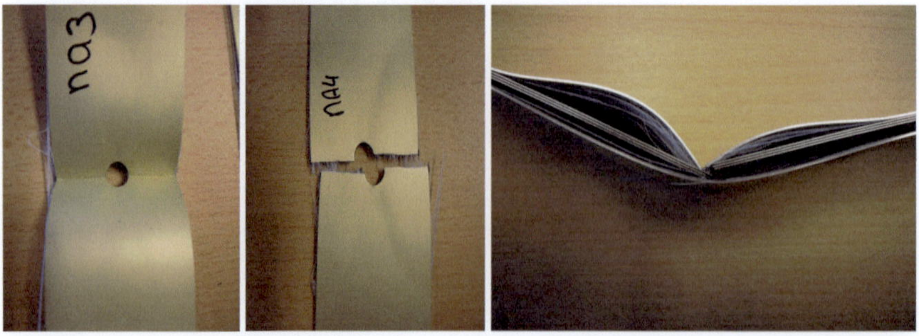

Fig. 5.9 Typical phenomena observed for the hybrid concept B1 blunt notch tests: hole ovalization, lateral aluminium layer contraction, delamination shapes as result of dynamic specimen failure and failure of inner steel-carbon plies while outer aluminium layers yield but remain intact [33]

5.2.7 Blunt Notch Strength and Ultimate Strength

Because blunt notch failure occurs when stresses and strains locally at the perimeter of the blunt notch reach the ultimate strength value, one may assume a relation between blunt notch strength and ultimate strength exists. For the standard GLARE laminates, the general experience is that the ratio between blunt notch strength and ultimate strength is in the range of 0.65–0.85.

van Rooijen illustrated this relationship by taking the blunt notch strength values of FMLs reported in literature and plotting them against the ultimate strength. The relation is given in Fig. 5.10, in which one data point has been added from a blunt notch strength test performed by Rensma [33], who investigated the hybrid laminate concept by varying constituents. The data point in Fig. 5.10 that has been added was determined for a laminate containing two metal types and two composite types, i.e. stainless steel and aluminium with glass fibre and carbon fibre epoxy layers ('Concept B1' in Fig. 2.5).

Apparent in Fig. 5.10 is that the hybrid concepts line up along two distinct trend lines. The FMLs containing aluminium as the metal constituent line up along the lower trend line, indifferent of the fibre type used. The laminates containing titanium as constituents line up along the other trend line indifferent of the fibre type applied.

It is interesting to observe that the hybrid concept containing both aluminium and stainless steel within the laminate, acts similarly as the laminates with titanium as metal constituent. Apparently, the relation between blunt notch strength and

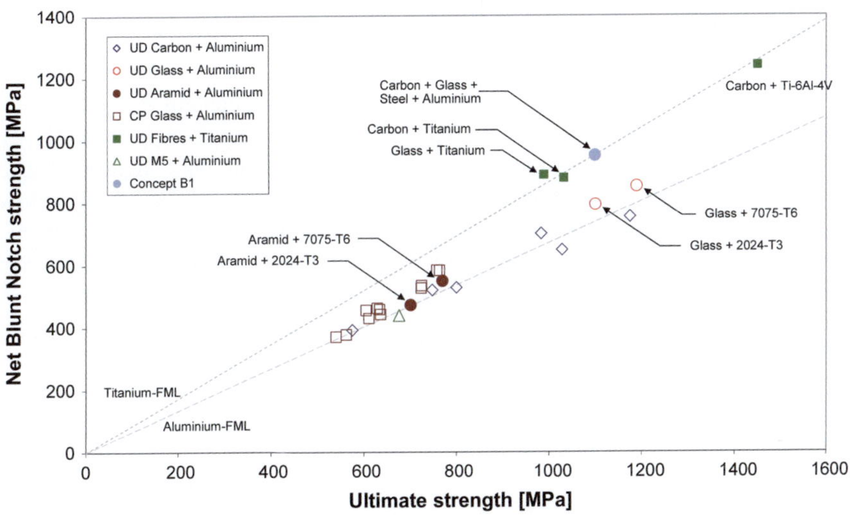

Fig. 5.10 Correlation between the blunt notch strength and ultimate strength for various FMLs (data taken from [34]) with one data point (concept B1) for an FML containing stainless steel, aluminium, glass fibre epoxy and carbon fibre epoxy (data from [33])

tensile failure strength of FMLs or arbitrary hybrid laminates is dictated by the metallic constituents.

5.3 Theoretical Approaches

In the literature, several theories describing blunt notch failure have been proposed. The methodologies are either based on the composite failure theories, for example Tsai–Hill and Norris, or on the metallic plastic failure criteria enhanced with additional composite failure criteria.

5.3.1 *Tsai–Hill/Norris Failure Criteria*

As mentioned in the previous chapter on the theories and criteria for failure, various failure criteria have been proposed over the past years. A theory that accounts for stress interaction, is the one employing the Tsai–Hill and Norris failure criteria, derived from the generalized Hill theory [35]. This theory was applied for blunt notch strength prediction for FMLs by Bosker et al. [36, 37]. The theory accounts for stress interaction by interpolating the stresses while assuming plane stress using an elliptical failure strength surface. The general equation is given by

$$\left(\frac{S_{11}}{S_{11ult}}\right)^2+\left(\frac{S_{11}}{S_{11ult}}\right)\left(\frac{S_{22}}{S_{22ult}}\right)+\left(\frac{S_{22}}{S_{22ult}}\right)^2+\left(\frac{S_{12}}{S_{12ult}}\right)^2=1 \tag{5.7}$$

where S_{11}, S_{22} and S_{12} are defined in Fig. 5.2b.

Whereas the criteria were meant to describe failure for a particular stress system, Bosker proposed to apply them to the blunt notch strength levels S_{BN}. Hence, $S_{11\mathrm{ult}}$ and $S_{22\mathrm{ult}}$ are then the measured blunt notch strengths in the longitudinal and transverse direction. These blunt notch strengths are nominal stresses calculated by considering the applied stresses to be homogeneously distributed over the net section of the specimens. The blunt notch strength is therefore a nominal or net strength property, as gross stresses may vary significantly depending on the specimen geometry. Even though the nominal stress formulation reduces the specimen geometry effect considerably, this effect has been observed to remain to some extent. For this reason, a standardized test procedure for blunt notch strength in FMLs has been formulated in which the specimen geometry has been prescribed [28, 29].

5.3.2 Point and Average Stress Criteria

Whitney and Nuismer [38, 39] introduced two stress criteria for predicting blunt notch failure of orthotropic composites. These criteria have in common that the criteria are based on a location away from the notch root causing the stress concentration, but they differ in the stress considered: the point stress or the average stress.

The major assumption underlying these criteria is that failure occurs once the point stress or average stress at a certain distance away from the notch root equals the un-notched strength of the laminate; see Fig. 5.11.

The substantiation of the proposed criteria relates to the observation that the blunt notch strength of composite laminates normalized for infinite width is dependent on the hole size. This relationship cannot be explained with the stress concentration factor K_t, and therefore requires the addition of a relation as function of the hole diameter.

For orthotropic laminates, K_t is given by [40]

$$K_t^{\infty} = 1 + \sqrt{2\sqrt{\frac{E_2}{E_1}} + \frac{E_2}{G_{12}} - 2\nu_{21}} \tag{5.8}$$

which is another way of writing Eq. (5.6).

The stress distribution along the axis perpendicular to the applied load is based on the solution for an infinite plate containing a circular hole subjected to remote uniaxial tensile loading [26]. This stress distribution is approximated for $x > R$ with

$$\sigma_y(x,0) = \frac{\sigma^{\infty}}{2}\left\{2 + \left(\frac{R}{x}\right)^2 + 3\left(\frac{R}{x}\right)^4 - (K_t^{\infty} - 3)\left[5\left(\frac{R}{x}\right)^6 - 7\left(\frac{R}{x}\right)^8\right]\right\} \tag{5.9}$$

The point stress criterion assumes that failure occurs when the stress at distance d_0 away from the notch root equals the failure strength

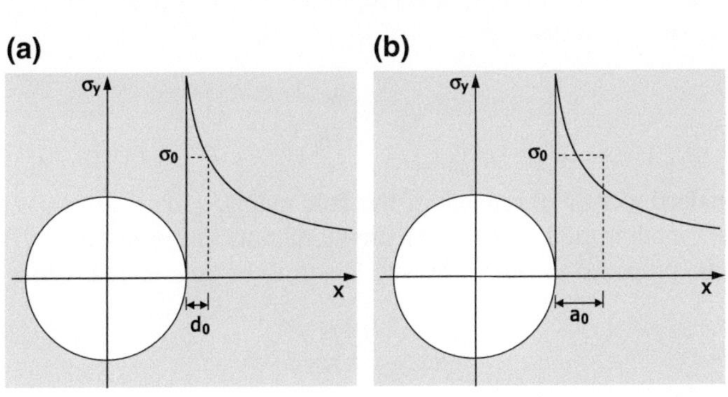

Fig. 5.11 Illustration of the point stress criterion (**a**) and the Average stress criterion (**b**)

$$\sigma_y(R+d_0,0)=\sigma_{\max}, \tag{5.10}$$

whereas the average stress criterion assumes that

$$\sigma_{\max}=\frac{1}{a_0}\int\limits_{R}^{R+a_0}\sigma_y(x,0)\mathrm{d}x \tag{5.11}$$

will lead to failure.

Combination of Eqs. (5.8), (5.9) and (5.10) gives for the ratio between notched and un-notched strength based on the point stress criterion

$$\frac{\sigma_N^\infty}{\sigma_0}=\frac{2}{2+\xi^2+3\xi^4-\left(K_t^\infty-3\right)\left(5\xi^6-7\xi^8\right)} \tag{5.12}$$

with

$$\xi=\frac{R}{R+d_0} \tag{5.13}$$

Similarly, combining Eqs. (5.8), (5.9) and (5.11) yields for the ratio between notched and un-notched strength

$$\frac{\sigma_N^\infty}{\sigma_0}=\frac{2(1-\xi)}{2-\xi^2-\xi^4+\left(K_t^\infty-3\right)\left(\xi^6-\xi^8\right)} \tag{5.14}$$

with

$$\xi=\frac{R}{R+a_0} \tag{5.15}$$

Karlak [41] modified the point stress criterion based upon his observations that the predicted blunt notch strength was proportional to the square root of the circular hole radius. The parameter

$$k_0=\frac{d_0}{\sqrt{R}} \tag{5.16}$$

appeared indeed to be independent of the hole radius.

A second modification to the point stress criterion was proposed by Pipes et al. [42, 43] who proposed the characteristic length to be determined with

$$d_0=\frac{\left(R/R_0\right)^m}{C} \tag{5.17}$$

in which C is the notch sensitivity factor and R_0 a reference notch root radius. For the ratio between notched and un-notch strength one can then obtain Eq. (5.12), but then with ξ defined as

$$\xi = \left(1 + R^{m-1} R_0^{-m} C^{-1}\right)^{-1} \tag{5.18}$$

Based upon various studies, it has been demonstrated that the blunt notch strength could be predicted using constant values for d_0, a_0 or k_0. However, because these parameters vary in value for each laminate lay-up, the parameters must be established experimentally first, before any prediction can be made.

This is immediately the weakness of both stress criteria (with or without the correction of Karlak); they do not provide a generic prediction method for FMLs, unless for each possible lay-up, experiments have first been performed to determine the values of d_0 and a_0. Because, for industrial application of the methodology, the statistical basis for the obtained values is required, an extensive number of blunt notch tests is necessary before predictions can be made. With the given amount of experimental data, more simple and straightforward methods could easily be adopted, such as for instance the MVF method earlier discussed in Sect. 4.11.1.

In addition, van Rijn [44] clearly showed that the assumption of the characteristic lengths d_0 and a_0 to be material parameters is not supported by the test evidence. The application of both stress criteria gave rather poor fits with the obtained experimental results. Although the modification by Pipes et al. appeared to give better results, the drawback of establishing the values for two parameters (C and m) with test data does not compete with the benefits of other, more easily applied, methods.

5.3.3 *Blunt Notch Factor to Ultimate Strength in Net Section*

Müller [45] initially proposed based on the work of Vermeeren [25] to calculate the gross blunt notch strength using

$$S_{\mathrm{BN,gross}} = K_{\mathrm{BN}} \frac{A_{\mathrm{net}}}{Wt} S_{\mathrm{ult}} \tag{5.19}$$

where K_{BN} is the so called blunt notch factor, S_{ult} he ultimate strength of the material, A_{net} the net section of the sheet with width W and thickness t. The ultimate strength of the FML at that time was established with the rules of mixtures using Eq. (4.35).

The disadvantage of this approach was the empirical nature of the method. For example, the values of K_{BN} where determined empirically by Müller [45] using the data of Vermeeren [25] for Glare3-3/2-0.3 and monolithic aluminium to be, respectively, 0.64 and 0.9. Despite that the data of Vermeeren was taken from open-hole blunt notch specimens, Müller successfully applied the approach and values to mechanically fastened joints where the holes are filled and loaded by fasteners.

5.4 Applicability to General Loading Conditions

The theoretical approaches discussed in the previous section use as input laminate properties or strengths which have been generated with uniaxial test specimens. In addition, the theories in most cases have been validated with tests using uniaxial loaded specimen in their principal material directions. Only a few experimental uniaxial off-axis and biaxial loading configurations have been reported [46–48]. This section provides an overview of the general observations during these tests and discusses the consequences for application of the theories.

5.4.1 *Uniaxial Off-Axis Loading*

The Tsai–Hill or Norris failure criterion in Eq. (5.7) describes the interaction of stresses that results in failure. This means that in case an FML specimen is tested in a uniaxial direction that is not aligned with the principal material direction, additional shear stresses may interact with the imposed uniaxial stresses as result of laminate orthotropy.

The stresses in the principal material direction can be calculated from the stresses in loading directions; see Fig. 5.12, with

$$\begin{bmatrix} S_{11} \\ S_{22} \\ S_{12} \end{bmatrix} = \begin{bmatrix} \cos^2\theta & \sin^2\theta & -2\sin\theta\cos\theta \\ \sin^2\theta & \cos^2\theta & 2\sin\theta\cos\theta \\ \sin\theta\cos\theta & -\sin\theta\cos\theta & \cos^2\theta - \sin^2\theta \end{bmatrix} \begin{bmatrix} S_{xx} \\ S_{yy} \\ S_{xy} \end{bmatrix} \tag{5.20}$$

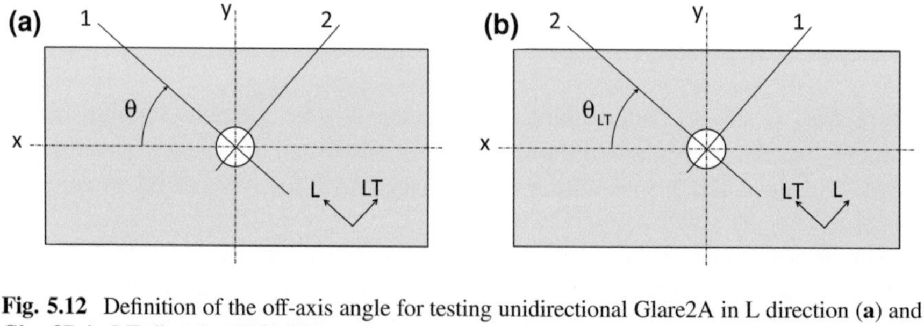

Fig. 5.12 Definition of the off-axis angle for testing unidirectional Glare2A in L direction (**a**) and Glare2B in LT direction [46] (**b**)

Here, one has to be aware that although the application of the load by the test machine is uniaxial, as a result of the off-axis loading of the orthotropic laminate, an imbalance will be created in loading direction. This effectively imposes an additional transverse load from the machine on the specimen. Here, the slenderness ratio of the test specimen determines whether the transverse force that can be calculated using CLT applies, or whether a smaller force must be considered [49].

Hooijmeijer [46] reported blunt notch tests performed on Glare2B-4/3-0.4 and Glare2B-7/6-0.4 under off-axis angles ranging between 0° and 15°. With reference to Fig. 5.12, one has to keep in mind that in case of FMLs often laminates have been considered with a specific application in mind. The difference between Glare2A and Glare2B is the orientation of the fibre layers with respect to the rolling direction of the aluminium. In case of Glare2B, the fibres are oriented in the direction transverse to the rolling direction, and hence the off-axis angle reported by Hooijmeijer was defined as the angle between the loading direction and the LT direction, hereafter referred to as θ_{LT}. The relation between the two angles obviously is

$$\theta = \frac{\pi}{2} - \theta_{LT} \tag{5.21}$$

for θ and θ_{LT} in (rad).

In case of uniaxial loading, the transverse and shear stresses, i.e. S_{yy} and S_{xy}, are zero. This reduces Eq. (5.20) to

$$\begin{aligned} S_{11} &= S_{xx}\cos^2\theta \\ S_{22} &= S_{xx}\sin^2\theta \\ S_{12} &= S_{xx}\sin\theta\cos\theta \end{aligned} \tag{5.22}$$

Substitution into Eq. (5.7) and rewriting then yields

$$S_{xx} = S_{\mathrm{BN}} = \sqrt{\frac{1}{\left(\frac{\cos^2\theta}{S_{11\mathrm{ult}}}\right)^2 - \left(\frac{\cos^2\theta}{S_{11\mathrm{ult}}}\right)\left(\frac{\sin^2\theta}{S_{22\mathrm{ult}}}\right) + \left(\frac{\sin^2\theta}{S_{22\mathrm{ult}}}\right)^2 + \left(\frac{\sin\theta\cos\theta}{S_{12\mathrm{ult}}}\right)^2}} \tag{5.23}$$

The correlation between this relation and the experimental data from [46] is presented in Fig. 5.13.

Here one should note that above-mentioned slenderness ratio of the test specimens is different for the specimens tested by Hooijmeijer [46], compared to the standard blunt notch test specimens [2]. Hooijmeijer tested 100 mm wide specimen, whereas the standard test specimens generally are 50 mm wide.

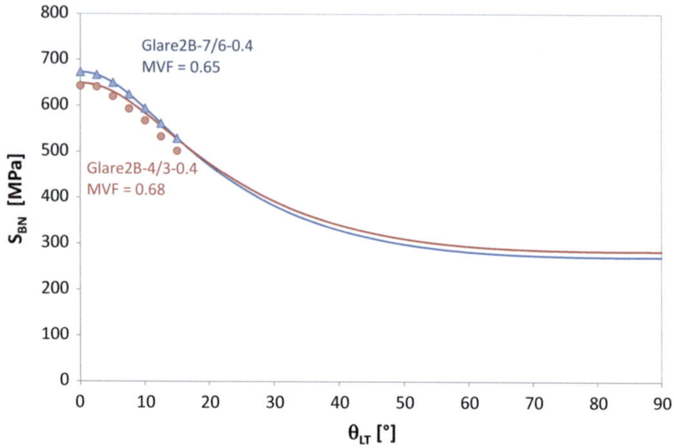

Fig. 5.13 Relation between the blunt notch strength of unidirectional Glare2B and the off-axis angle θ_{LT}

5.4.2 *Shear Loading*

Equation (5.7) uses an ultimate shear strength value $S_{12,\text{ult}}$, which is generally derived from uniaxial blunt notch tests under an off-axis angle of 45° using Eq. (5.20). For the off-axis angle of 45°, Eq. (5.20) reduces to

$$\begin{aligned} S_{11} &= 0.5\ S_{xx} \\ S_{22} &= 0.5\ S_{xx} \\ S_{12} &= 0.5\ S_{xx} \end{aligned} \tag{5.24}$$

and hence Eq. (5.7) can be rewritten to

$$S_{12\text{ult}} = \sqrt{\frac{S_{xx}^2}{4 - \left(\frac{S_{xx}}{S_{11\text{ult}}}\right)^2 + \left(\frac{S_{xx}}{S_{11\text{ult}}}\right)\left(\frac{S_{xx}}{S_{22\text{ult}}}\right) - \left(\frac{S_{xx}}{S_{22\text{ult}}}\right)^2}} \tag{5.25}$$

To determine whether this method is appropriate for establishing the input values for shear, Borgonje [47] tried to perform blunt notch tests under true shear. For this purpose, he modified the Iosipescu test specimen to include a circular notch in the centre, or to alter the edge notches with different radii with or without the V-notch.

Although the values for certain specimen geometries seem to correlate fairly well with the values obtained from the off-axis unidirectional tests (Fig. 5.14), the general conclusion was that further research was necessary to substantiate any assessment. Most failures for example were observed to not occur under pure shear. Instead, buckling deformations occurred near the support areas.

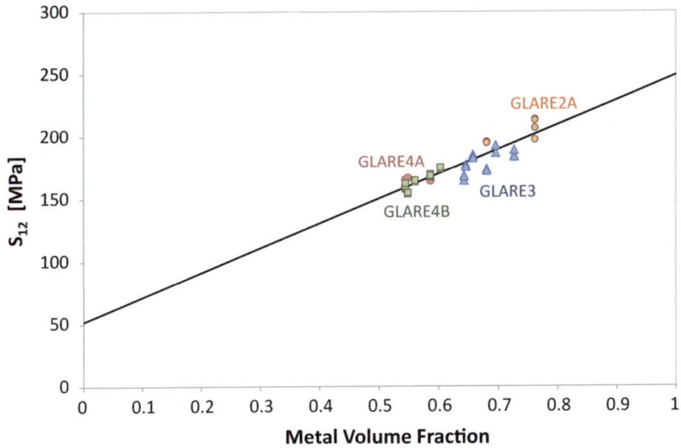

Fig. 5.14 Relation between the ultimate blunt notch shear S_{12} and the metal volume fraction; data from [50]

To improve such studies, Borgonje recommended the application of additional tabs near the support area to avoid the buckling deformation and to allow higher shear stresses in the centre cross section of the specimen. Additionally, alternative specimen geometries must be developed changing the geometry of the edge notches further.

5.4.3 Biaxial Loading

The biaxial blunt notch strength tests are complex tests to perform. The standardized geometry determined for FMLs in [2, 28, 29] is no longer applicable and dedicated specimen geometries must be designed. Especially the load introduction into the specimen adds to the complexity [51], because the constraint induced by specimen clamping should not affect the applied load in the other principal direction.

To validate the Tsai–Hill or Norris failure criterion given by Eq. (5.7), attempts have been made to perform biaxial blunt notch tests in a biaxial load frame. The first series of tests all failed in the load introduction area rather than in blunt notch [48]. A second series of tests, after improving the load introduction, resulted in one blunt notch failure out of three biaxial blunt notch tests [51]. Hence, the general conclusion was that the biaxial blunt notch tests were unsuccessful.

Therefore, the validation of the Tsai–Hill or Norris failure criterion has so far been made with only uniaxial tests under a range of off-axis loading angles [50].

5.5 Simple Methods for Design Purposes

The theoretical description of the blunt notch strength is rather complicated, because the composite laminate theories have to be combined with the elastic-plastic material response of the metallic constituents. A thorough and detailed description therefore not only requires fundamental understanding of the mechanisms, but often requires some assumptions to formulate a mechanistic model. In an engineering environment, especially those related to certification and material qualification, simple methods are therefore often preferred. Some of these simplified approaches are discussed in this section.

5.5.1 *Metal Volume Fraction*

Similar to the approach explained in Sect. 4.8.1, the blunt notch strength can be predicted with the metal volume fraction. This prediction approach applies to uniaxial loading in which the blunt notch strength is predicted based on the blunt notch strength of the metal (generalized for any orientation, because of the metal's isotropy) and the prepreg in that particular orientation

$$S_{\mathrm{BN,lam}} = MVF \cdot S_{\mathrm{BN,metal}} + (1 - MVF) S_{\mathrm{BN,prepreg}} \tag{5.26}$$

This means that the composite plies with fibres perpendicular to the applied load are assumed to not contribute to the blunt notch strength [37]; see also Table 5.2. Similarly, the shear blunt notch strength is predicted with the same rule based on the shear blunt notch strength of the constituents; see Table 5.2.

To demonstrate the ease of approximating blunt notch strength properties of arbitrary hybrid laminates using the MVF method, concept B1 (Fig. 2.5) is taken as example. The properties of the GLARE constituents are known and given in Table 5.2. Without having determined the corresponding properties of the stainless steel and carbon fibre prepreg, a first approximation can be made assuming that for both the same relation between blunt notch and ultimate strength of the constituents holds. The obtained blunt notch properties for the Concept B1 constituents are then given in Table 5.3 together with their corresponding volume fraction.

Equation (5.26) is reformulated for an arbitrary hybrid laminate as

$$S_{\mathrm{BN,lam}}(\varphi) = \frac{\sum n_i t_i}{t_{\mathrm{lam}}} \cdot S_{\mathrm{BN},i}(\varphi) \tag{5.27}$$

Table 5.2 Blunt notch strength of the GLARE constituents [37]

Material	$S_{\mathrm{BN,xx}}$ (MPa)	$S_{\mathrm{BN,yy}}$ (MPa)	$S_{\mathrm{BN,xy}}$ (MPa)
Aluminium 2024-T3	417	394	249
UD S2-glass/FM94 prepreg	1193	0	52

Table 5.3 Overview of the blunt notch strength contributions of the constituents of concept B1

Material	S_{BN} (φ = 0°) (MPa)	S_{BN} (φ = 90°) (MPa)	Volume fraction (%)
Aluminium 2024-T3	417	394	42.4
UD S2-glass/FM94 prepreg	1193	0	14.1
Stainless steel[a]	1879	1775	15.9
T700/FM943[b]	1269	0	27.6

[a]Estimated with the same S_{BN}/S_{ult} ratio as for aluminium
[b]Estimated with the same S_{BN}/S_{ult} ratio as for S2/FM94

Table 5.4 Blunt notch test results for concept B1

Concept B1	F_{max} (N)	W (mm)	D (mm)	t_{lam} (mm)	$S_{BN,gross}$ (MPa)	$S_{BN,net}$ (MPa)
Specimen 1	44,835	30	6	1.95	766.4	958.0
Specimen 2	44,672	24	6	1.95	763.6	954.5
Specimen 3	45,103	24.1	6	1.97	763.2	953.9
Specimen 4	44,116	24.1	6	1.96	750.3	937.8

To calculate the blunt notch strength of Concept B1, one additional calculation step needs to be made. The T700/FM layers are oriented under an angle of ±15° with respect to the principal axis of the laminate. To estimate the blunt notch strength under this angle, as a first approximation the $\cos(\varphi)^4$ contribution can be taken from the value for $S_{BN}(\varphi = 0)$, which gives $S_{BN}(\varphi = 15°) = 1105$ MPa.

With Eq. (5.27) and the volume fraction of the constituents given in Table 5.3, the blunt notch strength of the overall laminate can then be calculated. This gives $S_{BN,lam} = 949$ MPa. Considering the approximations made for the raw constituents and the simplicity of the procedure, the value seems quite accurate if compared to the values obtained in the tests, given in Table 5.4.

5.5.2 *Neuber's Postulate*

Another method to estimate the uniaxial blunt notch strength is based on the assumption that the linear elastic peak stress at the open hole is reduced due to plasticity in the metal layer, with as a consequence an increase in strain. With the elastic peak stress and strain described with the stress concentration factor, this implies [23]

$$\begin{aligned} \sigma_{\text{peak}} &< K_t \sigma_{\text{nom}} \\ \varepsilon_{\text{peak}} &> K_t \varepsilon_{\text{nom}} \end{aligned} \tag{5.28}$$

The postulate of Neuber states that for the initial deviation of the linear elastic material response, i.e. the initial portion of the elastic–plastic stress–strain curve, it can be assumed that

$$\sigma_{\text{peak}} \varepsilon_{\text{peak}} = K_t^2 \sigma_{\text{nom}} \varepsilon_{\text{nom}} = K_t^2 \frac{\sigma_{\text{nom}}^2}{E} \tag{5.29}$$

The blunt notch strength is defined as the net stress, which is equal to the definition of nominal stress. Therefore, failure may be assumed to occur when the peak stress equals the ultimate strength of the material, corresponding to the ultimate strain. Therefore, the blunt notch strength may be approximated with

$$S_{\text{BN}} = \frac{\sqrt{E S_{\text{ult}} \varepsilon_{\text{ult}}}}{K_t} \tag{5.30}$$

With this method, it can be discussed which stress concentration factor is most appropriate. Generally, the stress concentration factor for the FMLs is determined with the expression accounting for the orthotropic nature of FMLs, according to the theory explained in detail in Sect. 7.4. This includes the similarity approach to account for the finite dimensions of the blunt notch specimen, Eq. (7.3).

There may be an advantage to Eq. (5.30), because it is explicitly based on the stress concentration factor K_t, which accounts for the specimen geometry if the similarity approach for finite width is applied. In that case, one may assume that the actual dimensions of the specimen that may have some influence on the net blunt notch strength should be described by this approach. The methods mentioned earlier only implicitly address that, as they are based on a standardized specimen geometry [2].

However, the major disadvantage of Eq. (5.30) is that it violates the condition under which Neuber's postulate may be applied; the initial part of the stress–strain curve beyond first yielding. As a consequence of extrapolating the relations to the final failure of the materials, i.e. ultimate strength and strain, the method becomes highly inaccurate when FMLs are considered with limited fibre contributions to the mechanical properties.

High strain hardening, as observed in unidirectional FMLs, is favourable for the method, because the error in extrapolating Eq. (5.29) towards final failure then remains limited. The error becomes larger for cross-ply laminates, which have less strain hardening, and the error is extremely large for unidirectional FMLs perpendicular to the fibre direction. In that case, the strain hardening is negligible, and the error in applying Neuber's postulate to final failure is maximal.

References

1. Bosker OJ (1998) Blunt notch strength of aluminium 2024-T3, report B2v-98-33. Delft University of Technology, Delft
2. Brügemann V (2003) Test procedures for fibre metal laminates, report TD-R-03-005. Fibre Metal Laminates Centre of Competence, Delft, The Netherlands
3. ASTM Standard D5766/D5766M, Standard test method for open-hole tensile strength of polymer matrix composite Laminates1. ASTM International, 100 Barr Harbor Drive, PO Box C700, West Conshohocken, PA 19428-2959. United States
4. Beumler T (2004) Flying GLARE®, A contribution to aircraft certification issues on strength properties in non-damaged and fatigue damaged GLARE® structures. PhD dissertation, Delft University of Technology, Delft, The Netherlands
5. Bosker OJ (1998) Uni-axial blunt notch strength of GLARE 3 and GLARE 4, report B2v-98-28. Delft University of Technology, Delft
6. Bosker OJ (1998) Overview of existing literature on the blunt notch strength of GLARE, report B2v-98-35. Delft University of Technology, Delft
7. Ypma M (2000) Overview of tests concerning the influence of temperature and environmental exposure on GLARE, report B2v-00-41. Delft University of Technology, Delft
8. Hooijmeijjer PA (2000) Blunt notch strength of GLARE at elevated test temperatures, report B2v-00-68. Delft University of Technology, Delft
9. Mortier W, Pellenkoft FJ (2001) Blunt notch strength of open hole versus filled hole, report B2v-01-04. Delft University of Technology, Delft
10. Pellenkoft FJ, Mortier WJ (2001) Influence hole diameter on blunt notch strength, report B2v-01-05. Delft University of Technology, Delft
11. Hooijmeijjer PA (2001) Experimental project to assess the failure mechanisms during blunt notch of GLARE under 45°, report B2v-01-33. Delft University of Technology, Delft
12. Borgonje B (2009) GLARE outdoor exposure program: blunt notch strength—final issue, report B2v-02-40. Delft University of Technology, Delft
13. Hooijmeijer PA (1998) Blunt notch strength of spliced GLARE, inside doubler, preliminary thesis. Delft University of Technology, Delft
14. Mattousch A (1993) Open hole properties of some GLARE grades, report TD-R-93-014. Structural Laminates Company
15. van Rijn JCFN (1995) Blunt notch behaviour of GLARE3, NLR CR 95237 L. National Aerospace Laboratory NLR
16. Bosker OJ (1998) Finite element calculations to predict the uni-axial on-axis and off-axis blunt notch strength of GLARE and comparison with test results, report B2v-98-34. Delft University of Technology, Delft
17. Hagenbeek M (2000) Investigation into applicability of analytical methods for blunt notch strength prediction of FML's, report B2v-00-57. Delft University of Technology, Delft
18. Bosker OJ (2000) Progress on the development of a GLARE blunt notch strength prediction tool, report B2v-00-66. Delft University of Technology, Delft
19. Bosker OJ (1998) Development of a GLARE blunt notch strength prediction tool for the Airbus A3XX. MSc thesis, Delft University of Technology, Delft
20. Vermeeren CAJR (1995) The residual strength of fibre metal laminates. PhD dissertation, Delft University of Technology, Delft
21. van Rijn JCFN (1994) Stress distribution around a circular hole in a fibre metal laminate finite width plate specimen, NLR TP 94058 L. National Aerospace Laboratory NLR
22. van Rijn JCFN (1994) Preliminary model for the blunt notch behaviour of fibre metal laminates, NLR TP 94532 L. National Aerospace Laboratory NLR
23. Peterson RE (1953) Stress concentration design factors. Wiley, New York
24. Hagenbeek M (2001) Investigation of the finite width and notch size effect for the analytical blunt notch strength prediction method, report B2v-01-20. Delft University of Technology, Delft
25. Vermeeren CAJR (1990) The blunt notch behaviour of metal laminates: ARALL and GLARE, report LR-617. Delft University of Technology, Delft, The Netherlands

26. Lekhnitskii SG (1968) Anisotropic plates (Translated from the Second Russian Edition by S. W. Tsai and T. Cheron). Gordon and Breach Science Publishers, Inc., New York
27. Tan SC (1987) Laminated composites containing an elliptical opening. I. Approximate stress analyses and fracture models. J Compos Mater 21:925–948
28. Bosker OJ, de Kraker L (1998) Determination of optimum blunt notch specimen geometry, report B2v-98-37. Delft University of Technology, Delft
29. Mattousch A (1993) Test procedures for fiber metal laminates, report TD-R-93-003. Structural Laminates Company, Delft, The Netherlands
30. Wu G, Tan Y, Yang J-M (2007) Evaluation of residual strength of notched fiber metal laminates. Mater Sci Eng A 457:338–349
31. Yeh P-C, Chang P-U, Yang J-M, Wu PH, Liu MC (2011) Blunt notch strength of hybrid boron/glass/aluminum fiber metal laminates. Mater Sci Eng A 528:2164–2173
32. Meziere Y (2000) Experimental project to assess the failure mechanisms during blunt notch of GLARE, report B2v-00-47. Delft University of Technology, Delft
33. Rensma E (2007) Investigation of innovative concepts for hybrid structures. MSc thesis, Delft University of Technology, Delft
34. Rooijen RGJ (2006) Bearing strength characteristics of standard and steel reinforced GLARE. PhD dissertation, Delft University of Technology, Delft
35. Sun CT, Quinn BJ, Tao J, Oplinger DW (1996) Comparative evaluation of failure analysis methods for composite laminates, report DOT/FAA/AR-95/109. Department of Transportation, Federal Aviation Administration, Washington D.C
36. Bosker OJ (2000) GLARE uni-axial blunt notch test results, report B2v-99-27. Delft University of Technology, Delft
37. Bosker OJ (2001) Blunt notch strength. In: Vlot A, Gunnink JW (eds) Fibre metal laminates—an introduction. Kluwer Academic Publishers, Dordrecht, The Netherlands
38. Whitney JM, Nuismer RJ (1974) Stress fracture criteria for laminated composites containing stress concentrations. J Compos Mater 8:253–265
39. Nuismer RJ, Whitney JM (1975) Uniaxial failure of composite laminates containing stress concentrations. Fract Mech Compos ASTM-STP 593:117–142
40. Konish HJ, Whitney JM (1975) Approximate stresses in an orthotropic plate containing a circular hole. J Compos Mater 9:157–166
41. Karlak RF (1977) Hole effects in a related series of symmetrical laminates. In: Ornie JA, Crossman FW (eds) Failure modes in composites IV. The Metallurgical Society of AIME, Warundale, PA, pp 105–117
42. Pipes RB, Wetherhold RC, Gillespie JW (1979) Notched strength of composite materials. J Compos Mater 13:148–179
43. Pipes RB, Gillespie JW, Wetherhold RC (1979) Superposition of notched strength of composite laminates. Polym Eng Sci 19:1151–1155
44. van Rijn JCFN (1992) The use of composite fracture models to describe the blunt notch behaviour of metal laminates. NLR TP 92061 U, National Aerospace Laboratory NLR
45. Müller RPG (1995) An experimental and analytical investigation on the fatigue behaviour of fuselage riveted lap joints, the significance of the rivet squeeze force, and a comparison of 2024-T3 and GLARE 3. PhD dissertation, Delft University of Technology, Delft
46. Hooijmeijjer PA (2002) GLARE 2 under small off-axis angles, report B2v-02-01. Delft University of Technology, Delft
47. Borgonje B (2001) Blunt notch strength testing under shear loading, report B2v-01-12. Delft University of Technology, Delft
48. Bosker OJ (2000) GLARE bi-axial blunt notch test results, report B2v-00-30. Delft University of Technology, Delft
49. Herakovich CT, Aboudi J, Lee SW, Strauss EA (1988) Damage in composite laminates: effects of transverse cracks. Mech Mater 7(2):91–107
50. Bosker OJ (2001) Tsai-Hill failure criterion to predict the bi-axial blunt notch strength of GLARE, report TD-R-00-014. Structural Laminates Industries
51. Brügemann VP (2002) Validation of the load introduction for bi-axial blunt notch tests, report B2v-02-37. Delft University of Technology, Delft

Chapter 6
Bearing Strength

Abstract The bearing strength of FMLs is discussed with respect to the individual constituent phenomena. In particular the two main methods to experimentally obtain the bearing strength are discussed, addressing ply delamination buckling as an important phenomenon. The influence of the various characteristic specimen dimensions on the bearing strength is discussed, and theories are presented to predict the bearing strength of FMLs based on their constituent materials.

6.1 Introduction

The bearing strength is an important parameter for static strength evaluations of mechanically fastened joints. The load transfer from one sheet to another sheet through pin loading requires assessment of the strength of the pin–hole detail and evaluation of the related failure mechanisms.

Aside from the rivet or bolt failure mode, the sheet may fail in different modes. These failure modes are illustrated in Fig. 6.1. The type of failure mode that occurs often depends on the geometrical dimensions of the pin–hole joint, i.e. the diameter-to-width ratio (*D*/*W*) and the edge distance-to-diameter ratio (*e*/*D*). High values of the *D*/*W* ratio often result in net-section failure, whereas low values of *e*/*D* often result in shear-out failure.

This chapter primarily focuses on the bearing strength failure, Fig. 6.1c. Similar to the blunt notch strength, discussed in the previous chapter, the failure phenomena for bearing strength in FMLs differ from metallic or fibre reinforced polymer composite materials. Bearing failure is defined as the given amount of permanent deformation of the hole loaded by the pin. In metals, this deformation is primarily caused by the plastic deformation of the material, while for composite materials, this deformation may be the result of delamination buckling, fibre buckling, fibre splitting and other fracture mechanisms.

This chapter starts by describing the definition of bearing strength as it is considered for mechanical joining in FMLs. The failure phenomena reported in the literature will be described to provide a background for the evaluation methods for bearing strength in FMLs given thereafter.

R. Alderliesten, *Fatigue and Fracture of Fibre Metal Laminates*, Solid Mechanics and Its Applications 236, DOI 10.1007/978-3-319-56227-8_6

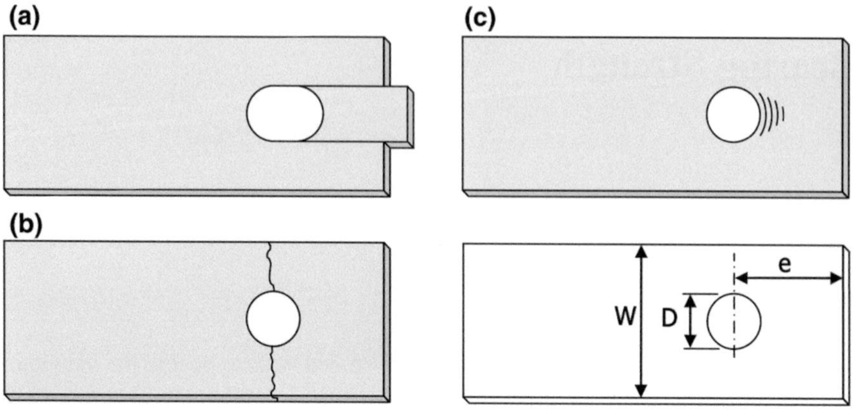

Fig. 6.1 Typical failure modes of pin-loaded holes; shear-out (**a**), net-section (**b**) and bearing failure (**c**)

6.2 Definition of Bearing Strength

The bearing strength is typically described as the pin load divided by the area defined by the pin diameter and sheet thickness, as illustrated in Fig. 6.2. The pin load required to cause failure is then referred to as the bearing ultimate strength (BUS).

Because the FMLs developed up until today have been certified as metallic structures, the bearing strength allowables are defined similar to the allowables for

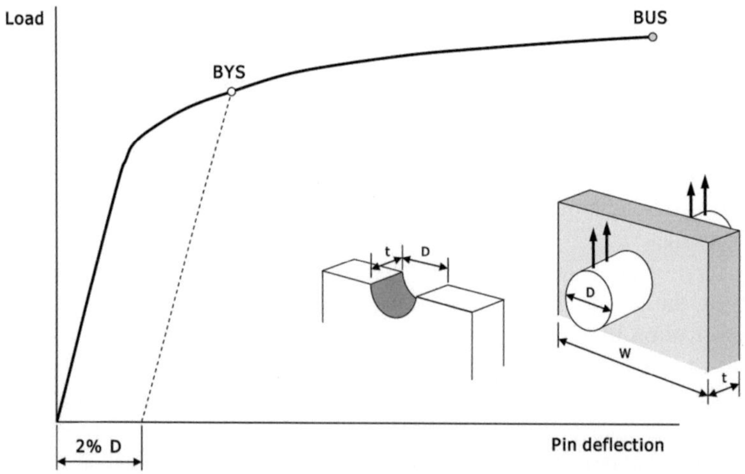

Fig. 6.2 Definition of the bearing allowables according to the ASTM standards [1]; the bearing strength is defined as the pin load divided by the area of pin diameter times thickness

monolithic metallic structures. These allowables are related to the bearing yield strength (BYS) and BUS illustrated in Fig. 6.2.

6.3 Failure Phenomena

6.3.1 Delamination Buckling

The failure modes that can occur during pin loading FMLs are either typical metallic or typical composite failure modes. The deformation indicated in Fig. 6.2 can be attributed to the plastic deformation of the metallic layers in the FML and to the composite fracture mechanisms at the same time. However, as a result of the laminated structure of FMLs, delamination buckling may occur as an additional failure mode when either the metallic layers are too thin, or when insufficient constraint is provided around the pin-loaded hole. The out-of-plane deformation of the laminate illustrated in Fig. 6.3 may induce interlaminar stresses of such magnitude that delamination between the layers occurs. This delamination implies that the layers are no longer joined and supported by each other, resulting in buckling of the individual metal layers.

As a consequence, the lack of hole constraint in the pin-bearing tests for metallic sheets [1] may be insufficient to observe bearing failure in the FML, making a pin-bearing test with anti-buckling constraints necessary [2]. This type of pin-bearing tests is often referred to as bolt bearing [3, 4], as it represents the clamping provided by the bolt head and washer. Typical BYS and BUS values for GLARE and ARALL laminates are given in Table 6.1.

The difference in typical load–deflection curves that one may obtain with either the ASTM E238-84 [1] or the ASTM D953-54 [2] test standard is illustrated in Fig. 6.4. Due to the lack of constraint in the first standard test, delamination buckling may occur at relatively low stress levels, whereas bearing failure modes can be observed when constraint is provided according to the latter standard test.

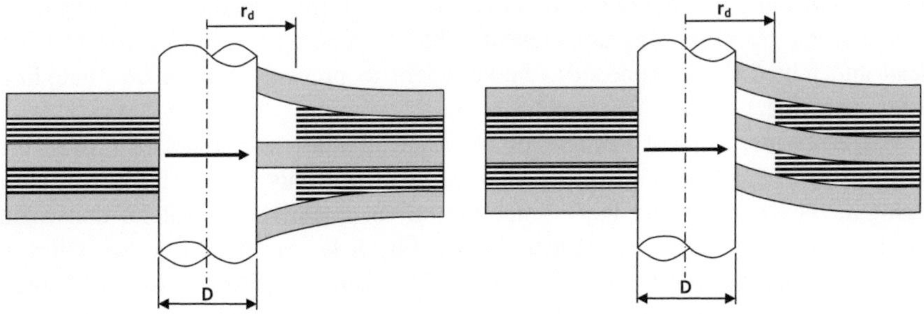

Fig. 6.3 Illustration of symmetric and asymmetric out-of-plane deformation due to pin loading [4]

Table 6.1 Standardized ARALL and GLARE grades [4]

Grade of laminate and lay-up	ASTM E238-84		ASTM D953-54	
	BYS[a] (MPa)	BUS[b] (MPa)	BYS (MPa)	BUS (MPa)
GLARE2-3/2-0.3	–	549	530	709
GLARE3-3/2-0.3	–	537	546	789
GLARE4-3/2-0.3	–	510	518	658
ARALL2-3/2-0.3	–	563	492	727
ARALL3-2/1-0.3	–	593	716	946
ARALL3-3/2-0.3	–	556	615	825

[a]Bearing yield strength could not be determined
[b]Failure by delamination buckling

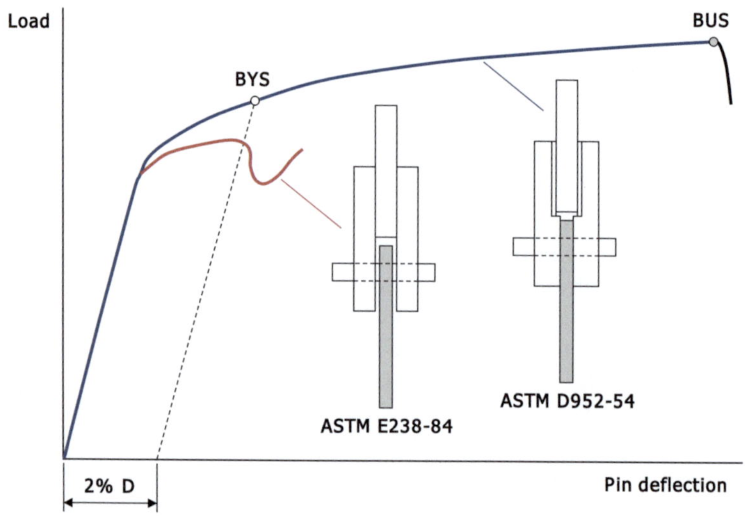

Fig. 6.4 Characteristics load–deflection curves for pin-bearing and bolt-bearing tests [4]

Slagter [4] investigated the strength of mechanically fastened joints in FMLs. He observed that in the numerous static failure tests performed on these types of joints, delamination buckling was not present. The constraint provided by the fastener head and joined sheets appears to be sufficient to prevent delamination buckling from occurring.

For assessment of the strength of mechanically fastened joints in FMLs, the bearing type of failure is therefore of prime interest. This means that for bearing strength, the characteristic load–deflection curve obtained with lateral constraint (ASTM D953-54 standard—illustrated in Fig. 6.4), is the curve of interest. Therefore, delamination buckling is not considered in the remainder of this chapter, and only bearing failure is discussed. More information on delamination buckling due to pin loading of holes is given by Slagter [4].

Another aspect often not explicitly addressed, but worth to note is the influence of the laminates Young's modulus on the determination of the BYS. This is illustrated for unidirectional and cross-play FMLs in Fig. 6.5. The offset definition of 2%*D* in the determination of BYS implicitly incorporates an influence of the elastic stiffness. In addition, depending of the amount of strain hardening after yielding FMLs with relatively higher BYSs may end up with relatively lower BUSs.

6.3.2 Bearing Failure

In experiments, the different failure modes illustrated in Fig. 6.1 may be chosen by carefully selecting the geometry of the specimens. Limiting the edge distance (*e*/*D*) too much will cause shear-out, while taking the diameter too large compared to the specimen width, most likely will give net-section failure.

However, although a standardized test may be defined for which the bearing type of failure mode is most common [7], the other failure modes may not be fully excluded. In fact, depending on the FML constituents considered, the specimen geometry may need to be redefined in order to avoid net-section failure or shear-out [8]. That fact even applies to the lateral constraint applied to the specimen [9]; depending on the thickness of the individual metallic sheets, even the slightest delamination may induce buckling of the metallic sheets.

In addition to that, the bearing failure mode may occur together with crack formation in width direction (initiation of net-section failure) and in loading direction (initiation of shear-out). This was observed by Buczynski [10] who tested

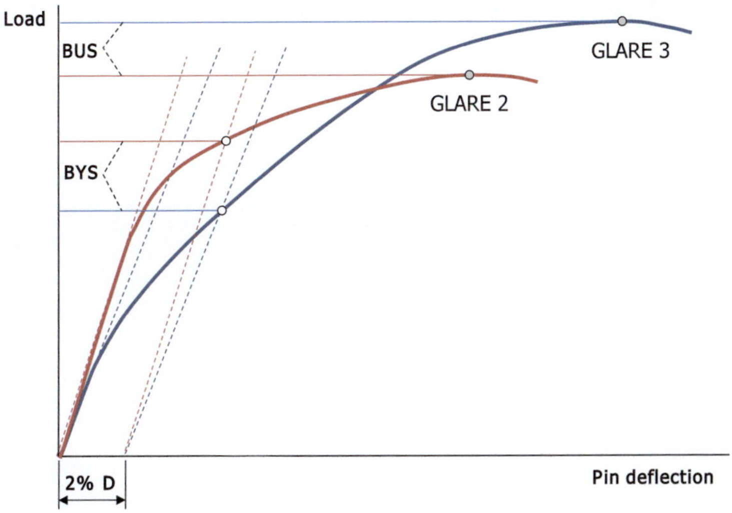

Fig. 6.5 Influence of Young's modulus on the measured BYS and BUS [5, 6]; note the similarity with the influence illustrated in Fig. 2.2

FMLs made of 0.1 mm thin stainless steel sheets reinforced by carbon fibre epoxy layers. The very thin sheets easily buckled due to the lack of lateral support, but the typical bearing failure often occurred together with shear-out cracking (Fig. 6.6) and net-section onset cracking (Fig. 6.7).

The observations of Buczynski are not exclusively related to the application of thin stainless steel foils, as demonstrated by the experimental observations of Frizzell et al. [11].

Here one should bear in mind that the definition of bearing strength, related to the 2%*D* deflection, may be easily obtained with, for example, ARALL and GLARE, containing ductile aluminium alloys, but requires more damage (mechanisms) before similar deflections or deformations are obtained with high stiffness stainless steel alloys and carbon fibre epoxy systems.

Cross sections made by Buczynski revealed a high amount of buckling deformation in the individual stainless steel sheets before 2% permanent deformation, see Fig. 6.8. The observations seem in agreement with the observations reported by

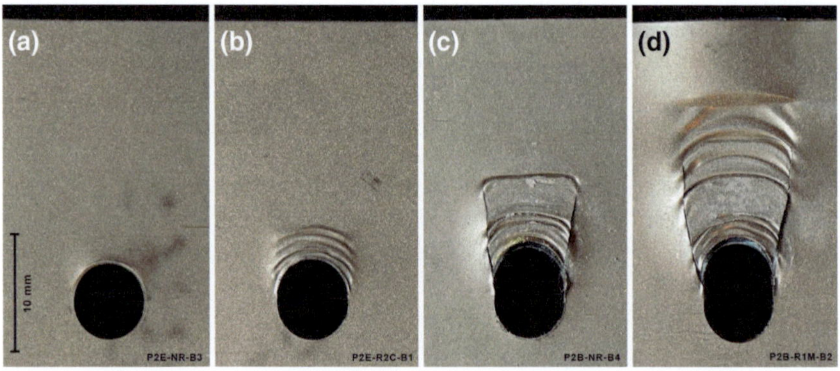

Fig. 6.6 Typical stages of bearing failure combined with buckling and shear-out cracking in stainless steel–CFRP laminates observed in four different specimens [10]

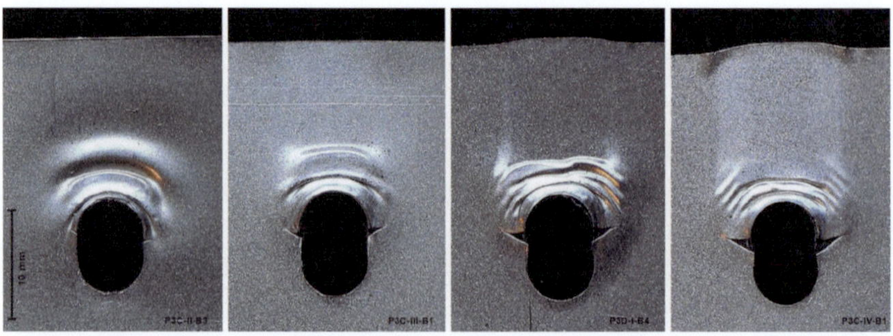

Fig. 6.7 Typical combinations of buckling and bearing failure and net-section onset cracking in laterally constrained stainless steel–CFRP laminates [10]

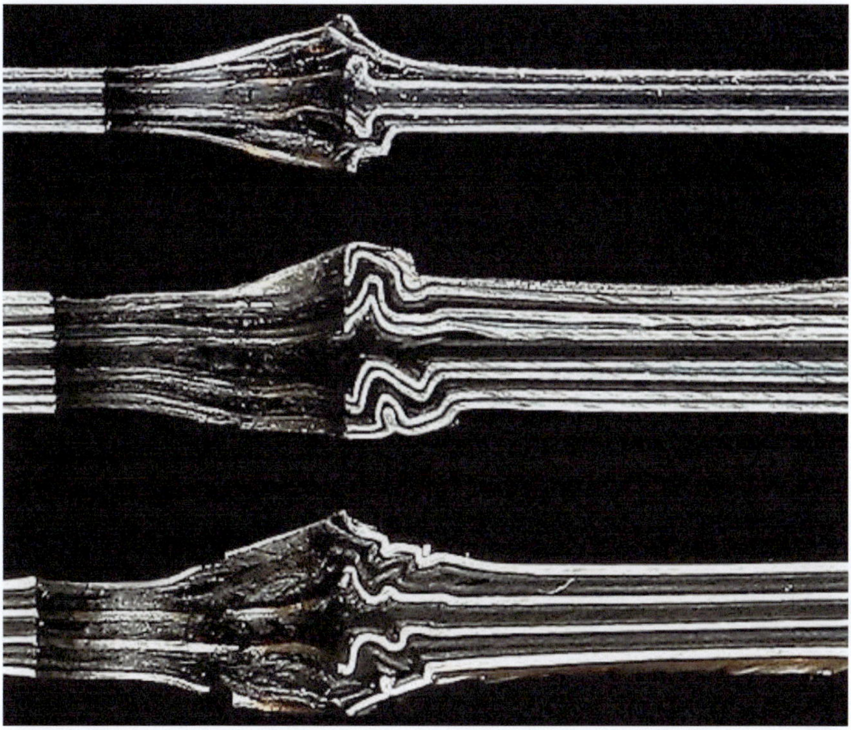

Fig. 6.8 Typical out-of-plane deformations as illustrated in Fig. 6.3 in stainless steel–CFRP laminates with 0.1 mm thin steel sheets [10]

Yamada et al. [12] for FMLs using thin titanium sheets together with carbon fibre epoxy layers.

A significant increase in bearing strength values could be obtained by improving the stability of the individual stainless steel sheets by increasing the thickness. However, Buczynski illustrated that even bonding two 0.1 mm sheets together may not be sufficient to increase the buckling stability, as the flexibility of the adhesive would not sufficiently support the individual plies, see the centre specimen in Fig. 6.8.

6.4 Diameter-to-Thickness Ratio

Hakker [5, 6] studied the bearing strength of unidirectional GLARE laminates. Aside from the edge distance-to-diameter (*e*/*D*) and diameter-to-width (*D*/*W*) ratio, he investigated the effect of diameter-to-thickness (*D*/*t*) ratio. The results of his experiments are plotted in Fig. 6.9. These data have been presented before by Wu et al. in [13].

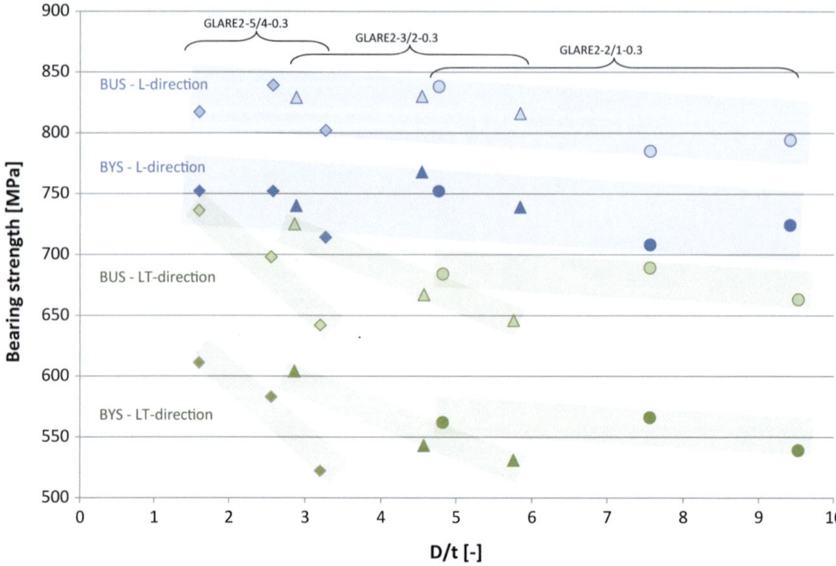

Fig. 6.9 Relation between the bearing strength and the diameter *t* thickness ratio *D*/*t*, data from [6]. Three diameters were tested: 4, 6.35 and 8 mm

Although Hakker concluded that there is not a clear or significant effect of increasing the *D*/*t* ratio on the bearing yield and BUS, a slight trend still can be observed. Figure 6.9 illustrates that increasing the *D*/*t* ratio decreases the bearing strength slightly. This trend should be attributed to the effect of the pin diameter itself, as was also pointed out by Hakker. Increasing the pin diameter decreases the bearing strength, which implies that the largest pin diameter of 8 mm, most commonly applied, yields the lowest strength. In any case, the results presented by Collings [14] seem to suggest that with sufficient lateral constraint, the effect of *D*/*t* will disappear.

6.5 Influence of the Diameter-to-Width Ratio

The diameter-to-width ratio (*D*/*W*), or *W*/*D* ratio, of the specimens used to establish the bearing strength influences the failure mode. As mentioned before, too low values of *W*/*D* will induce net-section failure, which may result in lower strength values. This aspect was evaluated by among others Meola et al. [15] who tested a modified GLARE laminate at several diameter-to-width ratios.

A key observation is that for values of *W*/*D* above 2, the measured bearing strength remained constant (Fig. 6.10). This implies that in order to ensure bearing type of failure mode, a width of twice the diameter should be sufficient. Comparing this value with glass fibre reinforced polymer laminates [16], this value is very low [15].

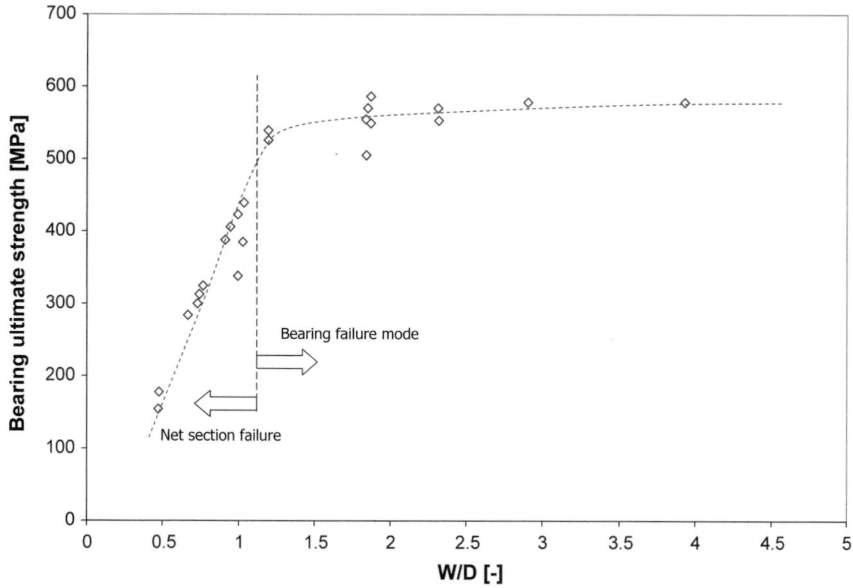

Fig. 6.10 Relation between measured bearing strength and diameter-to-width ratio (data from [15])

6.6 Influence of Edge Distance

The edge distance-to-diameter ratio (*e*/*D*) may also influence which failure mode occurs [17]. Hence, the static strength evaluation of mechanically fastened joints should consider the influence of the edge distance. Where in monolithic aluminium, the edge distance of, for example, $e/D = 2$ may be sufficient to obtain bearing type of failure, selection of the edge distance in FMLs may require some further elaboration.

To investigate the influence of edge distances on the bearing strength and failure modes, Wu and Slagter [18], Broest [19] and Meola et al. [15] performed bearing strength tests at different values for the edge distance.

The results for unidirectional GLARE2B and cross-ply GLARE3 and GLARE4B are consistent between the experiments performed by Wu and Slagter and the ones reported by Broest. It is therefore interesting to observe the difference between the standardized GLARE laminates reported by these authors and the modified lay-up tested by Meola et al., both illustrated in Fig. 6.11. Apparently, the edge distance required to obtain pure bearing strength failures with corresponding bearing strength values is lower for the modified lay-up compared to the standardized laminates reported by Wu, Slagter and Broest, as illustrated in Fig. 6.12. The transition takes place near $e/D \sim 1$ instead of $e/D \sim 2.5$.

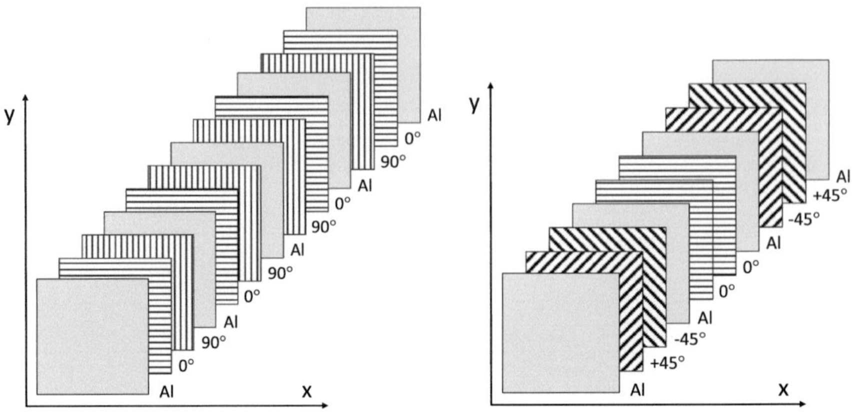

Fig. 6.11 Illustration of the GLARE3-5/4-0.3 lay-up tested by Broest [19], and the lay-up of the improved GLARE laminate tested by Meola et al. [15]

In other words, one may be able to adapt the minimum edge distance in design by changing the lay-up of the laminate, rather than changing any of the constituent materials.

6.7 In-Axis Versus Off-Axis Loading

The directionality of loading is an important parameter for design. Mechanically fastened joints, for which the bearing strength forms an important parameter, may not be perfectly loaded in their major material axes. This has led to testing FML lap joints in a uniaxial test set-up where the joint was placed under an angle, as reported by De Rijck [20].

To evaluate the influence of the load orientation with respect to the major material axes, bearing strength tests have been reported by Broest and Nijhuis [21] at loading angles of 0°, 45° and 90°. The tests were performed on unidirectional GLARE2B laminates and cross-ply GLARE3 and GLARE4B laminates.

Due to the presence of metal layers in FMLs, the influence of the fibre layers on the bearing strength remains limited, though not negligible. The results by Broest [19] and Broest and Nijhuis [21] show that unidirectional FMLs have a lower ultimate bearing strength compared to their cross-plied counterparts. The fibre layers in unidirectional FMLs only contribute with the adhesive's resistance to shear deformation, whereas in cross-ply laminates, the combination with transverse layers improves the resistance. This is schematically illustrated in Fig. 6.13.

Obviously, this resistance to shear deformation can be further improved by adding fibre layers under ±45° angles, as demonstrated by Meola et al. with their improved lay-up [15]. This improvement seems to outperform the improvements that potentially can be obtained by adding thin high-strength metal inserts at the edges as investigated by Van Rooijen [22].

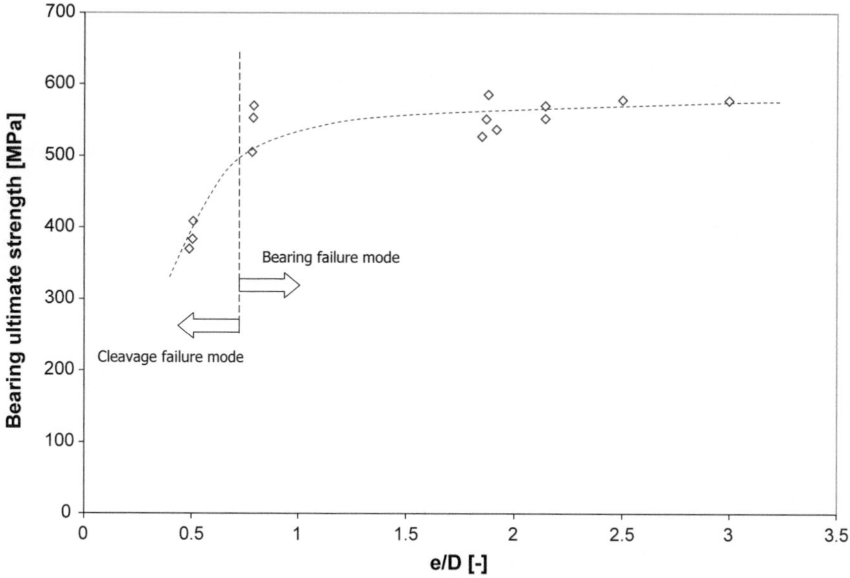

Fig. 6.12 Relation between measured bearing ultimate strength and edge distance-to-diameter ratio for laterally unconstrained bearing tests (data from [15]). A similar transition was reported by Broest [19] at $e/D \sim 2.5$)

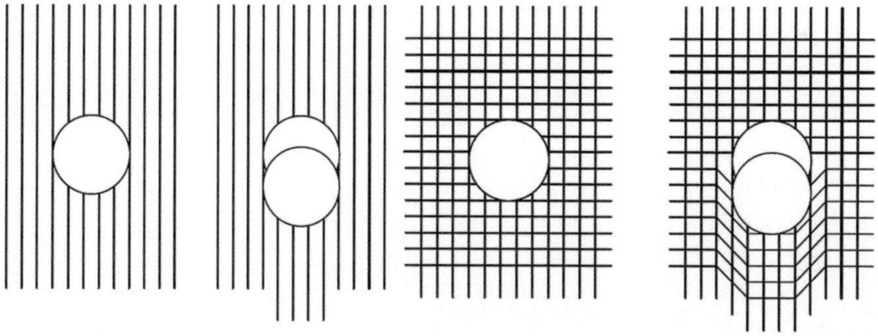

Fig. 6.13 Illustration of the influence of fibre orientation in unidirectional and cross-ply FMLs under bearing loading [19]

An interesting observation reported by Broest and Nijhuis [21] is that the dependence of the measured bearing strength on the so-called off-axis angle is fairly small, even for the unidirectional FML GLARE2A. Hence, they conclude that performing bearing strength tests under off-axis angles for determining design allowables seems irrelevant. This observation and conclusion are supported by the numerical analysis of Van Rooijen [22] who investigated the off-axis bearing properties of unidirectional GLARE2 and cross-ply GLARE3. A comparison between the numerical results of Van Rooijen with the experimental results obtained by Broest and Nijhuis is given for both laminate types in Figs. 6.14 and 6.15.

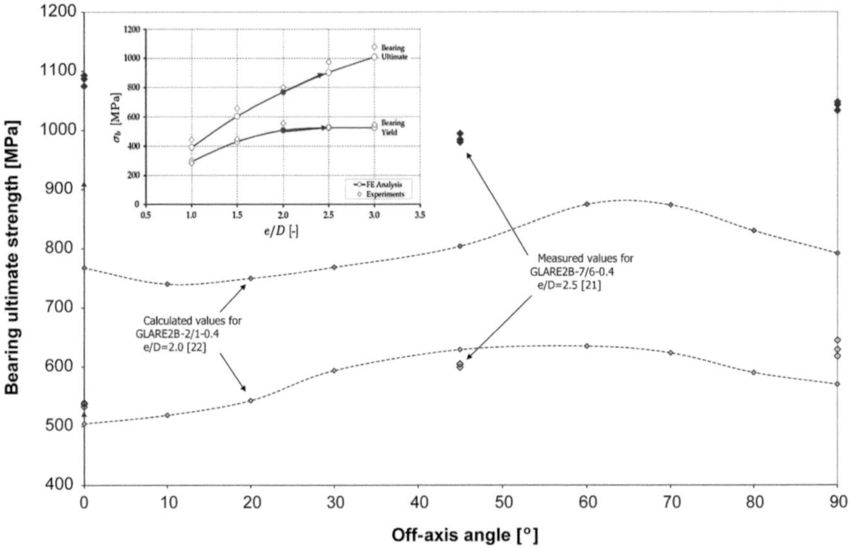

Fig. 6.14 Relation between measured [21] and calculated [22] bearing ultimate strength and applied off-axis angle for GLARE2B (insert from [22] to indicate correlation between *e*/*D* ratios)

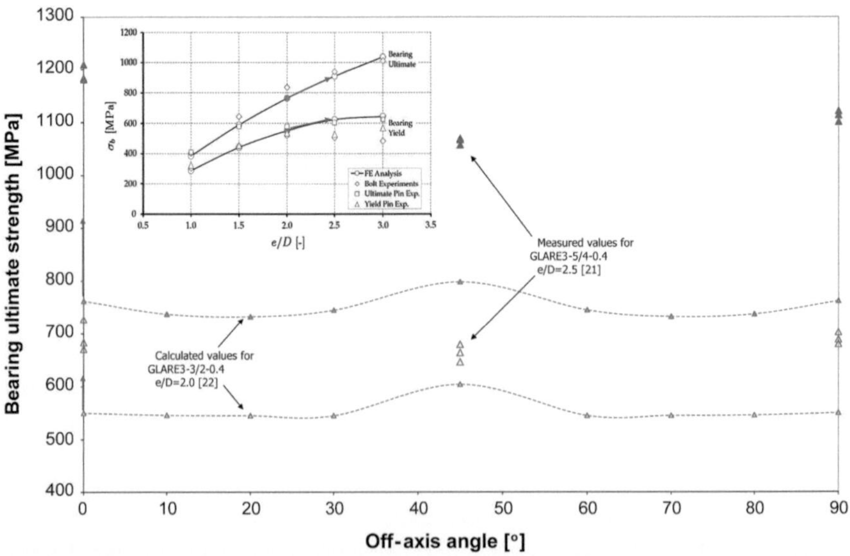

Fig. 6.15 Relation between measured [21] and calculated [22] bearing ultimate strength and applied off-axis angle for GLARE3 (insert from [22] to indicate correlation between *e*/*D* ratios)

6.8 Analysis and Prediction Methods

To evaluate the bearing strength of FMLs, various analysis methods have been proposed and reported in the literature. These methods vary from simplified analytical models to three-dimensional finite element analyses (FEA). This section provides a brief description of the most important methods for describing the bearing strength of FMLs.

6.8.1 *Bilinear Constituent Representation with Rules of Mixtures*

The load–deflection curve typical for FMLs can be approximated with a bilinear curve, as proposed by Slagter [4] and later again confirmed by Holleman [23] and Krimbalis et al. [24]. Slagter developed a simplified method for predicting the bearing yield and BUS. The assumption is that if the bearing strength tests were conducted on the metallic and fibre layers separately (with ASTM D953-54 for the fibre layers), one would obtain two hypothetical curves as illustrated in Fig. 6.16. In this representation, K_m and K_f represent the slopes of the linear elastic part of the load deflection curve, referred to as the elastic foundation modulus.

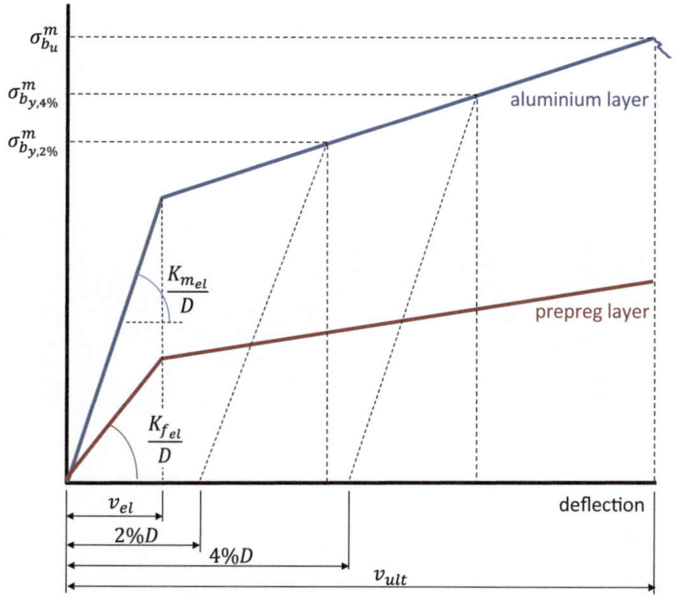

Fig. 6.16 Schematic pin-loading stress–deflection behaviour of the metal and fibre layer of an FML [4, 23]

Slagter evaluated the elastic foundation modulus K for a monolithic aluminium plate with a pin-loaded hole by making an energy analysis for the displacement of the pin in the hole. He assumed the pin to be a circular cylindrical beam resting on an elastic foundation without friction and clearance. The rigidity of the pin assured that the foundation modulus only depended on the stiffness of the foundation and not on the bending stiffness of the pin. Slagter assumed for all considered foundations an isotropic material with Young's modulus E and a constant Poisson's ratio ν.

The elastic foundation modulus was calculated with

$$K = \frac{P}{v \cdot t} \tag{6.1}$$

where t is the nominal sheet thickness, and v is the deflection of the pin under the application of bearing load P [23]. Dimensional analysis reveals that the elastic foundation modulus K is proportional to the Young's modulus E with

$$K = E \cdot f\left(\frac{e}{W}, \frac{e}{D}\right) \tag{6.2}$$

which explains the name 'foundation modulus'.

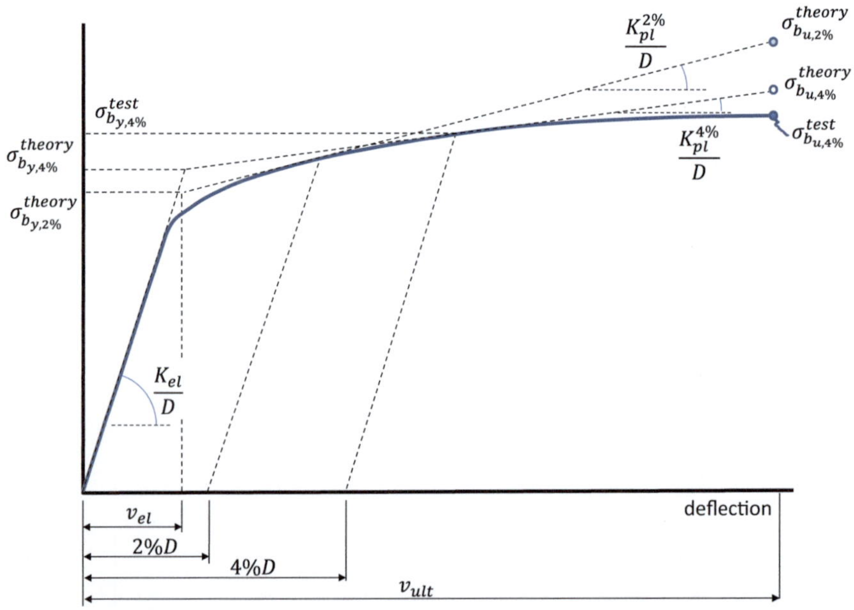

Fig. 6.17 Illustration of the idealized bilinear bearing stress–deflection behaviour in consideration of both the 2 and 4% yield point [23]

According to Slagter, bearing yield and BUS of FMLs can then be predicted with the two curves in Fig. 6.16 using the MVF approach. To enable a straightforward calculation using the MVF approach, Slagter had to assume

- that yielding of the fibre layer under the pin loading occurs at the same deflection as that of the metallic layer (v_{el} in Fig. 6.16)
- that both bearing yield and BUS of the FML correspond to the same deflections at which the metal layers reach their bearing yield and BUS.

The BYS and BUS of the FML (Fig. 6.17) are then determined with, respectively,

$$\sigma_{b_y}^{\text{lam}} = \begin{cases} \sigma_{b_{y,2\%}}^{m} \cdot \text{MVF} + \frac{K_{f,el}}{K_{m,el}} \left[\sigma_{b_{y,2\%}}^{m} + \frac{2K_{m,el}\left(\kappa_m^{2\%} - \kappa_f^{2\%}\right)}{100\left(\kappa_m^{2\%} - 1\right)} \right] (1 - \text{MVF}) \\ \sigma_{b_{y,4\%}}^{m} \cdot \text{MVF} + \frac{K_{f,el}}{K_{m,el}} \left[\sigma_{b_{y,4\%}}^{m} + \frac{4K_{m,el}\left(\kappa_m^{4\%} - \kappa_f^{4\%}\right)}{100\left(\kappa_m^{4\%} - 1\right)} \right] (1 - \text{MVF}) \end{cases}$$

$$\sigma_{b_u}^{\text{lam}} = \begin{cases} \sigma_{b_u}^{m} \cdot \text{MVF} + \frac{K_{f,el}}{K_{m,el}} \left[\left(\sigma_{b_u}^{m} - \sigma_{b_{y,2\%}}^{m} \right) \frac{\kappa_f^{2\%}}{\kappa_m^{2\%}} + \sigma_{b_{y,2\%}}^{m} + \frac{2K_{m,el}\left(\kappa_m^{2\%} - \kappa_f^{2\%}\right)}{100\left(\kappa_m^{2\%} - 1\right)} \right] (1 - \text{MVF}) \\ \sigma_{b_u}^{m} \cdot \text{MVF} + \frac{K_f}{K_m} \left[\left(\sigma_{b_u}^{m} - \sigma_{b_{y,4\%}}^{m} \right) \frac{\kappa_f^{4\%}}{\kappa_m^{4\%}} + \sigma_{b_{y,4\%}}^{m} + \frac{4K_{m,el}\left(\kappa_m^{4\%} - \kappa_f^{4\%}\right)}{100\left(\kappa_m^{4\%} - 1\right)} \right] (1 - \text{MVF}) \end{cases} \tag{6.3}$$

in which

$$\kappa_m^{2\%} = \frac{K_{m,pl}^{2\%}}{K_{m,el}}; \quad \kappa_m^{4\%} = \frac{K_{m,pl}^{4\%}}{K_{m,el}}; \quad \kappa_f^{2\%} = \frac{K_{f,pl}^{2\%}}{K_{f,el}}; \quad \kappa_f^{4\%} = \frac{K_{f,pl}^{4\%}}{K_{f,el}} \tag{6.4}$$

and $\sigma_{b_{y,2\%}}^{m}$, $\sigma_{b_{y,4\%}}^{m}$ and $\sigma_{b_u}^{m}$, respectively, the bearing yield at 2 and 4% deflection and BUS of the metal layers, often available in handbooks or provided by metal producers. In these equations, K_m, K_f, κ_m and κ_f must be determined with experiments, in particular for highly orthotropic prepreg layers [4].

According to Slagter [4], experiments support that for GLARE grades, one can assume E_f/E_m to approximate $K_{f,el}/K_{m,el}$, and that for cross-ply FML grades (GLARE3) $\kappa_m = \kappa_f$, and that the unidirectional prepreg behaves in bearing as elastic-perfectly plastic, or $\kappa_f = 0$.

Predictions with Eq. (6.1) provide excellent results for both bearing yield and BUS as illustrated in Table 6.2. All experimental results lie within 10% of the model predictions, except for the cross-ply GLARE4 laminate, of which the fibre layer contribution should be expected to lie in-between the unidirectional GLARE2 and cross-ply GLARE3 laminates. However, one has to keep in mind that the values of the bearing strength in Table 6.2 are obtained for an edge distance of $e/D = 2.0$, see also Fig. 6.18. Holleman [23] and Mattousch [25] presented significantly higher values for edge distances of $e/D = 3.0$ for some of the laminates in

Table 6.2 Comparison between bearing yield and ultimate strength obtained by bolt-type bearing test at e/D = 2.0 and the analytical model from [4]

Grade and lay-up	MVF	BYS			BUS		
		Test[a] (MPa)	Theory[b] (MPa)	Error (%)	Test[a] (MPa)	Theory[b] (MPa)	Error (%)
GLARE2-3/2-0.3	0.64	530	530	0.0	709	742	4.7
GLARE3-2/1-0.2	0.62		495		810	754	−6.9
GLARE3-2/1-0.3	0.71		526		856	802	−6.3
GLARE3-3/2-0.2	0.55		470		702	717	2.1
GLARE3-3/2-0.3	0.64	546	504	−7.7	789	769	−2.6
GLARE3-4/3-0.4	0.68		517		778	789	1.4
GLARE3-4/3-0.5	0.73		534		832	814	−2.2
GLARE3-5/4-0.4	0.67		513		769	781	1.6
GLARE4-3/2-0.3	0.55	518	498	−3.8	658	760	15.5

[a]Bolt-type bearing test values determined in accordance with ASTM D953-54
[b]Equation (6.1) with $\kappa_m = \kappa_f = 0.1$ (cross-ply) or $\kappa_m = 0.1$, $\kappa_f = 0$ (unidirectional), $K_f/K_m = E_f/E_m$, and $\sigma^m_{b_y}$ = 629 MPa and $\sigma^m_{b_u}$ = 959 MPa

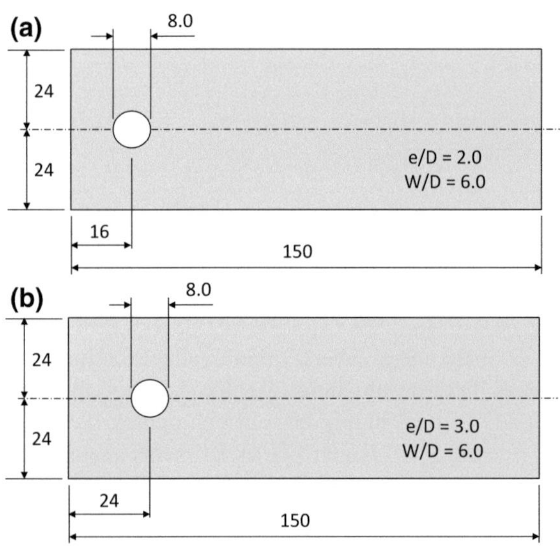

Fig. 6.18 Illustration of the dimensions of the specimens tested by Slagter [3, 4] (**a**) and the dimensions for the specimens tested by Holleman [23] and Mattousch [25] (**b**)

See Tables 6.2 and Table 6.3. Predicting the theoretical values with Eq. (6.3) then requires the bearing yield and ultimate strength for the metal layers at the same edge distance, which are given by Mattousch [25], see Table 6.4.

Table 6.3 Comparison between bearing yield and ultimate strength obtained by bolt-type bearing test at e/D = 3.0 [23, 25] and the analytical model from [4, 23]

Grade and lay-up	MVF	Test[a]			Theory ($\kappa_{al} \neq \kappa_f$)[b]				Theory ($\kappa_{al} = \kappa_f$)		
		$\sigma^m_{b_{y,2\%}}$ (MPa)	$\sigma^m_{b_{y,4\%}}$ (MPa)	$\sigma^m_{b_u}$ (MPa)	$\sigma^m_{b_{y,2\%}}$ (MPa)	$\sigma^m_{b_{y,4\%}}$ (MPa)	$\sigma^m_{b_{u,2\%}}$ (MPa)	$\sigma^m_{b_{u,4\%}}$ (MPa)	$\sigma^m_{b_{y,2\%}}$ (MPa)	$\sigma^m_{b_{y,4\%}}$ (MPa)	$\sigma^m_{b_u}$ (MPa)
GLARE2-2/1-0.3 L	0.70	724		1011	571	657	915	919	623	730	1073
GLARE2-2/1-0.3 LT	0.70	539		1008	493	575	837	837	502	588	864
GLARE2-3/2-0.3 L	0.64	739		991	549	629	863	868	613	717	1055
GLARE2-3/2-0.3 LT	0.64	531		976	455	530	769	769	466	546	802
GLARE3-3/2-0.3 L	0.64	657		1004	564	627	1003	918	539	632	928
GLARE4-2/1-0.4 L	0.68	667	763	1020	582	664	1009	970	576	675	992
GLARE4-3/2-0.3 L	0.55	627	719	896	543	612	946	890	535	627	922
GLARE4-3/2-0.4 L	0.62	671	759	983	563	639	978	931	535	627	922
GLARE4-5/4-0.3 L	0.50	617	710	903	530	594	924	863	521	611	898
GLARE4-5/4-0.4 L	0.57	653	759	945	550	622	957	905	535	627	922

[a]Bolt-type bearing test values determined in accordance with ASTM D953-87
[b]Bilinear representation based on empirical bearing strain hardening ratios from [23]

Table 6.4 Comparison between measured prepreg bearing yield and ultimate strength obtained by bolt-type bearing test at e/D = 3.0 [23, 25] and the values used in the MVF method [26]

Material/lay-up	t (mm)	Experiments[a]			MVF method	
		$\sigma^m_{b_{y,2\%}}$ (MPa)	$\sigma^m_{b_{y,4\%}}$ (MPa)	$\sigma^m_{b_u}$ (MPa)	$\sigma^m_{b_{y,2\%}}$ (MPa)	$\sigma^m_{b_u}$ (MPa)
[0]6	0.85	b	b	160	794	585
[0]12	1.77	b	b	159		
[90]6	0.85	b	b	82	231	565
[90]12	1.77	b	b	81		
[0/90/90/0]	0.58	171	207	373	657	689
[0/90/90/0]3	1.8	175	225	446		
[0/90]3	0.88	183	216	301		
[90/0/0/90]	0.59	152	191	373		
[90/0/0/90]3	1.78	199	237	433		
[0/90/0]	0.45	181	202	389	583	608
[0/90/0]2	0.88	212	255	469		
[0/90/0]4	1.75	196	233	427		
[90/0/90]	0.44	183	208	391	434	689
[90/0/90]2	0.89	187	230	532		
[90/0/90]4	1.76	211	244	420		
2 × 0.28 Al	0.56	689	804	1078	693	1200
Monolith Al	0.66	688	804	1194		
Monolith Al	1.3	640		1200		
Monolith Al	1.6	639	754	1200		

[a]Bolt-type bearing test values determined in accordance with ASTM D953-87
[b]Elastic until failure

6.8.2 Simplified MVF Method

The bearing strength prediction method represented by Eq. (6.3) relates both laminate bearing yield and bearing ultimate to the contribution of the bearing strength of the metal constituent and the volume fraction of the metal layers in the total laminate thickness. This is visible in Eq. (6.3) with the MVF and the (1-MVF), of which the latter is the fibre volume fraction.

To reduce the complexity of the method, Eq. (6.3) can be simplified to the rule of mixtures, discussed before for the static properties in Sect. 4.11.1 and for the blunt not strength in Sect. 5.8.1. Similar to Eq. (5.24), one could rewrite Eq. (6.3) into the form

$$\begin{aligned} S^{\text{lam}}_{b_y} &= \text{MVF} \cdot S^m_{b_y} + (1 - \text{MVF}) S^f_{b_y} \\ S^{\text{lam}}_{b_u} &= \text{MVF} \cdot S^m_{b_u} + (1 - \text{MVF}) S^f_{b_u} \end{aligned} \quad (6.5)$$

This rule of mixtures has been studied by Roebroeks [26], who developed the MVF approach for bearing strength at an edge distance of $e/D = 3.0$. The bearing yield, defined by the 2%D deflection (see definition in Fig. 6.16), and the BUS were calculated with the linear interpolation rule. For this calculation, the bearing yield and BUS for the metal layers were obtained with a bearing strength test on monolithic aluminium with an edge distance of $e/D = 3.0$. These results were reported by Mattousch [25], which were, respectively, $\sigma_{b_y}^m = 640$ MPa and $\sigma_{b_u}^m = 1200$ MPa, though Roebroeks [26] used for the BYS $\sigma_{b_y}^m = 693$ MPa instead, see Table 6.4. Note that these values are higher than the values given in Table 6.2, because the edge distance here is $e/D = 3.0$ instead of the $e/D = 2.0$ in Table 6.2.

Obviously, reducing the complexity by reducing Eqs. (6.3) and (6.4) to Eq. (6.5) requires empirical assessment of the fibre layer contribution. In the attempt to evaluate the elastic and plastic foundation moduli and the κ ratios of Eq. (6.4) for the standardized GLARE grades, Holleman [23] has illustrated that for any standard GLARE grade, an assessment must be made for the longitudinal and transverse direction of the stack of fibre layers between two aluminium layers. As a result, the corresponding bearing yield and ultimate strength values of the prepreg layers must be tuned in order to obtain best fits with the bearing strength test data, as, for example, presented by Slagter [4], Holleman [23] and Mattousch [25].

Here, one has to keep in mind that testing the prepreg lay-up alone may yield different values, because the failure mechanisms differ from the mechanisms of failure of the same stack within an FML lay-up. This is illustrated in Table 6.4 by comparing the bearing yield and ultimate strength values measured by Holleman [23] and Mattousch [25] with the values determined by Roebroeks [26] for the MVF method.

6.8.3 *Finite Element Analyses*

An approach towards describing the bearing strength behaviour alternative to the analytical formulation of Slagter [3, 4] is through the use of FEA. Various authors have proposed FEA adopting different failure criteria or methods to define bearing failure.

For example, Van Rooijen et al. [27] presented FEA using solid elements to describe the individual layers of the FML. For the metal, he assumed elastic-perfectly plastic behaviour, while for the prepreg layers, he adopted a degradation factor for various in plane properties after failure initiation criteria were met.

Krimbalis et al. [28] suggested a different approach using FEA based on a revised compression characteristic dimension, defined as the distance between the hole edge and a point in the stress profile where the compressive stress equals the yield strength of the metal layer. In the literature, this approach is also often referred to as the critical distance method. Their definition is in agreement with their observations in slightly modified bearing strength tests they performed [29], and the observation by Caprino et al. [9] discussed above that the aluminium layers within GLARE play a dominant role in determining the true bearing strength of a joint.

Similar to Van Rooijen, Hundley et al. [8] introduced FEA incorporating a three-dimensional progressive failure constitutive model for the prepreg layers. The interlaminar properties were described using cohesive zone formulations, while with a user material subroutine, the onset of damage in the prepreg layers was evaluated at each integration point in the mesh, similar to the failure initiation criteria adopted by Van Rooijen et al. [27].

Recently, Garg et al. [30] presented FEA using a multi-scale approach incorporated into the existing GENOA software. The multi-scale approach calculates stresses and strains at the micro-scale that are derived from the lamina scale using micro-stress theory. Although this approach is often presented as a significant advancement, one may argue that this approach is scientifically not more sophisticated than the analyses of Van Rooijen et al. [27] and Hundley et al. [8]. The reported calibration exercise with data from Hagenbeek [31] transfers the macroscopic properties established with experiments to the microscopic level; these properties are then again recalculated to macroscopic levels for predictions. The reported correlations claiming accuracy in prediction therefore merely represent a verification that this calibration is performed consistently, rather than a true validation.

6.9 Additional Studies

The studies discussed in this chapter on the bearing strength of FMLs approached the problem from various perspectives, either investigating the bearing strength experimentally, or developing predictive capabilities with analytical or numerical methods. What all these studies have in common is that they consider the subject of bearing strength of FMLs as a separate property and only for laboratory-air and room temperature environments.

Only a few studies report putting the bearing strength into context with, for example, the blunt notch strength discussed in Chap. 5, or addressing the influence of environmental conditions. This section highlights two of those studies.

6.9.1 Bearing/ByPass Diagrams

The fasteners in mechanically fastened joints often transfer portions of the load, while the remainder of the load goes around the fastener. These loads are often referred to as bypass loads. In monolithic aluminium, interaction between bypass loads and bearing loads is assumed to be small, because the ductility of the alloys effectively redistributes the stresses. This is similar to the blunt notch strength. However, in composites the bypass loads have a significant effect on the bearing strength, requiring the assessment of the so-called bearing/bypass interaction.

A first attempt to evaluate the bearing/bypass interaction for Fibre Metal Laminates was made by Ypma [32]. Ypma tested a number of specimens, varying the number of fasteners over which the load is transferred. The approach is not according to the procedure prescribed in [33], but Ypma explains that the incentive for the exploratory study was to identify whether the interaction could be demonstrated with a simple test specimen and test set-up.

Some of the specimen configurations Ypma tested are illustrated in Fig. 6.19. He assumed the single fastener joint, illustrated in Fig. 6.19a, to represent the pure bearing strength case, while the joints with two or more fasteners represent various levels of bearing/bypass interaction.

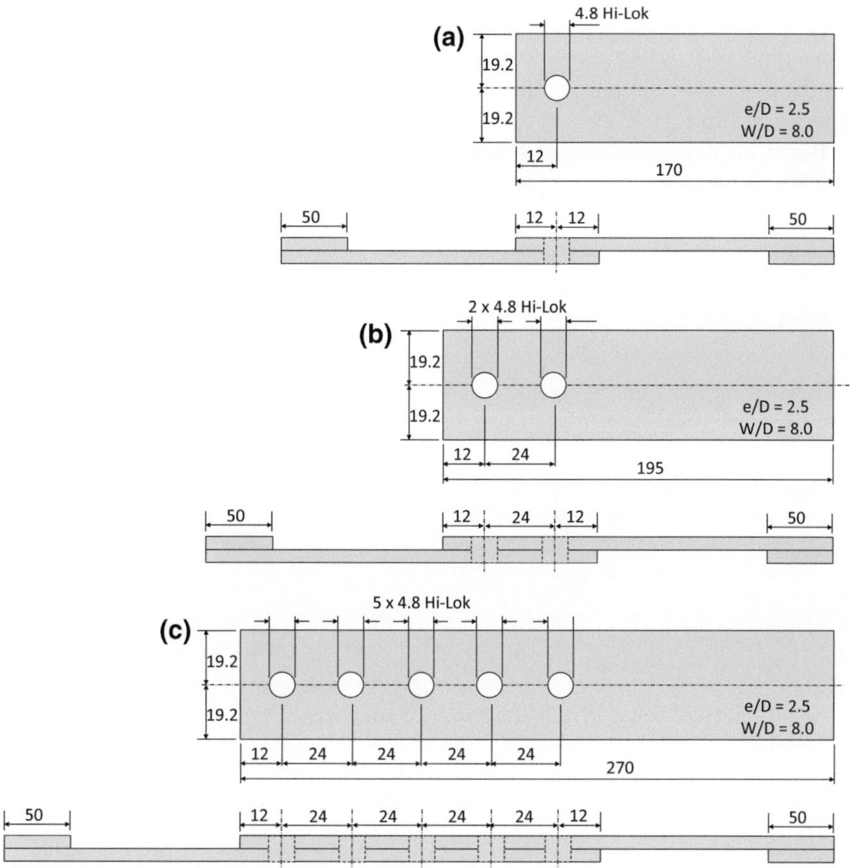

Fig. 6.19 Illustration of specimen dimensions tested by Ypma [32] with a single Hi-Lok (**a**), two Hi-Loks (**b**) and five Hi-Loks (**c**). Ypma tested in a similar fashion specimens with three and four Hi-Loks

The fastener load F_{fastener} was determined by dividing the total measured load by the number of fasteners in the specimen. Obviously, this is a crude approach, because the intermediate fasteners are known to bear less load than both outer fasteners [34]. The bearing stress was then calculated with

$$S_b = \frac{F_{\text{fastener}}}{D_{\text{Hi-Lok}} t_{\text{lam}}} \tag{6.6}$$

whereas the bypass stress was determined at the first fastener, exhibiting the highest bypass stress, with

$$S_{\text{by-pass}} = \frac{F_{\text{total}} - F_{\text{fastener}}}{W t_{\text{lam}}} \tag{6.7}$$

Because bearing/bypass diagrams are generally plotted in the form of bearing stress versus bypass strain, Ypma used the stress–strain relationships for the corresponding FML using the software developed and reported by Hagenbeek [35]. For the monolithic aluminium specimens Ypma, used the stress–strain relationships reported in [36].

Based on the experimental results, the bearing/bypass diagram was presented, which is given in Fig. 6.20. Ypma concluded that although the performed tests and the specimen geometry are far from ideal for studying bearing/bypass interaction, the experiments did demonstrate an interaction is visible for the FML GLARE. He recommended to further study this interaction using proper experiments, as, for example, proposed in [33].

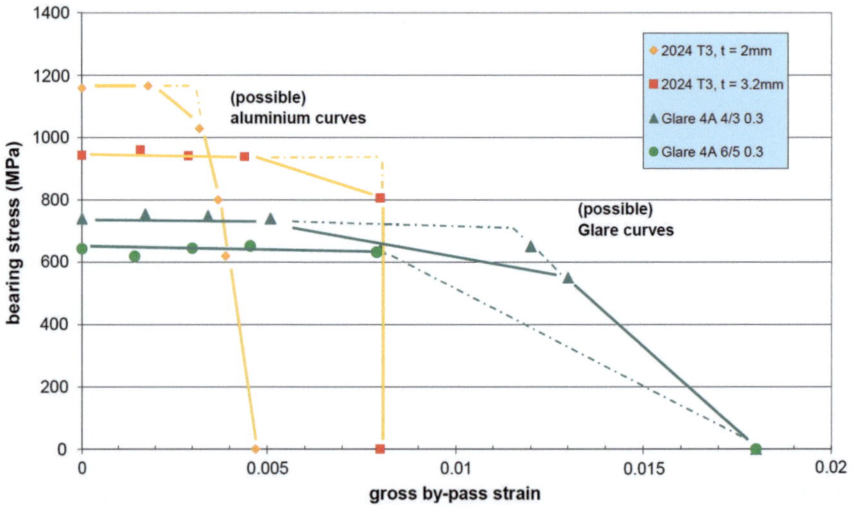

Fig. 6.20 Bearing/bypass interaction curves for aluminium 2024-T3 and GLARE4A-4/3-0.3 [32]

6.9.2 Environmental Exposure

Borgonje [37] investigated the influence of realistic hot/wet environment on the bearing strength properties, testing specimens that were exposed at a site in Australia for 2 and 6 years. He compared the observations to specimens tested without exposure at both room temperature and at 70 °C. Unexposed bearing strength tests at 70 °C were previously performed by Broest and Nijhuis [21]. In addition, specimens were tested after accelerated exposure at 70 °C and 85% RH, both for 1500 and 3000 h.

The experiments revealed that the exposure to the hot/wet environments deteriorates the BYS and BUS of GLARE. The observed reduction, however, was still less than the tests without exposure tested at 70 °C. Because the bearing strength tests were tested according to the test specification specifying clearance between specimen and its support, the obtained bearing strength values appeared to be unrealistically low. The results did seem to correlate with the bearing strength tests with and without accelerated exposure performed within the material qualification programme [38], but as Ypma already reported in [39], these were tested at an edge distance of $e/D = 2$, while Broest and Nijhuis used $e/D = 3$. For that reason, Borgonje recommended to further study the environmental influences, but to harmonize the specimen geometry and test specification, preferably using the bolt-bearing test specification.

References

1. ASTM Standard E238-84, Standard method of test for bearing strength of metallic materials, Annual Book of ASTM Standards. American Society for Testing and Materials, Philadelphia, USA
2. ASTM Standard D953-54, Standard test method for bearing strength of plastics, Annual Book of ASTM Standards. American Society for Testing and Materials, Philadelphia, USA
3. Slagter WJ (1992) On the Bearing Strength of Fibre Metal Laminates. J Compos Mater 26:2542–2566
4. Slagter WJ (1994) Static strength of riveted joints in fibre metal laminates. PhD dissertation, Delft University of Technology, Delft
5. Hakker M (1992) Geometry effects on the bearing strength of GLARE laminates. Preliminary thesis, Faculty of Aerospace Engineering, Delft University of Technology
6. Hakker M (1993) Bearing strength of GLARE laminates. Master's thesis, Faculty of Aerospace Engineering, Delft University of Technology
7. Brügemann V (2003) Test procedures for fibre metal laminates, Report TD-R-03-005. Fibre Metal Laminates Centre of Competence, Delft, The Netherlands
8. Hundley JM, Hahn HTh, Yang J-M, Facciano AB (2010) Three-dimensional progressive failure analysis of bolted titanium-graphite fiber metal laminate joints. J Compos Mater 45 (7):751–769
9. Caprino G, Squillace A, Giorleo D, Nele L, Rossi L (2005) Pin and bolt bearing strength of fibreglass/aluminium laminates. Compos A 36:1307–1315

10. Buczynski A (2009) Development of a carbon-fiber/stainless steel laminate concept—improving the bearing characteristics of carbon-fiber composites by addition of steel strip reinforcements. MSc thesis, Delft University of Technology, Delft
11. Frizzell RM, McCarthy CT, McCarthy MA (2008) An experimental investigation into the progression of damage in pin-loaded fibre metal laminates. Compos B 39:907–925
12. Yamada T, Nakatani H, Ogihara S (2011) Evaluation of bearing damage behaviour in thin titanium films-CFRP hybrid laminate, ICCM19?
13. Wu HF, Wu LL, Slagter WJ (1994) An investigation on the bearing test procedure for fibre-reinforced aluminium laminates. J Mater Sci 29:4592–4603
14. Collings TA (1977) The strength of bolted joints in multi-directional CFRP laminates. Aeronautical Research Council, Current Papers No. 1380
15. Meola C, Squillace A, Giorleo G, Nele L (2003) Experimental characterization of an innovative GLARE® fiber reinforced metal laminate in pin bearing. J Compos Mater 37 (17):1543–1552
16. Caprino G, Giorleo G, Nele L, Squillace A (2002) Pin-bearing strength of glass mat reinforced plastics. Compos A Appl Sci Manuf 33:779–785
17. Meesters (2000) Edge distance and rivet pitch in joints, a literature survey, Report B2v-00-13. Delft University of Technology, Delft
18. Wu HF, Slagter WJ (1992) Parametric studies of bearing strength for fiber/metal laminates, SLC-report SL-019-E. Structural Laminates Company, Delft
19. Broest P (2000) The influence of the edge distance on the joint strength in GLARE laminates, Report B2v-00-67. Delft University of Technology, Delft
20. de Rijck R (2003) Residual strength prediction of XJ-advanced specimens, Report B2v-03-07. Delft University of Technology, Delft
21. Broest P, Nijhuis P (2000) Determination of bearing strength, Report B2v-00-39. Delft University of Technology, Delft
22. Rooijen RGJ (2006) Bearing strength characteristics of standard and steel reinforced GLARE, PhD dissertation. Delft University of Technology, Delft
23. Holleman E (1994) Bearing strength prediction for some GLARE grades (BE2040 subtask 5.1-a), Memorandum M-683. Delft University of Technology, Delft
24. Krimbalis PP, Poon C, Fawaz Z, Behdinan K (2007) Experimental characterization of the bearing strength of fiber metal laminates. J Compos Mater 41:3109–3131
25. Mattousch AC (1994) Bearing properties of GLARE 2-2/1-0.3, GLARE 2-3/2-0.3 and GLARE 3-3/2-0.3, Report TD-R-94-010. Structural Laminates Company
26. Roebroeks G (2000) The metal volume fraction approach, Report: TD-R-00-003. Structural Laminates Industries, Delft, The Netherlands
27. Rooijen RGJ, Sinke J, de Vries TJ, van der Zwaag S (2006) The bearing strength of fiber metal laminates. J Compos Mater 40(1):5–19
28. Krimbalis PP, Poon C, Fawaz Z, Behdinan K (2007) Prediction of bearing strength in fiber metal laminates. J Compos Mater 41(9):1137–1157
29. Krimbalis PP, Poon C, Fawaz Z, Behdinan K (2008) On the pin bearing behavior of orthotropic fiber metal laminates. J Compos Mater 42:1547–1566
30. Garg M, Falugi M, Abdi F, Abumeri G (2011) Predicting bearing strength of fiber metal laminates via progressive failure analysis. 52nd AIAA/ASME/ASCE/AHS/ASC structures, structural dynamics and materials conference 19th, 4–7 April 2011, Denver, Colorado
31. Hagenbeek M (2005) Characterisation of fibre metal laminates under thermo-mechanical loadings, PhD dissertation. Delft University of Technology, Delft
32. Ypma M (2002) Exploratory investigation on the effect of bearing/bypass interaction on the static joint strength of GLARE, Report B2v-02-05. Delft University of Technology, Delft
33. MIL HDBK—17—1E. Department of Defense U.S.A
34. Schijve J (2009) Fatigue of structures and materials. Springer Science + Business Media, B.V.
35. Hagenbeek M (2000) Estimation tool for basic Material properties, ETMP_release 2, Report B2 V-00-29. Delft University of Technology, Delft, The Netherlands
36. MIL HDBK—5H, December 1998. Department of Defense U.S.A

37. Borgonje B (2002) GLARE outdoor exposure program: bearing strength, Report B2v-02-36. Delft University of Technology, Delft
38. Hoeven W. van der, Nijhuis P (2002) Qualification testing of semi-finished flat GLARE products, NLR-CR-2002-146. NLR National Aerospace Laboratory
39. Ypma M (2000) Overview of tests concerning the influence of temperature and environmental exposure on GLARE, Report B2v-00-41. Delft University of Technology, Delft

Chapter 7
Fatigue Initiation

Abstract This chapter explains how fatigue initiation in FMLs can be evaluated and predicted, assuming that the phenomenon primarily affects the fatigue property of the metal constituents. It is illustrated how definition of initiation in the theory enables the utilization of Wöhler failure life curves obtained for the monolithic metal, in combination with stress concentration factors and laminated plate theory. The accuracy of prediction is discussed in relation to the match between the conditions for the adopted Wöhler curves and the predicted conditions. In the end, the theory is related to the application of mechanically fastened joints, where a combination of tension and bending impose different fatigue initiation lives in each individual metal layer.

7.1 Introduction

Fatigue initiation in hybrid materials in general and FMLs in particular is considered a damage phenomenon primarily related to the metallic constituents. The methods to assess fatigue initiation are as a consequence similar to the methods for metals. It is assumed that the initiation of a crack under cyclic loading in a metallic layer of an FML is equal to the initiation behaviour of a monolithic sheet of the same metal, provided that the cyclic stresses at the location of initiation are identical for both cases.

7.2 Definition of Initiation

Traditionally, the transition from initiation to crack growth in monolithic metals has been related to the transition from nucleation and formation of micro-cracks towards propagation of macro-cracks [1]. The initiation phase is predominantly affected by the surface conditions, whereas the subsequent crack growth phase is determined by the bulk material properties.

Indeed, because of the argument that the initiation behaviour in the metal layers of the FML is determined solely by the metal layer [2], there seems to be no scientific reason for taking a different definition for fatigue initiation in FMLs.

R. Alderliesten, *Fatigue and Fracture of Fibre Metal Laminates*, Solid Mechanics and Its Applications 236, DOI 10.1007/978-3-319-56227-8_7

For the thin sheets in FMLs (commonly between 0.3 and 0.5 mm), it can be argued that the transition length, i.e. the crack length at which transition is assumed to take place, thus would be in the same order of magnitude as the sheet thickness.

However, from an engineering perspective a different transition length is preferred. A first suggestion for a different transition length was given by Homan [3], who has shown that the initiation behaviour is predominantly described by the actual nominal stress cycle and the peak stress cycle at the notch root. He derived a theory in which these stresses are calculated for the individual metal layers of an FML using the Classical Laminate Theory. These nominal stresses, together with the corresponding stress concentration factor K_t at the notch in these layers, could be related to S–N_i (fatigue life until initiation) curves, obtained from fatigue tests on monolithic metal sheets.

The problem here is that the available fatigue data for metallic materials consist either of fatigue life until failure data (S–N) or of crack resistance data (da/dN vs. ΔK). In engineering practice, the fatigue initiation in monolithic metals is often related to the S–N data, because the growth period in these tests is relatively small; see Fig. 7.1. In FMLs, however, the growth of cracks from the transition length of 0.1–0.3 mm onwards is more substantial, because fibres start to bridge these cracks. Hence, working with such a small transition lengths requires generation of S–N_i curves defined as life until the formation of that transition length.

This means that statistically significant number of monolithic metal samples must be fatigue-tested to generate S–N_i = curves, for which already S–N curves to failure have been generated, or another engineering approach should be adopted to use the available fatigue life data. Here, the hypothesis is that if the transition length is selected appropriately, the vast amount of experimental S–N data, available in the open literature, can be used to predict the initiation behaviour of FMLs.

Figure 7.1 illustrates the evolution of cracks length as percentage of the life until failure for monolithic aluminium 2024-T3. From this figure, it is evident that once a transition crack length of a_i = 1 mm is taken; the error with the total fatigue life is rather small, in the order of a few per cent. However, if initiation would be defined up to a transition crack length of about 0.1 mm the error would be significant, up to 60%. Based upon this figure, it is suggested to take a transition crack length of at least 1 mm, but preferably greater, to further limit the error with the S–N data taken from the literature.

On the other hand, Fig. 7.1 also illustrates the crack growth for two GLARE grades for crack lengths up to 5 mm. From this figure, it can be observed that the fibres start having an effect on the crack growth once the crack is still in the order of a tenth of millimetres. In other words, this graph suggests that a transition length should be defined as small as possible—preferably less than 1 mm—in order to exclude any effect of fibre bridging on the initiation life.

In the past, when discussing the transition from initiation to crack growth, the selected transition length of 1 mm was often justified with the argument that after 1 mm the fibre bridging became effective. In other words, fatigue initiation was related to fibres being ineffective, while crack growth was related to effective fibre bridging. This distinction seems to be supported by the illustration in Fig. 7.2, presented by Homan [3].

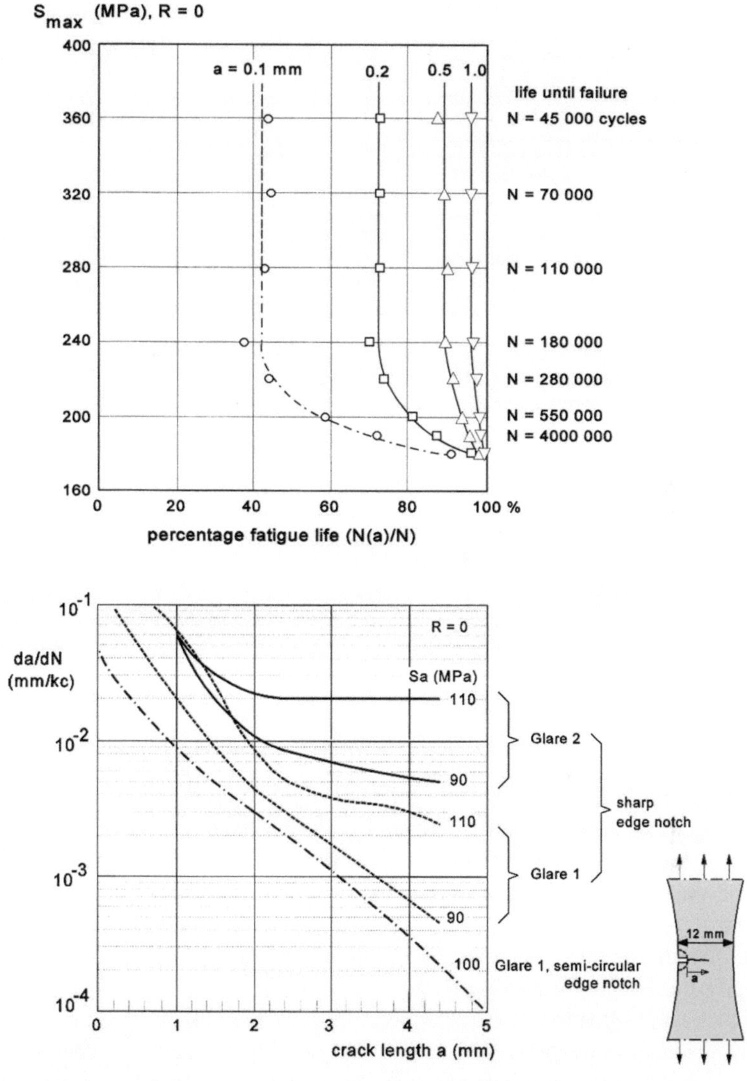

Fig. 7.1 Length of small cracks as percentage of the fatigue life for un-notched Al2024-T3 [1, 4] (*upper*) and the growth of small cracks in GLARE [1, 5] (*lower*)

Homan explains that there are two transitions in phenomena in FMLs: the transition between micro-crack and macro-crack lengths, and the transition from ineffective to effective fibre bridging. He explicitly notes that there is no physical relation between these two transitions. Where in monolithic metals commonly the transition between micro-crack to macro-crack length is considered [1], in FMLs the transition between ineffective and effective fibre bridging is considered dominant.

However, Fig. 7.1 illustrates that the fibres start influencing the crack growth immediately. Nevertheless, based upon other work [2] it is known that the length at

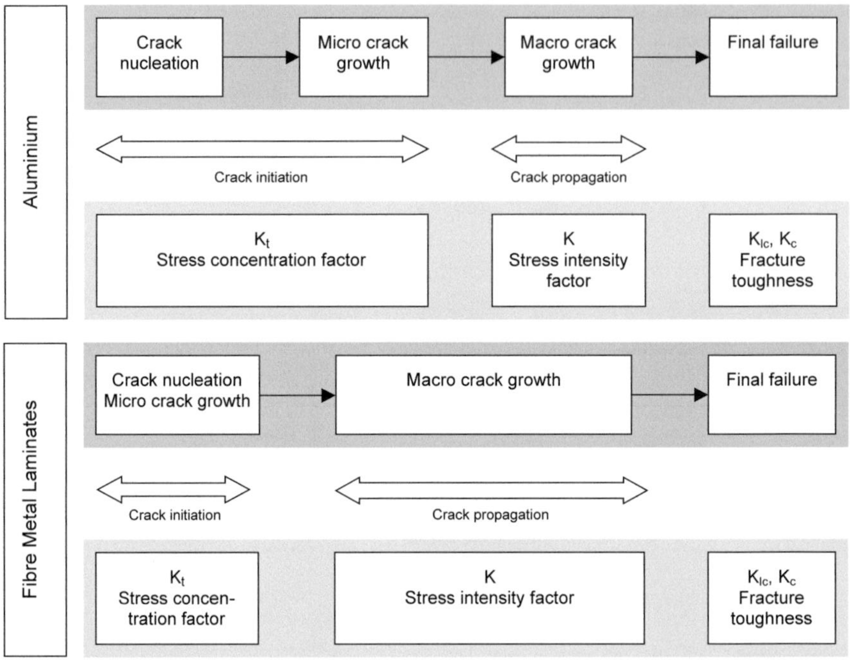

Fig. 7.2 Different phases of the fatigue life in FMLs compared to monolithic aluminium together with relevant factors

which fibre bridging becomes fully effective might range between tenth of millimetres up to several millimetres depending on the laminate, notch and loading.

To conclude, the transition crack length of 1 mm is not directly related to any physical phenomenon. It is a somewhat arbitrary, but convenient engineering selection to enable application of *S–N* data to the prediction of fatigue initiation in FMLs. This means that when discussing fatigue initiation in FMLs, one always has to specify the selected transition length.

A practical implication of the selection of 1 mm as crack initiation length is that various inspection methods for open holes could be used for detection. For example, Airbus has demonstrated that the 1 mm crack length can be detected online at the mating surfaces of a single shear joint using either crack wires or by comparative vacuum monitoring.

The method originally presented by Kieboom [6] and Homan [3] has been further developed and corrected by Spronk et al. [7]. Although the overall methodology remains similar, they justified their paper by arguing that the original method [3, 6] did not describe how the obtained laminate stresses were recalculated to nominal stresses, which is common practice in fatigue life evaluation of metallic materials when working with the stress concentration factor.

They also argued that although it was mentioned that the residual stresses affect the stress concentration factor, making it dependent on the applied load, it was unclear which stress concentration factor should be taken to characterize the peak stresses in comparison with the reference S–N data. Moreover, the stress ratio that has to be considered for correlating to S–N data was also not well addressed. For this reason, the explanation in this chapter follows the description of the work by Spronk et al. [7].

7.3 Definition of Stresses

In the methodology for predicting fatigue crack initiation, different types of stresses are considered. Starting with the stresses applied to the laminate, equivalent far field stresses can be computed using the Classical Laminate Theory (CLT). Here one has to be aware that the far field stresses in each layer consist of a mechanical stress component, induced by the applied laminate stress, and a thermal stress component, induced by laminate curing. The method to calculate these stress components is explained in Chap. 4.

Fatigue life evaluations are performed using stress concentration factors that are defined as

$$K_t = \frac{\sigma_{\text{peak}}}{S_{\text{nom}}} \tag{7.1}$$

where the nominal stress S_{nom} is defined in Fig. 7.3. The peak stress σ_{peak} is the maximum local stress in the gradients illustrated in Fig. 7.3.

With Fig. 7.4a, the nominal stress in case of an open hole in a uniaxial loading condition is defined as

$$S_{\text{nom}} = S_{\text{applied}} \frac{W}{W - D} \tag{7.2}$$

Here one should note that the nominal stress is a function of the applied mechanical stress, but that the local stress is influenced by the thermal stresses induced by curing; see Fig. 7.4b. These residual stresses include any additional residual stress induced by, for example, a post-stretching procedure, as explained in Sect. 4.8.

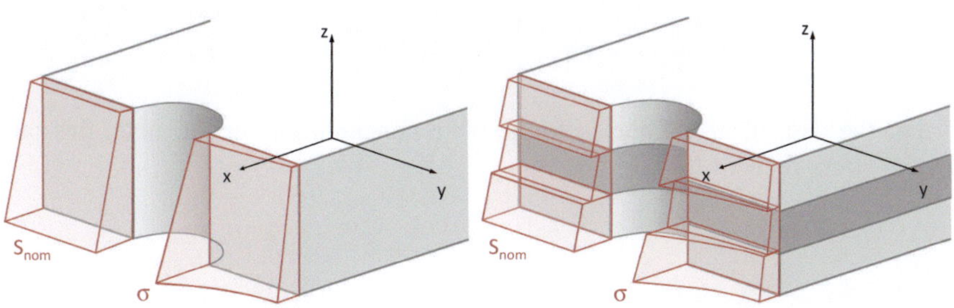

Fig. 7.3 Definition of nominal stresses S_{nom} and the local stress σ at corresponding z-coordinates; monolithic plate (*left*) and FML (*right*)

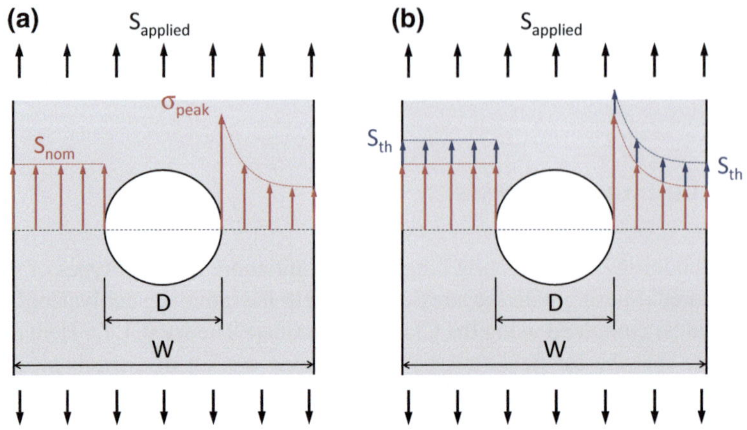

Fig. 7.4 Definition of nominal stresses S_{nom} and the peak stress σ_{peak} (**a**) and the influence of thermal stresses on these nominal and peak stresses (**b**)

7.4 Stress Concentration in a Uniaxial Stress Field

For the case of an orthotropic plate with finite dimensions containing an open hole, the stress concentration generally can be approximated with a similarity principle in relation to isotropic plates following

$$\left.\frac{K_{t,\,\mathrm{ortho}}}{K_{t,\,\mathrm{iso}}}\right|_{\mathrm{finite}} = \left.\frac{K_{t,\,\mathrm{ortho}}}{K_{t,\,\mathrm{iso}}}\right|_{\mathrm{infinite}} \tag{7.3}$$

Similarly as discussed in Chap. 5, the problem of an open hole in an infinite orthotropic plate can be evaluated with the solutions of Lekhnitskii [8]. When the uniaxial stress field is aligned with the laminate's principal material axes, Lekhnitskii's equation for the tangential stress around the perimeter of the circular hole can be simplified to [7]

$$\left.\frac{\sigma_{\theta,\,\mathrm{ortho}}}{S}\right|_{\varphi=0^\circ,\,\psi=90^\circ} = 1 + \sqrt{2\left(\sqrt{\frac{E_x}{E_y}} - \nu_{xy}\right) + \frac{E_x}{G_{xy}}} \tag{7.4}$$

For the case of an infinite isotropic plate containing a circular hole, this tangential stress reduces to

$$\left.\frac{\sigma_{\theta,\,\mathrm{iso}}}{S}\right|_{\varphi=0^\circ,\,\psi=90^\circ} = 3 \tag{7.5}$$

which implies for isotropic plates $K_t = 3$.

For finite isotopic plates under an uniaxial loading, often Heywood's equation [9] is applied

$$K_{t,\,\mathrm{iso}} = 2 + \left(1 - \frac{D}{W}\right)^3 \tag{7.6}$$

with D and W defined in Fig. 7.4. Peterson [10] provides another solution for the stress concentration factor as function of the ratio D/W

$$K_{t,\,\mathrm{iso}} = 2 + 0.284\left(1 - \frac{D}{W}\right) - 0.6\left(1 - \frac{D}{W}\right)^2 + 1.32\left(1 - \frac{D}{W}\right)^3 \tag{7.7}$$

Taking either Eqs. (7.6) or (7.7) divided by $K_t = 3$ provides the ratio to be substituted into Eq. (7.3), which in rewritten form is

$$K_{t,\,\mathrm{ortho}} = K_{t,\,\mathrm{ortho}}^{\infty} \frac{K_{t,\,\mathrm{iso}}}{K_{t,\,\mathrm{iso}}^{\infty}} \tag{7.8}$$

Here, it must be noted that the methodology described here is not limited to the given equations for the stress concentration factors. In principle, different equations can be adopted, which, for example, have been derived directly for anisotropic plates. For example, an alternative finite width correction has been proposed by Tan [11].

$$\frac{K_t^{\infty}}{K_t} = \frac{1}{2}\left[2 - \left(\frac{D}{W}\right)^2 - \left(\frac{D}{W}\right)^4\right] + \frac{1}{2}\left(\frac{D}{W}\right)^6 (K_t^{\infty} - 3)\left[1 - \left(\frac{D}{W}\right)^2\right] \tag{7.9}$$

with

$$K_t^{\infty} = \sqrt{\frac{2}{A_{66}}\left(\sqrt{A_{11}A_{22}} - A_{12} + \frac{A_{11}A_{22} - A_{12}^2}{2A_{66}}\right)} \tag{7.10}$$

where A_{ij} denote the laminate in-plane stiffness.

The above equations have been derived for circular central cut-outs in orthotropic or isotropic plates. However, the methodology applies to any cut-out with a shape different than circular. For example, the expressions proposed by Tan [11] apply in principle to any elliptical shape. Similarly, Wu [12] proposed a higher-order theory to assess the stress concentration in an orthotropic plate containing an arbitrary elliptical cut-out.

The experience here is, however, that accurate theoretical methods are not a necessity. Similar to the approximation of Eq. (7.3), one could adopt a first-order approximation such as

$$\left.\frac{K_{t,\text{ortho}}}{K_{t,\text{iso}}}\right|_{\text{arbitrary shape}} = \left.\frac{K_{t,\text{ortho}}}{K_{t,\text{iso}}}\right|_{\text{circular}} \tag{7.11}$$

For example, for an elliptical central cut-out, illustrated in Fig. 7.5, one could approximate the stress concentration factor for orthotropic material with

$$K_{t,\text{ortho}}^{\text{ellipse}} = K_{t,\text{ortho}}^{\text{circle}} \frac{K_{t,\text{iso}}^{\text{ellipse}}}{K_{t,\text{iso}}^{\text{circle}}} = K_{t,\text{ortho}}^{\text{circle}} \frac{1+2\sqrt{\frac{a}{b}}}{3} \tag{7.12}$$

7.5 Peak Stresses at Locations Other Than $\psi = \pm 90°$

Equations (7.4) and (7.5) specify the location of the peak stress at an angle of $\psi = \pm 90°$. In case of orthotropic laminates with ply orientations other than 0/90°, or the applied loading is at an off-axis angle with respect to the material's principal directions, the highest peak stress may occur at an angle different from $\psi = \pm 90°$. For that case, in theory the complete solution from Lekhnitskii [8] must be taken to calculate the magnitude and location of the peak stress.

Here, two aspects must be taken into consideration. First, the definition of the net section, which is subjected to the nominal stresses considered in the fatigue life analysis. Second, the location of the peak stresses in the stress concentration factor solutions provided in handbooks for isotropic plates.

Concerning the definition of the net section, Spronk et al. [7] discuss the arguments in favour of either option 1 or 2 in Fig. 7.6. One of the arguments is that close to the peak stress location option 2 is most appropriate, whereas further away the net section should be considered transverse to the applied load, thus similar to option 1. Spronk et al. conclude that the correct net section is positioned between the two options illustrated in Fig. 7.6.

There are two reasons, however, why one may consider the exact location of peak stress or net section less relevant, which are not well discussed in the paper. First, the

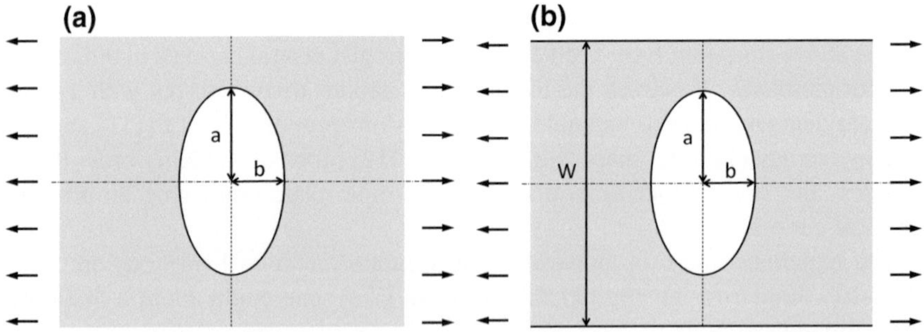

Fig. 7.5 Definition of the major axes of an elliptical cut-out in a uniaxial loaded infinite plate (**a**) and similar definition for a finite plate with a central ellipse (**b**)

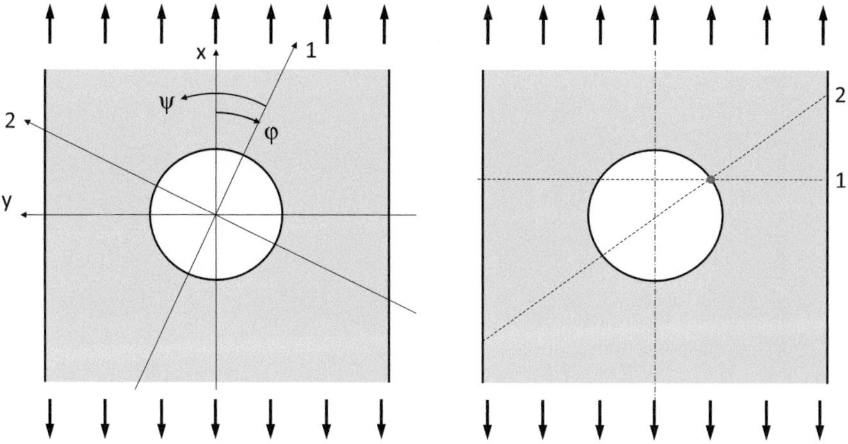

Fig. 7.6 Definition of angle ψ (*left*) and definition of net section, in case the maximum stress concentration occurs at a location different from $\psi = \pm 90°$ (*right*)

determination of peak stress and nominal stress, as discussed here, aims at assessing the fatigue initiation life of FMLs containing stress concentrations. One should acknowledge that fatigue life assessment is never an exact science, illustrated by the number of approximations one can make. So in case the diameter of the circular cut-out is relatively small, for example the diameter of a rivet or fastener in a mechanically fastened joint, one could argue that the exact location is unimportant and that the error made by ignoring the influence of the location remains very small.

Secondly, even for cases where the theoretical stress concentration is positioned at $\psi = \pm 90°$ the general observation is that cracks may not initiate exactly at that position. At a mesoscopic or even microscopic level, the stress concentration even for those cases may be at an angle different from $\psi = \pm 90°$.

This can be illustrated by the observations of Kieboom [6], illustrated in Fig. 7.7, and Vašek et al. [13], illustrated in Fig. 7.8. It seems that because fibres

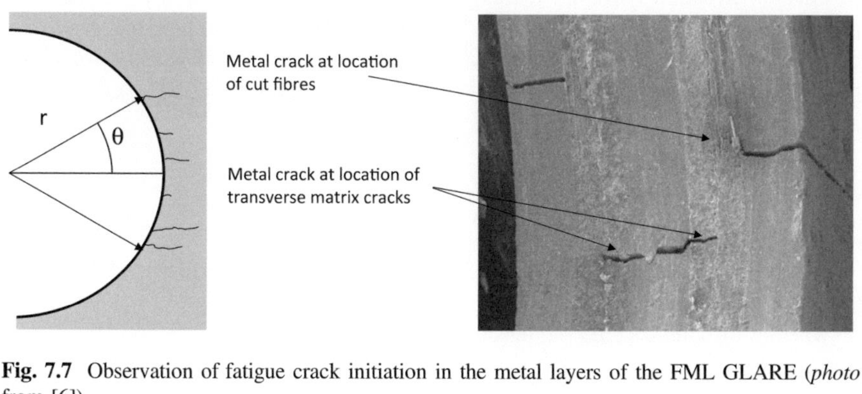

Fig. 7.7 Observation of fatigue crack initiation in the metal layers of the FML GLARE (*photo* from [6])

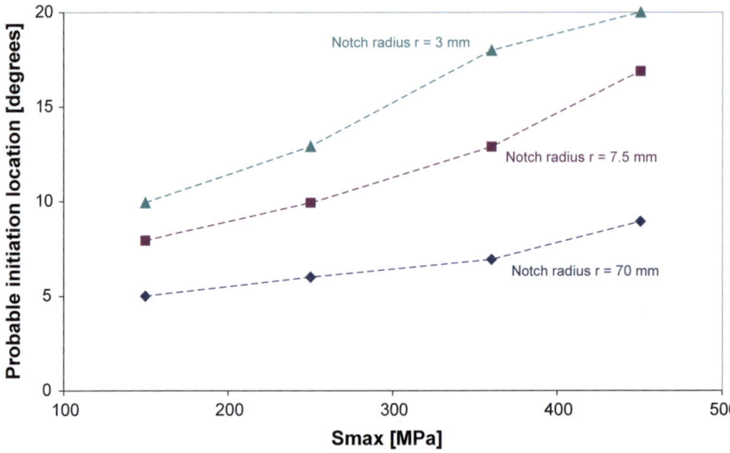

Fig. 7.8 Probable initiation location in notched FMLs as function of the notch radius [13] (initiation location defined in Fig. 7.7)

are cut by drilling the hole, locally additional stress concentrations are introduced that are not considered in any of the theories, which assume homogeneous material behaviour.

7.6 Stress Concentration in a Biaxial Stress Field

In the previous sections, uniaxial loading was considered, where the loading axes were either oriented in the material's principal directions or, in case of off-axis loading for example, where the loading axes are rotated with respect to the material's directions.

Another form of interaction with loading may come from loading in two perpendicular directions. Biaxial loading is known to influence the stress concentrations, because stresses may be superimposed. Take, for example, the case of a circular cut-out in an infinite isotropic plate. For each individual loading direction, the tangential stress at the location $\psi = \pm 90°$ is then given by Eq. (7.5). The stress at the location $\psi = 0°$ or $\psi = 180°$ is equal to

$$\left.\frac{\sigma_{\theta,\mathrm{iso}}}{S}\right|_{\varphi=0°,\,\psi=0°} = -1 \tag{7.13}$$

A biaxial stress field implies a superposition of two perpendicular load fields and may therefore yield stress concentrations ranging between $K_t = 2$ and $K_t = 4$ depending on whether the stresses are both tension or whether one is in compression [1].

For the case where both loads are acting in the laminate's material axes, the tangential stress at the locations $\psi = 0°$ or $\psi = 180°$ is given by [7]

$$\sigma_{\theta,\,\mathrm{ortho}} = S_x\left(1+\sqrt{2\left(\sqrt{\frac{E_x}{E_y}}-v_{xy}\right)+\frac{E_x}{G_{xy}}}\right)+S_y\left(-\sqrt{\frac{E_x}{E_y}}\right) \tag{7.14}$$

and at the locations $\psi = \pm 90°$

$$\sigma_{\theta,\,\mathrm{ortho}} = S_y\left(1+\sqrt{2\left(\sqrt{\frac{E_y}{E_x}}-v_{yx}\right)+\frac{E_y}{G_{xy}}}\right)+S_x\left(-\sqrt{\frac{E_y}{E_x}}\right) \tag{7.15}$$

In the case that both loads are similar in magnitude, the exact location of the highest peak stress may be at a location different from $\psi = 0°$ or $\psi = 90°$, and hence, the considerations discussed in the previous section may apply.

7.7 Other Load Cases

Theoretically, multi-axial load cases may occur which are not acting in-phase. If that is the case, the above methodology may not apply and for each location of the notch perimeter the stress sequences must be established. This could be done algebraically or numerically using the full stress solutions for orthotropic plates containing a cut-out. From such an analysis, the most severe case at the notch root could be taken for an assessment of the fatigue initiation life.

If the frequencies are different, then the peak stress cycles at the notch root are no longer constant in amplitude. The assessment should then adopt either an equivalent constant amplitude load spectrum analysis or perform a full variable amplitude load spectrum analysis. These options are discussed briefly in Sect. 7.12.

7.8 Fatigue Stresses at the Notch Root

The residual stresses in FMLs that exist after curing make the fatigue analysis a bit less straightforward. The key aspect here is that the stress concentration factors discussed in previous sections apply to the mechanical load cycles only. Spronk et al. [7] explain that the in-plane stresses that exist after curing, induced by the mismatch in coefficients of thermal expansion, remain as they are in the laminate, even after removing a piece of material.

This means that an FML containing a hole will experience thermal stresses within the laminate that remain constant irrespective of the hole diameter and shape.

Thus, the peak stresses that occur at the notch root are a function of the mechanical nominal stress and thermal stress, but with only the first multiplied by the stress concentration factor.

Hence, the peak stresses at the notch root in layer k are calculated with

$$\sigma_{k,\,\mathrm{peak}} = K_{t,\,\mathrm{FML}} S_{k,\,\mathrm{nom}}^{\mathrm{mech}} + S_k^{\mathrm{th}} \tag{7.16}$$

where the nominal stress is calculated without the influence of the thermal stresses, i.e.

$$S_{k,\,\mathrm{nom}}^{\mathrm{mech}} = \left(S_{k,\,\mathrm{farfield}} - S_k^{\mathrm{th}}\right) \frac{W}{W - D} \tag{7.17}$$

Here $S_{k,\ \mathrm{farfield}}$ are the far field stresses in layer k that are calculated with the CLT.

It is important to understand this aspect, because whereas traditionally the stress concentration factor is considered a constant describing the influence of the geometry, the stress concentration factor in FMLs is no longer a constant depending on the geometry only. The mechanical load applied to the laminate changes the actual value of the stress concentration factor.

So for the layer k described in Eqs. (7.16) and (7.17), the effective stress concentration factor at the notch root is described by

$$K_{t,\,k} = \frac{\sigma_{k,\,\mathrm{peak}}}{S_{k,\,\mathrm{nom}}} \tag{7.18}$$

with the nominal stress $S_{k,\ \mathrm{nom}}$ including the thermal stresses

$$S_{k,\,\mathrm{nom}} = S_{k,\,\mathrm{nom}}^{\mathrm{mech}} + S_k^{\mathrm{th}} \tag{7.19}$$

Hence, throughout the applied fatigue load cycle, the stress concentration factor at the notch root varies between a minimum value at minimum applied load and maximum value at maximum applied load. The corresponding stress ratio R then has to be calculated with

$$R_{k,\,\mathrm{peak}} = \frac{\sigma_{k,\,\mathrm{peak}}^{\mathrm{min}}}{\sigma_{k,\,\mathrm{peak}}^{\mathrm{max}}} \tag{7.20}$$

which is different from the stress ratio applied to the specimen, i.e. $R = S_{\mathrm{min}}/S_{\mathrm{max}}$, but also different from the nominal stress ratio

$$R_{k,\,\mathrm{nom}} = \frac{S_{k,\,\mathrm{nom}}^{\mathrm{min}}}{S_{k,\,\mathrm{nom}}^{\mathrm{max}}} \tag{7.21}$$

7.9 Fatigue Initiation Life Estimation

In general, the stress concentration factor is used to determine fatigue lives of metallic materials and components. The approach is based on correlating the applied amplitude stress to a life until failure for a given geometry, i.e. for a given stress concentration factor. An extensive amount of data has been generated in past decades for many metals and alloys to provide this *S–N* correlation [14–17].

Therefore, it certainly would be beneficial if these data could be used for estimating fatigue lives in FMLs. However, the contribution of (what is considered) the fatigue initiation phase to the total fatigue life until failure is considerably smaller for FMLs as compared to monolithic metals. The *S–N* data therefore cannot be applied to metals in FMLs if life until failure is being considered.

But because of the large contribution of the initiation phase to the total fatigue life for metals, the *S–N* data can be used to determine the fatigue initiation life in FMLs. It requires an engineering definition of fatigue initiation life in FMLs that approximates the total fatigue life in metals. This subject was discussed in Sect. 7.2. Defining initiation as the life until crack length of 1 mm has been generated, the *S–N* data for monolithic aluminium can be used as S–N_i data for the aluminium layers in FMLs [18].

The key question in this approach is which stresses to take in correlating these *S–N* curves to the case of initiation in FMLs. The previous section illustrates that in fact three different sets of stresses could be considered: the far field stresses, the nominal stresses and actual peak stresses. *S–N* curves for monolithic metals generally are defined in terms of nominal amplitude and mean stresses with curves for particular values of K_t.

Because initiation is a surface phenomenon, the actual peak stress cycle should determine the initiation behaviour. The earlier discussed peak stresses cannot be used, however, because the reference data are expressed in nominal stresses. The nominal stresses in each layer on the other hand do not cycle with the stress ratio of the peak stress at the notch root. Hence, it is advised to take the nominal amplitude stress in a layer, Eq. (7.19), together with the stress ratio of the peak stresses at the notch root in that layer, i.e. Eq. (7.20).

One has to keep in mind that the approach presented here is only applicable to high cycle fatigue. In case of low cycle fatigue, the assumption of fibre layers remaining intact under the occurring load cycles may no longer be valid. Even if the fibre layers remain intact, the amount of plasticity created may impair the accuracy of the prediction. As a consequence, the theory to calculate the stresses is not applicable.

7.10 Adapting Reference Data for S_m and K_t

The *S–N* data taken from the literature have been produced for specific mean stresses and stress concentration factors. This implies that in the fatigue initiation analysis the calculated reference stresses must be adapted to the corresponding mean stress and stress concentration factor of the *S–N* data.

Hence, the nominal reference stresses in the FML are calculated with

$$\begin{aligned} S_{m,\,\text{nom}} &= \frac{1 + R_{k,\,\text{nom}}}{2} S_{k,\,\text{nom}}^{\max} \\ S_{a,\,\text{nom}} &= \frac{1 - R_{k,\,\text{nom}}}{2} S_{k,\,\text{nom}}^{\max} \end{aligned} \tag{7.22}$$

With these stresses, Spronk et al. [7] explain how the Goodman relation can be used twice to establish the amplitude stress for the equivalent *S–N* data. Obviously, if for the metal used in the FML data are available that suggests use of other relationships, for example the Gerber parabola, then the approach allows for that, but because Goodman generally yields conservative results it is considered the best first step in the fatigue initiation evaluation.

The fatigue response of metal containing a notch is dependent not only on the peak stresses at the notch root, but also on the distribution of the stresses in the net section [1]. Therefore, another translation of the amplitude stress is required related to the stress concentration factor. HSB 62131-01 describes a method to predict the fatigue life for cases with a stress concentration factor different from the *S–N* data [19].

The methodology makes a distinction between the case where only *S–N* data with a single K_t value are used for translation and the case where one can interpolate between multiple K_t values. The first method implies that the amplitude stress is multiplied with a load factor f to account for the different stress concentration factor. Obviously, the accuracy of this method is determined by how close the reference stress concentration factor is to the target value of the *S–N* data. The load factor is determined with

$$f = \begin{cases} K_t & K_{t,S-N} = 1 \quad K_t \geq 1 \\ \frac{K_t}{K_{t,S-N}} & K_{t,S-N} > 1 \quad K_t > K_{t,S-N} \\ 1 & K_{t,S-N} > 1 \quad K_t \leq K_{t,S-N} \end{cases} \tag{7.23}$$

and the corresponding stress amplitude is obtained with

$$S_{a,\,\text{corr}} = f\, S_{a,\,S-N} \tag{7.24}$$

In case more *S–N* curves are available for different stress concentration factors but the same stress ratio, the second method can be adopted. With that method, the number of cycles until fatigue crack initiation N_i can be established with

$$\log(N_i) = \log(N_2) + (\log(N_2) - \log(N_1)) \frac{\log(K_t) - \log\left(K_{t,\,S-N2}\right)}{\log\left(K_{t,\,S-N2}\right) - \log\left(K_{t,\,S-N1}\right)} \tag{7.25}$$

Here, the subscripts 1 and 2 indicate the *S–N* curves with, respectively, the lower and higher stress concentration factor.

The applicability of both the amplitude stress correction methods is limited to high cycle fatigue, that is, $N > 5 \times 10^3$ cycles [19].

7.11 Accuracy of Predictions

The accuracy of the methodology explained above depends on various aspects. The most important aspect relates to the reference *S–N* data used. Spronk et al. [7] illustrate that by comparing their predictions with fatigue initiation data obtained from Homan and Schra [20] using *S–N* curves for different stress concentration factors from [21]. The two most extreme cases are presented in Fig. 7.9.

It is important to understand the message that Fig. 7.9 conveys. In particular because very few studies on fatigue initiation in FMLs [3, 22] show a correlation between presented model and experimental data to demonstrate the validity and accuracy of the model. None of these papers well addresses the fact that most of the accuracy is achieved by taking reference *S–N* data for stress concentration factors close to the case considered in the FML.

Depending on these cases, the method utilizing a single *S–N* curve and the method interpolating between two *S–N* curves may prove to achieve the same level of accuracy, as illustrated by Spronk et al. [7]. In general, considering the stress concentration factor at the maximum level of the applied load cycle seems to provide the best results in comparison with experimental data. Furthermore, it is observed that once reference *S–N* data are taken with stress concentration factors close to, or at least less than the stress concentration for which the prediction is made, predictions generally are rather accurate. It is expected that this can be attributed to Eq. (7.23).

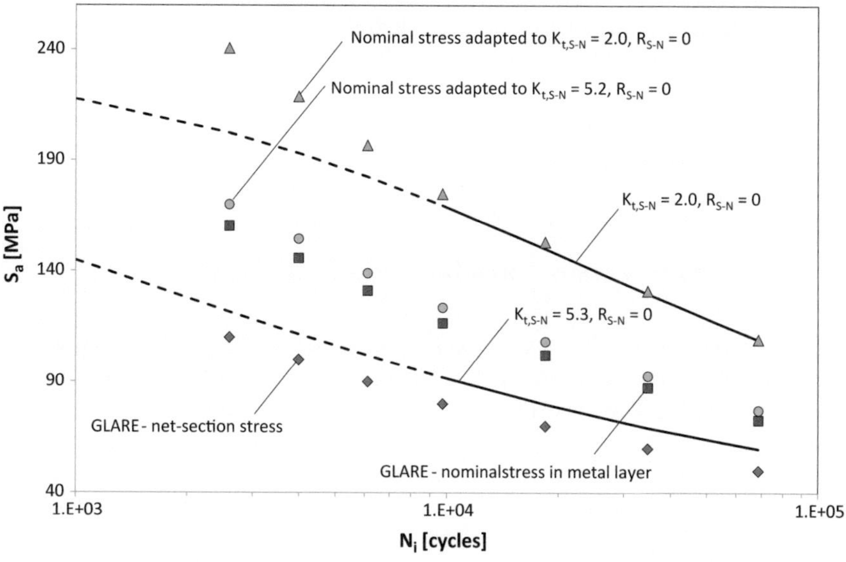

Fig. 7.9 Comparison of crack initiation life measurements of GLARE 4B-3/2-0.3 [20] to *S–N* curves with $K_t = 2.0$ and $K_t = 5.2$ ($R = 0$), valid above $N = 10^4$ cycles [21]

Table 7.1 Information on specimen geometry tested by Oldersma [23]

W (mm)	L (mm)	D (mm)	$K_{t,iso}^{\infty}$	$K_{t,iso}$
50	500	5	3	2.73
100	500	10	3	2.73

7.12 Size Effects

The initiation of fatigue cracks at blunt notches is not only affected by the actual value of the peak stress at the notch root, or the stress concentration factor, but also by the surface area in the notch root exposed to these high stresses. This is generally referred to as the size effect [1]. Two cases with different notch root radii may experience the peak stress and stress concentration factor, but because of the difference in area in the notch radius, different initiation lives are obtained.

A similar observation can be made for fatigue initiation in FMLs. This is illustrated by the work of Oldersma [23], who tested two open hole tension specimens geometries with different hole diameters and specimen widths; see Table 7.1.

Oldersma tested ARALL2, ARALL3 and GLARE2 with the MiniTwist and F27 load spectra and recorded the growth of cracks that initiated over the first few millimetres. For both load spectra, GLARE2 outperformed the other two laminates. Interestingly enough, ARALL2 performed better than ARALL3 with the F27 load spectrum, but worse for the MiniTwist.

Oldersma reported the observation of fibre failure during the MiniTwist load spectrum in ARALL2, which did not occur with the F27 specimens. ARALL3 represents the equivalent to ARALL2, except that post-stretching has been applied. This prevents fibre failure under the MiniTwist load spectrum, giving the better performance.

Important here is the observation that despite the similar stress concentration factor and stress conditions, the fatigue initiation life of the small diameter, $D = 5$ mm, is greater than the larger diameter of 10 mm. This is illustrated in Fig. 7.10. For all three laminate grades, the S–N_i curves for $D = 10$ mm are consistently positioned left of the curves for $D = 5$ mm.

7.13 Constant Versus Variable Amplitude Loading

The intention of the work by Oldersma [23], discussed in the previous section, was to develop S–N_i curves for ARALL2, ARALL3 and GLARE2 for the two load spectra mentioned. The concept was that in monolithic aluminium the difference between the time to initiate a crack and the time until a crack has reached a detectable length is fairly small and hence may be ignored. Because of the intrinsic high crack resistance of FMLs, however, this difference may be more significant. Technically, this argument applies to both constant and variable amplitude loading, but the MiniTwist and F27 load spectrum were adopted in Oldersma's research, which are both variable amplitude load spectra.

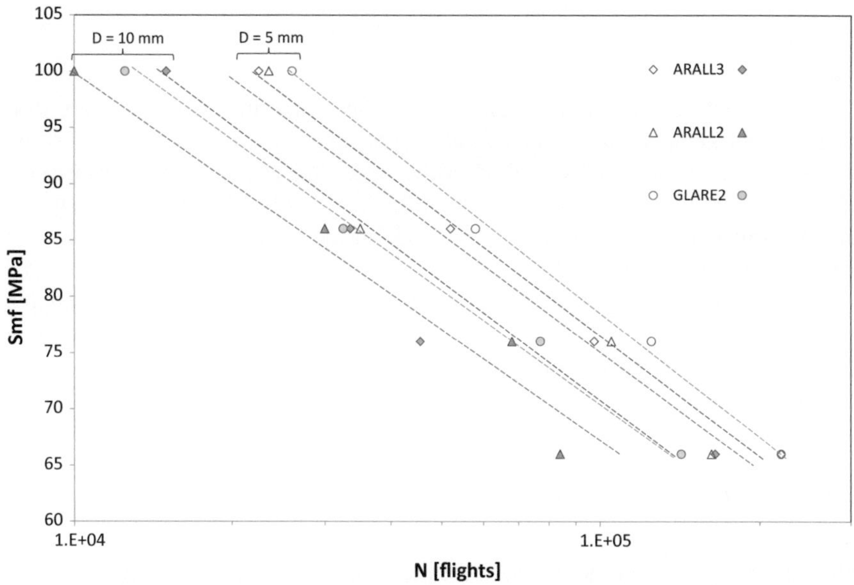

Fig. 7.10 Illustration of the effect of hole diameter on the fatigue initiation life (a_i = 1 mm) of ARALL2, ARALL3 and GLARE2 tested with the MiniTwist load spectrum [23]

The key finding of Oldersma's work is that the initiation life is entirely dependent on the load spectrum applied, indifferent of whether the mean stress in flight S_{mf} is kept constant. This seems rather obvious with hindsight, but it made clear that developing S–N_i curves for specific FML lay-ups requires the knowledge of the load spectrum beforehand. Hence, to predict the fatigue initiation life, or the life until a detectable crack length is reached, Oldersma concluded that the load spectrum should be known beforehand in order to generate reference data for these predictions. Because this obviously is not the case most of the time, other solutions should be adopted.

Rather than testing the fatigue initiation characteristics of the FML (which is also FML grade and lay-up dependent), the concept of the current chapter is that the constituent metal layer should be tested. Hence, fatigue initiation life curves S–N_i or fatigue life curves obtained from tests on monolithic metals sheets should be used to predict the life.

There are two ways to adopt the methodology discussed in this chapter for variable amplitude loading; either an equivalent stress cycle is determined and used following the same procedure as for constant amplitude loading, or the life is estimated on a cycle-by-cycle basis. The first method may be applied if experiments with representative load spectra are available.

The latter method may be performed using the Miner rule, which in its general form may be written as

$$\sum \frac{n_i}{N_i} = 1 \tag{7.26}$$

The general observation for monolithic metals is that this method is very inaccurate, if not incorrect. A survey by Schütz [24] indicated that the scatter in Miner values had a mean value close to 0.85, which makes the Miner rule un-conservative. However, similarly, it is known that the Miner rule can be highly conservative, if the effect of plasticity induced by high loads in the load spectrum on subsequent load cycles is not considered [1].

Beumler [25] investigated the application of the Miner rule for fatigue initiation in FMLs using various fuselage load spectra, observing that the value may be slightly less than 1. Hence, he concluded that predictions using Eq. (7.26) can be considered valid.

7.14 Mechanically Fastened Joints

The calculation methods described in the previous sections apply to generic cases of fatigue initiation in FMLs. However, for fatigue evaluation of actual FML structures the assessment will most often be applied to mechanically fastened joints. For that purpose, the stress similarity concept may provide the means to accurately predict the initiation life.

This concept is based on the assumption that the stresses at the hole perimeter can be related to the initiation life, rather than the nominal stresses. This implies the assumption that two different joints with the same peak stress cycle at the hole edge have identical initiation lives. Fatigue initiation prediction of arbitrary joints under given fatigue loading may then be performed using reference joints for which the initiation life has been experimentally determined and related to the peak stress cycle.

The peak stresses at the hole in the metal layers of a mechanically fastened FML joint can be derived including bending, geometry (net stress in the presence of holes) and stress concentration with

$$\sigma_{\text{peak}} = K_t \frac{s}{s - D} S_m \tag{7.27}$$

with s and D defined in Fig. 7.11 and

$$K_t = (1 - \gamma) K_{t,\text{open hole}} + \gamma K_{t,\text{pinloaded hole}} + k_{\text{bending}} K_{t,\text{bending}} \tag{7.28}$$

Here, γ defines the load transfer ($0 < \gamma < 1$), i.e. the amount of load bypassing the fastener hole and the amount transferred through the fastener in bearing. The stress S_m can be calculated similarly to the method explained in Sect. 7.8, while the bending factor k_b is defined as

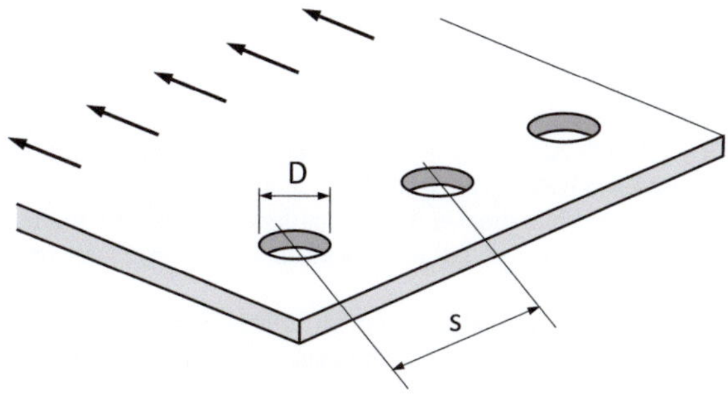

Fig. 7.11 Definition of diameter D and rivet pitch s

Fig. 7.12 Determination of bending factor for each individual layer based on laminate bending factor

$$k_b = \frac{\sigma_{\text{bending}}}{\sigma_{\text{normal}}} \tag{7.29}$$

Figure 7.12 illustrates how the bending factor for each layer can be related to the laminate bending factor k_b by considering the layer's through-thickness position in the laminate.

7.15 Influence of Post-stretching

The procedure of post-stretching has been discussed in Sect. 4.8, which presents a method to approximate the stress levels in the individual layers of the FML after post-stretching. Conceptually, the methodology presented in this chapter is not different in case post-stretching is considered. The residual stresses induced by the mismatch in coefficients of thermal expansion after the curing process, are altered by post-stretching.

This implies that once post-stretching is applied, the stresses S_k^{ps} in the individual layers must be estimated with the method from Sect. 4.8, after which the same methodology is applied as discussed before, except that instead of S_k^{th} the stresses S_k^{ps} are used.

References

1. Schijve J (2009) Fatigue of structures and materials. Springer, B.V
2. Alderliesten RC (2007) Analytical prediction model for fatigue crack propagation and delamination growth in GLARE. Int J Fatigue 29(4):628–646
3. Homan JJ (2006) Fatigue initiation in fibre metal laminates. Int J Fatigue 28:366–374
4. Schijve J (1967) Significance of fatigue cracks in micro-range and macro-range. Fatigue Crack Propag ASTM-STP 415:415
5. Papakyriacou M, Schijve J, Stanzl-Tschegg SE (1997) Fatigue crack growth behaviour of fibre-metal laminate GLARE-1 and metal Laminate 7475 with different blunt notches. Fatigue Fract Eng Mater Struct 20:1573
6. Kieboom O (2000) Fatigue crack initiation and early crack growth prediction model for the Fibre Metal Laminate GLARE. MSc thesis Delft University of Technology
7. Spronk SWF (2015) Predicting fatigue crack initiation in fibre metal laminates based on metal fatigue test data. Int J Fatigue 70:428–439
8. Lekhnitskii SG (1968) Anisotropic plates. Gordon and Breach
9. Heywood RB (1962) Designing against fatigue. Chapman & Hall, London
10. Peterson R (1974) Stress concentration factors. Wiley
11. Tan SC (1988) Finite-width correction factors for anisotropic plates containing a central opening. J Compos Mater 22:1080–1097
12. Wu XJ (2002) A higher-order theory for fiber-metal laminates. In: Proceedings of the 23rd international congress on aeronautical sciences, Toronto, Canada
13. Vašek A, Polák V, Kozák J (1997) Fatigue crack initiation in fibre-metal laminate GLARE 2. Mater Sci Eng A234–236:621–624
14. Metallic materials properties development and standardization (MMPDS) handbook
15. Handbuch Struktur Berechnung (HSB)
16. Fatigue data book: light structural alloys (1995), ASM International, Materials Park, OH 44073-507-9, USA
17. Kaufman J (2008) Properties of aluminium alloys, fatigue data and the effects of temperature, product form, and processing. ASM International, Materials Park, OH 44073-0002, USA
18. Alderliesten RC (2009) Fatigue & damage tolerance of hybrid materials & structures—some myths, facts and fairytales. In: Proceedings of the 25th ICAF symposium—Rotterdam
19. Homan JJ (2009) Handbuch Struktur Berechnung (HSB) 62131-01 issue a: guidelines for the prediction of the fatigue life for Kt values different to available databases, Technical Report
20. Homan JJ, Schra L (2002) Application of aluminium alloy 2024-T3 fatigue life data to GLARE laminates, NLR report NLR-CR-2002-185, National Aerospace Laboratory of the Netherlands
21. Hoang, V, Schwarmann L (1986) Handbuch Struktur Berechnung (HSB) 63111-01 issue D: Zeitfestigkeit 3.1354-T3, Technical Report
22. Chang P-Y, Yeh P-C, Yang J-M (2008) Fatigue crack initiation in hybrid boron/glass/aluminium fiber metal laminates. Mater Sci Eng A 496:273–280
23. Oldersma O (1994) S-N curve construction for fatigue crack initiation in Fibre-Metal Laminates under flight simulation loading, Report NLR-CR-94427 C. National Aerospace Laboratory NLR, Amsterdam, The Netherlands
24. Schütz W (1979) The prediction of fatigue life in the crack initiation and crack propagation stages. A state of the art survey. Eng Fracture Mech 11:405–421
25. Beumler Th (2004) Flying GLARE®, A contribution to aircraft certification issues on strength properties in non-damaged and fatigue damaged GLARE® structures. PhD dissertation, Delft University of Technology, Delft, The Netherlands

Chapter 8
Static and Fatigue Delamination

Abstract The delamination between metal and composite plies in FMLs is discussed for the two major crack opening modes and combinations thereof. The strain energy release rate is introduced as parameter to describe the static and fatigue behaviour, which is influenced by the interface geometry and fibre volume fraction of the composite plies. The relation between constant and variable amplitude loading is discussed. Furthermore it is explained how plasticity of the metal layers affects the quasi-static delamination fracture toughness. Some observations concerning delamination buckling and residual stress state are discussed.

8.1 Introduction

An important fracture mechanism in fibre-reinforced composites in general and fibre metal laminates (FMLs) in particular is disbonding of individual layers, often denoted as delamination. Delamination may occur in some cases as a single fracture type, but it is observed to occur in FMLs in most cases in balance with crack growth in metal layers. Interesting is that whereas delamination is usually considered detrimental to the material performance, especially for fibre-reinforced composites, it is to some extent considered beneficial in combination with crack growth in FMLs.

Once a crack in the metal layer has initiated, it tends to propagate further under (cyclic) loading, as will be discussed in later chapters. Opening of that crack under applied loads induces a stress singularity at the interface with the adjacent intact fibre layers. These layers are forced to strain significantly with the opening of the crack, risking fibre failure unless delamination occurs. The stress singularity induces mode II delamination (Fig. 8.1) initiation at the interface perpendicular to the crack and in line with the applied load. This enables the intact fibre layers to strain over a longer length, locally reducing the stress in the fibres.

Hence, to be able to describe how damage evolves under either quasi-static loading or fatigue loading, one needs to understand and be able to predict delamination growth. This chapter describes the results and observations of various studies addressing the delamination growth in FMLs.

R. Alderliesten, *Fatigue and Fracture of Fibre Metal Laminates*, Solid Mechanics and Its Applications 236, DOI 10.1007/978-3-319-56227-8_8

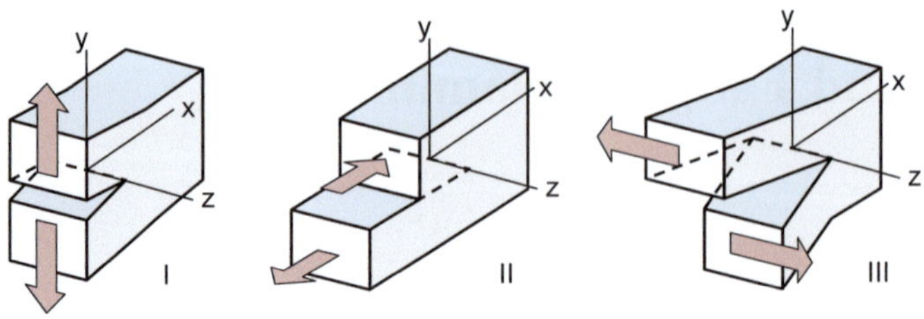

Fig. 8.1 Illustration of the three opening modes: mode I (tension), mode II (shear) and mode III (transverse shear)

8.2 Strain Energy Release Rate

One approach often used to describe delamination growth in laminated materials applies the strain energy release rate (SERR) G, which is based on linear elastic fracture mechanics (LEFM). This parameter is analogous to the stress intensity factor, but rather than describing a stress field in the vicinity of the crack, it describes the energy dissipation in the surrounding material around growing delaminations. The reason for using G instead of K in this case is that the stress field locally at the crack at the interface of two dissimilar and inhomogeneous materials is hard to describe [1]. Taking the energy upstream and downstream of the crack tip in the surrounding material seems then a straightforward engineering approach.

However, although the application of the SERR is rather common, it seems that consensus on the formulation of the effective SERR under cyclic loading has not yet been reached [2]. In particular where it concerns the description of fatigue delamination growth, the range ΔG may be taken as the difference between $G_{\max}$ and $G_{\min}$, or closer to the definition of ΔK, as the difference $\sqrt{G_{\max}} - \sqrt{G_{\min}}$. Hence, this section recapitulates the principles of LEFM concerning the SERR.

For a linear elastic system, there is no interaction between the different modes of deformation (modes I, II and III). For a given system, the total SERR can therefore be obtained with

$$G_{tot} = G_{I} + G_{II} + G_{III} \tag{8.1}$$

However, when considering the superposition of SERRs resulting from the same deformation mode, the rules for superposition are [3]

$$\begin{aligned} G_{I} &= \left[\sqrt{G_{I(1)}} + \sqrt{G_{I(2)}} + \sqrt{G_{I(3)}} + \cdots\right]^2 \\ G_{II} &= \left[\sqrt{G_{II(1)}} + \sqrt{G_{II(2)}} + \sqrt{G_{II(3)}} + \cdots\right]^2 \\ G_{III} &= \left[\sqrt{G_{III(1)}} + \sqrt{G_{III(2)}} + \sqrt{G_{III(3)}} + \cdots\right]^2 \end{aligned} \tag{8.2}$$

Thus, when expressing the SERR range for given deformation mode it should be formulated as

$$\begin{aligned}\Delta G_{\mathrm{I}} &= \left(\sqrt{G_{\mathrm{Imax}}} - \sqrt{G_{\mathrm{Imin}}}\right)^2 \\ \Delta G_{\mathrm{T}} &= \Delta G_{\mathrm{I}} + \Delta G_{\mathrm{II}} + \Delta G_{\mathrm{III}}\end{aligned} \tag{8.3}$$

Another way of looking at this formulation is by looking at the similitude with the stress intensity factor range. Because both parameters follow from LEFM, they can be related to each other by $G \propto K^2$.

Another aspect that advocates for the formulation of Eq. (8.2), and that appears to be important for the analysis in FMLs, is the presence of residual stresses after curing, discussed in Sect. 4.4. Because if G is broken up into a component due to applied load and a component due to residual stresses, it is easily shown that the SERR range becomes

$$\begin{aligned}\Delta G &= \left(\sqrt{G_{\mathrm{max}}} - \sqrt{G_{\mathrm{min}}}\right)^2 \\ &= \left[\left(\sqrt{G_{\mathrm{applied,max}}} + \sqrt{G_{\mathrm{r}}}\right) - \left(\sqrt{G_{\mathrm{applied,min}}} + \sqrt{G_{\mathrm{r}}}\right)\right]^2 \\ &= \left(\sqrt{G_{\mathrm{applied,max}}} - \sqrt{G_{\mathrm{applied,min}}}\right)^2\end{aligned} \tag{8.4}$$

In other words, despite the presence of residual stresses, the range of G remains unaffected as expected based on the selected similitude principle similar to ΔK.

The delamination growth rate $\mathrm{d}b/\mathrm{d}N$ can be related to the SERR similarly to the way the Paris equation relates the crack growth rate to the stress intensity factor range, i.e.

$$\frac{\mathrm{d}b}{\mathrm{d}N} = C_{\mathrm{d}}(\Delta G_{\mathrm{eff}})^{n_{\mathrm{d}}} = C_{\mathrm{d}}\left(\sqrt{G_{\mathrm{max}}} - \sqrt{G_{\mathrm{min}}}\right)^{2n_{\mathrm{d}}} \tag{8.5}$$

For an FML containing n_{m} number of metal layers, see Fig. 8.2, the SERR can be written as [4]

$$G_{\mathrm{II}} = \frac{S_{\mathrm{lam}}^2}{2jE_{\mathrm{m}}}\left[\gamma^2(n_{\mathrm{m}} - n_{\mathrm{cr}})t_{\mathrm{m}} - \lambda^2 n_{\mathrm{m}} t_{\mathrm{m}} + \frac{E_{\mathrm{f,0}}}{E_{\mathrm{m}}} n_{\mathrm{f,0}} t_{\mathrm{f,0}}\left(\gamma^2 - \lambda^2\right) + \frac{E_{\mathrm{f,90}}}{E_{\mathrm{m}}} n_{\mathrm{f,90}} t_{\mathrm{f,90}}\left(\gamma^2 - \lambda^2\right)\right] \tag{8.6}$$

with

$$\gamma = \frac{t_{\mathrm{lam}}}{(n_{\mathrm{m}} - n_{\mathrm{cr}})t_{\mathrm{m}} + \frac{E_{\mathrm{f,0}}}{E_{\mathrm{m}}} n_{\mathrm{f,0}} t_{\mathrm{f,0}} + \frac{E_{\mathrm{f,90}}}{E_{\mathrm{m}}} n_{\mathrm{f,90}} t_{\mathrm{f,90}}};\ \lambda = \frac{t_{\mathrm{lam}}}{n_{\mathrm{m}} t_{\mathrm{m}} + \frac{E_{\mathrm{f,0}}}{E_{\mathrm{m}}} n_{\mathrm{f,0}} t_{\mathrm{f,0}} + \frac{E_{\mathrm{f,90}}}{E_{\mathrm{m}}} n_{\mathrm{f,90}} t_{\mathrm{f,90}}} \tag{8.7}$$

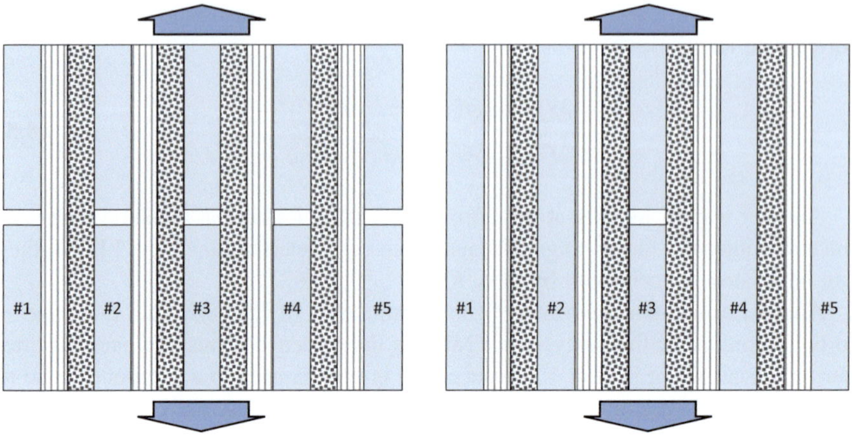

Fig. 8.2 Definition of the geometry of an FML containing n_m metal layers of which n_{cr} contain a crack; *left* $n_m = n_{cr} = 5$; *right* $n_m = 5$, $n_{cr} = 1$ (both cases: $n_{f,0} = n_{f,90} = 5$)

8.3 Interface Geometry

8.3.1 Resin-Rich Layers

It was mentioned before that the delamination growth at the interface between metal and fibre layers depends on the delamination resistance of that interface. This resistance is related not only to the adhesive type itself, but to some extent also to the geometry of the interface. The shear stresses that result from the load transfer from one layer to another induce a shear deformation at the interface. This means that the shear stiffness is an important parameter. This parameter depends on the material—solely adhesive in a resin-rich layer or fibres evenly distributed throughout the adhesive layer—and thus on the interface geometry.

To account for the effect of shear deformation in the case of fatigue delamination in ARALL, Verbruggen [5] and Marissen [6] developed a linear model utilizing an element breakdown in thickness direction of the fibre/adhesive layer, see Fig. 8.4. In this model, it was assumed that the aluminium and fibre elements carry normal loads only and the adhesive elements only shear loads.

This seems a reasonable assumption considering the fact that the fibre layers in ARALL consist of concentrated fibres in the middle of the fibre layers, while resin-rich layers were present at both interfaces. This geometry corresponds to the manufacturing method; woven fibres were cured between two adhesive films. Although the impregnation of the fibres was good, a resin-rich layer remained at each side of the fibre layer.

For standard GLARE, however, the through-the-thickness distribution of the fibres is different. Because GLARE is made of pre-impregnated layers, the adhesive is evenly distributed between the fibres, which results after curing in a fibre layer without resin-rich layers at each side. The difference between typical distributions for both ARALL and GLARE is shown in Fig. 8.3.

The interface geometry in GLARE no longer supports the original assumption by Verbruggen and Marissen, but requires another model. Although such model is somewhat different, the original approach of breaking down the fibre layer in elements through the thickness that deform linearly still applies [8, 9], see Fig. 8.4.

There is, however, a difference between the two concepts illustrated in Fig. 8.4. In the resin-rich layer configuration, the adhesive will crack along with the metallic layer, because the tensile strength of the adhesive is very low. In case of fibres evenly distributed throughout the full fibre layer, this cracking is prevented by the intact continuous fibres. The crack in the resin means that in the resin-rich layer configuration, even for zero delamination an adhesive shear deformation may be present greater than zero, while in the absence of a crack, shear deformation will be zero.

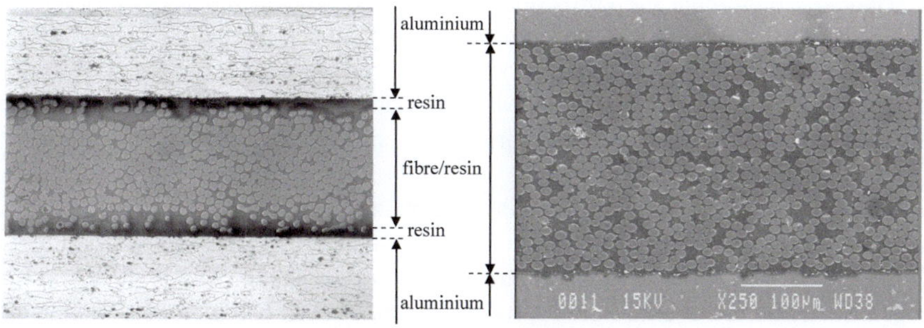

Fig. 8.3 Example of a through-the-thickness fibre distribution in ARALL (*left*) [7] and GLARE (*right*) [8]. The resin-rich layer at the interface in ARALL is not present in GLARE [1]

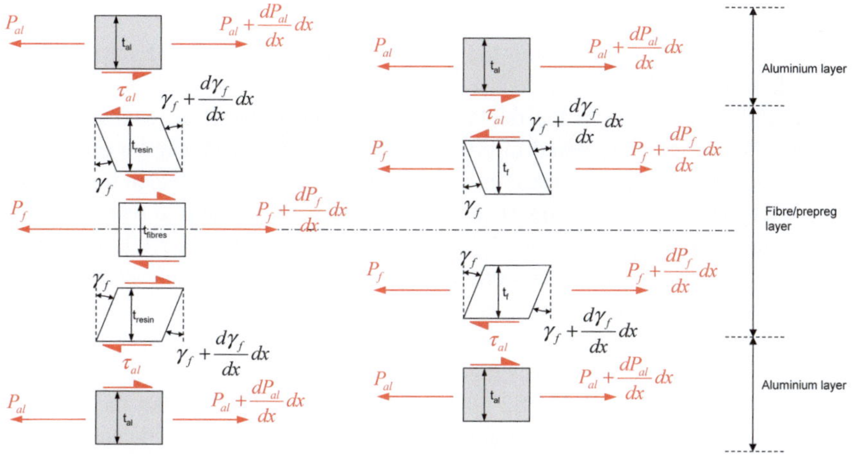

Fig. 8.4 Breakdown of the elements in thickness direction of the deformed fibre layers in ARALL [5] (*left*) and the unidirectional prepreg in GLARE [9] (*right*)

Marissen therefore described the crack opening due to adhesive shear deformation with [8, 9]

$$v(x) = S_{\mathrm{al}} n_{\mathrm{al}} t_{\mathrm{al}} \sqrt{\frac{1}{jF_{\mathrm{ad}}} \left(\frac{1}{F_{\mathrm{al}}} + \frac{1}{F_{\mathrm{f}}} \right)} \tag{8.8}$$

in which F_{ad} represents the adhesive stiffness. With the concept for GLARE, where shear deformation occurs within the fibre layers itself, the crack opening due to shear deformation is calculated with [9]

$$v(x) = C_{\mathrm{b}} S_{\mathrm{al}} t_{\mathrm{al}} \sqrt{\frac{t_{\mathrm{f}}}{G_{\mathrm{f}}} \left(\frac{1}{F_{\mathrm{al}}} + \frac{1}{F_{\mathrm{f}}} \right)} \tag{8.9}$$

Here, C_{b} represents the correction of the shear deformation for small delamination lengths. Equation (8.8) is valid for any delamination length, because at zero delamination length the resin-rich layer has cracked along with the metal layer and can deform. However, in case of shear deformation of the entire fibre layer in the absence of a resin-rich layer, the deformation is prevented in case the delamination length is close to zero.

In case the delamination length is close to zero, a balance must be obtained between the shear deformation building up towards the delamination tip and the decay of that deformation to the centre of the crack opening. This is illustrated in Fig. 8.5. For unidirectional fibre layers, an expression can be obtained with a linear breakdown of elements through the thickness as illustrated in Fig. 8.4, which yields [9]

$$C_{\mathrm{b}} = 1 - \left(\cosh \sqrt{2\alpha_{\mathrm{UD}}} b - \tanh \sqrt{2\alpha_{\mathrm{UD}}} b \sinh \sqrt{2\alpha_{\mathrm{UD}}} b \right) \tag{8.10}$$

with α_{UD} a parameter comprising the thickness, and the normal and shear stiffness of the fibre layers [8, 9].

The difference between the delamination resistance of ARALL laminates and GLARE laminates is presented in [10], see also Fig. 8.6. What can be observed is that considering the observed scatter in the data, the difference between the different adhesive systems is not very significant. The comparison is coloured by the fact that the interface geometry of FM94/S2-glass prepreg and the BSL-312-UL adhesive is different. Hence, the fact that both curves coincide may be interpreted as the advantage of the resin-rich layer is compensated by the poorer resistance of the BSL-312-UL adhesive itself.

On the other hand, the interface geometry of the 3 M adhesive AF163-2-based prepreg is similar to the FM94/S2-glass prepreg. Hence, the fact that the curve for the AF163-2 prepreg is lower than the curve for standard GLARE indicates the 3 M adhesive system has an inherent higher resistance to delamination.

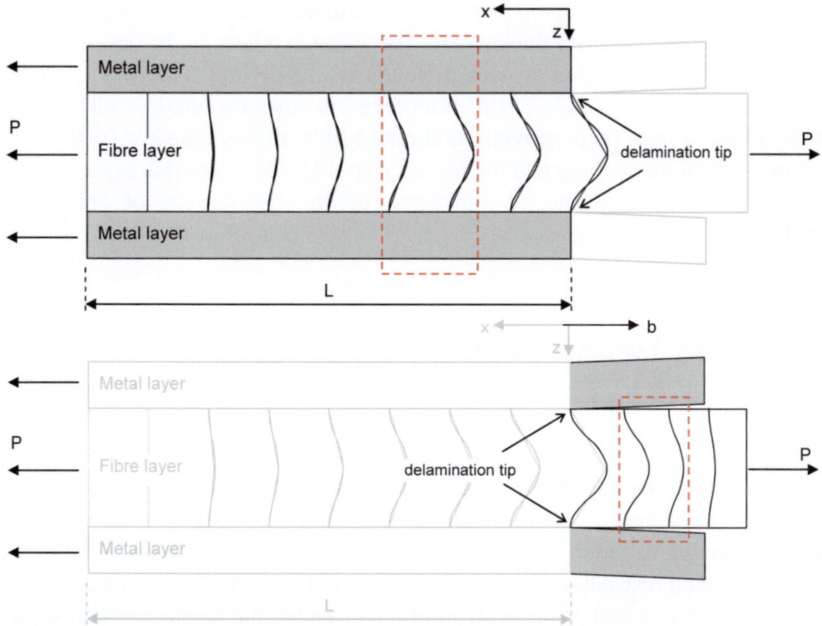

Fig. 8.5 Deformation of the unidirectional prepreg under tensile loading in the fibres and shear loading at the metal/fibre interface before and in the delaminated area

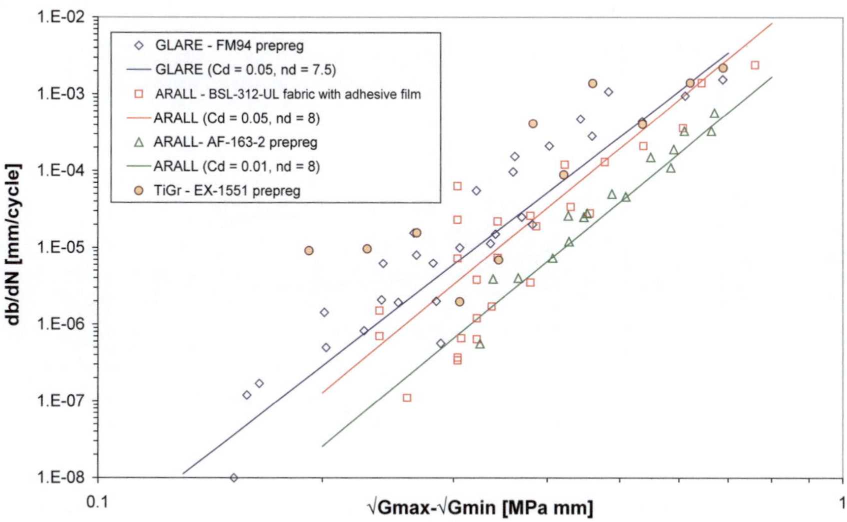

Fig. 8.6 Comparison between the delamination resistance of GLARE and ARALL interfaces [10]

Wilson [11] evaluated the delamination resistance of hybrid laminates containing different resin-rich layer thicknesses. The research related to the development of a resin-rich prepreg system is named Bondpreg, for hybrid laminates containing thicker metal layers [12, 13]. Bondpreg is manufactured with standard FM94/S2-glass fibre prepreg with additional FM94 K0.03 adhesive (without carrier) in between the prepreg and the metal layer [11]. This resin-rich layer improves the delamination resistance, as illustrated by the experiments of Wilson, see Fig. 8.7.

8.3.2 *Tapes Versus Weaves*

Another aspect concerning the interface geometry that must be considered is the difference between unidirectional tapes as applied in standard GLARE and the woven fabrics applied in ARALL laminates tested by Marissen [6], see Fig. 8.8.

With woven fabrics, the delamination propagates along an interface, where it faces continuous variation in interface topology. This has an evident influence on the delamination resistance. For example, Vugt [14] reported delamination growth at the interface in a composite scarf joint repair where the delamination resistance clearly increased when changing the orientation from 0°/90° to ±45°. Similarly, the difference between the interface in GLARE with UD prepreg and the woven fabrics applied in ARALL yields different delamination resistance curves.

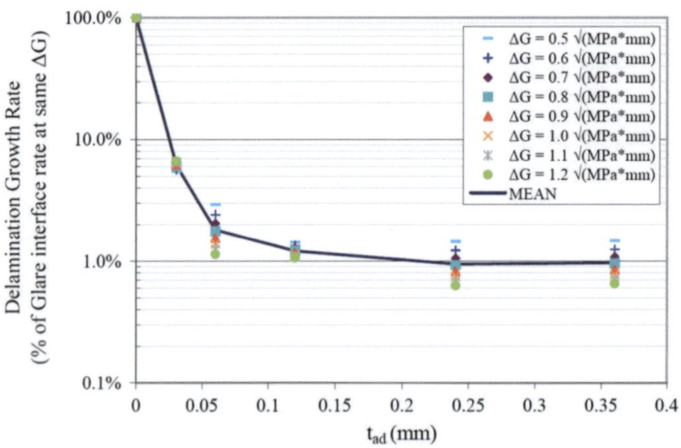

Fig. 8.7 Observed decrease in delamination growth rates with addition of adhesive in the bondline relative to the growth rates in GLARE [11]

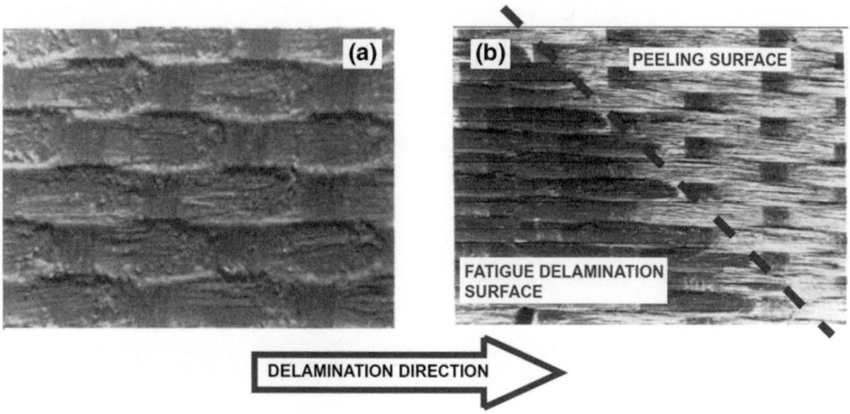

Fig. 8.8 Fatigue delamination surface in ARALL; observed fabric imprint in the adhesive (**a**) and delamination front at the fibre side (**b**) [6, 15]

8.4 Modes I, II and Mixed Mode

8.4.1 Mode I

Van der Hoeven and Frijns [16] evaluated the static delamination resistance of ARALL. They performed mode I fracture toughness tests using DCB tests. The results are presented in Fig. 8.9. The critical SERR G_{Ic} was calculated using

$$G_I = \frac{P^2}{2B}\frac{\delta C}{\delta a} \tag{8.11}$$

taking for P the load at which crack instability occurred. In Eq. (8.11), B is the specimen width and the derivative of compliance with respect to crack length $\delta C/\delta a$ was calculated using the classical beam theory.

The critical SERR G_{Ic} initially increases, which is commonly observed in mode I fracture toughness tests in unidirectional composites. The fibre bridging constitutes a crack-shielding mechanism, taking up a major portion of the applied loading. Hence, the SERR calculated with Eq. (8.11) increases with increasing crack length, until a plateau level has been reached.

8.4.2 Mode II

Mode II delamination at the interfaces between metal and fibre layers in FMLs has received significantly more attention than the mode I. Mostly, this attention is related to the occurrence of mode II-driven delaminations in fatigue crack

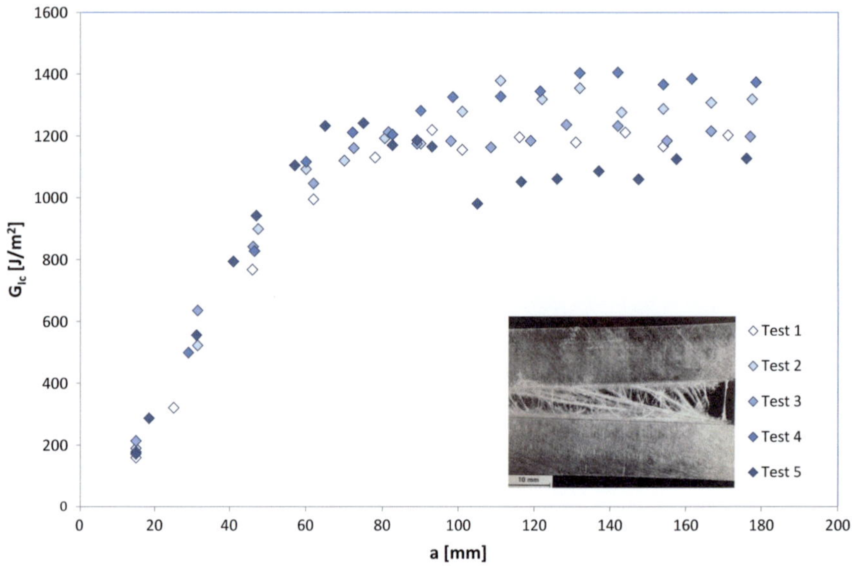

Fig. 8.9 Fracture toughness tests results obtained with DCB tests on ARALL2/1 laminates bonded to 10-mm-thick aluminium 2024-T3 beams using AF163-2 adhesive; inset: observation of fibre bridging [16]

configurations, the subject of Chap. 9. Hence, some results are presented in Figs. 8.6 and 8.7 for the S2/FM94 prepreg system applied in GLARE.

In relation to the topic of interface geometry (see Sect. 8.3), Mangkoesoebroto [17] investigated the delamination behaviour of ARALL laminates with fibre/adhesive layers that contained different levels of fibre volume.

Mangkoesoebroto used the effective cyclic SERR as proposed previously by Marissen [6]

$$G_{\mathrm{eff}} = \left(\sqrt{G_{\max}} - \sqrt{G_{\min}}\right)\sqrt{G_{\max}} \tag{8.12}$$

It should be noted that the adhesive thickness does not occur in the formulations that Marissen and Mangkoesoebroto used for the SERR. However, the effect is apparent in, for example, the adhesive shear deformation in Eq. (8.8). The stiffness parameter F_{ad} is defined as $F_{ad} = G_{ad}/t_{ad}$, which implies that with increasing thickness of the resin-rich layers, the shear deformation increases. Hence, the deformation field in the direct vicinity of the crack tip is affected, but not accounted for in the calculation of the SERR. The consequence is that when plotting the delamination growth rate against the effective SERR, the effect is revealed by different curves, see Fig. 8.10.

This observation is important, because if comparisons are made between different adhesive or prepreg systems, as shown, for example, in Fig. 8.6, one has to be aware that the presented curves do not relate straightforwardly to the inherent

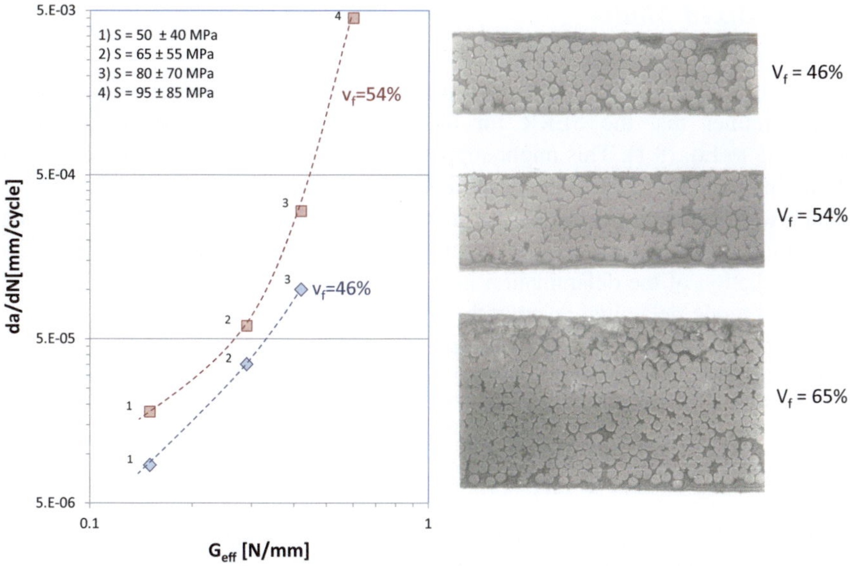

Fig. 8.10 Fatigue delamination growth rate as a function of the effective cyclic SERR for delamination (*left*) and cross-sectional ARALL specimens with the different fibre volume fractions (*right*) [17]

resistance of the adhesive system. Because of the SERR taken as a parameter describing the similitude, and the fact that interface geometry is not accounted for explicitly in the formulation of the SERR parameter, the effect appears in the position of the delamination growth curves.

The implication is that the adhesive system's resistance and the interface geometry may both shift the delamination growth rate curve in the same direction, but they might also have opposite effects.

Another aspect concerning mode II delamination growth tests is the difference between the test set-ups applied. In the FML community, it seems that the ply-interrupt specimen is generally adopted to obtain the mode II delamination characteristics [4, 6, 9, 17], whereas in the composites community people tend to work with the end-notched flexure tests [18].

Delgrange reported the correlation of both test set-ups for the interfaces in GLARE [19], highlighting the difficulty to correlate both results with each other. A major difference between both tests is that in the ENF tests a single delamination is tested, but at continuously changing levels of the SERR, whereas in the ply-interrupt test generally four interfaces are tested at the same time, but at constant SERR.

In most cases, the interfaces do not propagate exactly at the same rate and with the same length [4], which requires some averaging approaches in the data evaluation. Here, case has to be taken, because for ply-interrupt specimens, the differences in delamination lengths may impose secondary bending deformations that actually change the effective opening mode (mix). This is illustrated by the work of Pascoe [20].

8.4.3 Mixed Mode

According to LEFM, the different opening modes do not interact with each other, which implies that the SERR for the different modes can be superimposed according to Eq. (8.1). This might support the idea that for delamination extension at any interface in an FML the total SERR should be calculated and subsequently related to delamination growth with a relation like Eq. (8.5). However, one may have to be careful following such an approach, as it may lead to significant over-prediction of the delamination growth.

Mixed-mode delamination growth is generally approached in a very empirical and phenomenological manner. In quasi-static analyses often the B-K criterion is adopted [21]

$$G_c = G_{Ic} + (G_{IIc} - G_{Ic})\left(\frac{G_{II}}{G_I + G_{II}}\right)^{\alpha} \tag{8.13}$$

However, the question rises how mode-mixity should be considered in fatigue loading. For example, one may apply Eq. (8.1) at both the minimum and maximum of the load cycle (in-phase opening is considered only) and substitute the corresponding total SERR range in Eq. (8.5).

Such an approach implicitly assumes that a single delamination resistance characteristic exists that relates to a total SERR range indifferent from the contribution of individual opening modes. However, microscopic evidence seems to illustrate that decohesion features typically differ for the three opening modes [22]. Hence, one may also expect the delamination resistance characteristic to be different for the different opening modes, as illustrated in Fig. 8.11.

In that case, superposition of the different opening modes may have to be considered with

$$\frac{da}{dN} = C_1(\Delta G_I)^{m_1} + C_2(\Delta G_{II})^{m_2} + C_3(\Delta G_{III})^{m_3} \tag{8.14}$$

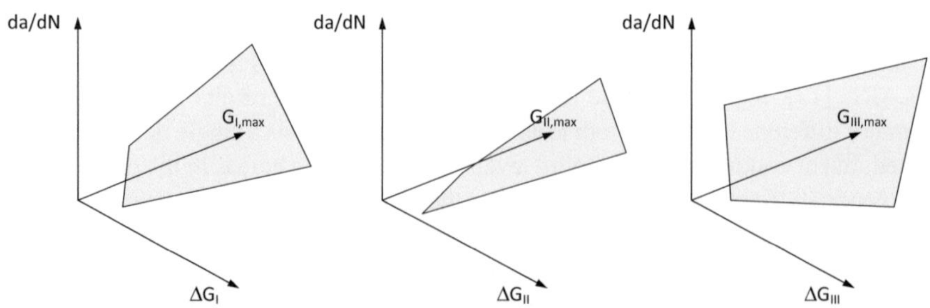

Fig. 8.11 Illustration of possible delamination of different resistance planes for the three opening modes; the delamination plane captures both the influence of ΔG and of G_{max}

The problem probably is that both methods of superposition may be proven to work only if additional curve fitting is applied; Eqs. (8.5) and (8.14) are both entirely empirical. The truth may lie in between both solutions.

However, to incorporate this in prediction methods may be very challenging, in particular if one has to consider that the opening mode ratio may change throughout the process of damage propagation. Take, for example, the observations reported by Roebroeks [23] who investigated the delamination fracture surfaces of ARALL laminates using microscopy.

Roebroeks discusses the different fracture surface features of these delaminations. The initial delamination appearance differs from the major part of the delamination surface, which Roebroeks explained with the changing opening mode. This aspect is illustrated in Fig. 8.12; the delamination initiates predominantly by a mode II opening, because bending deformation of the cracked metal layer is initially constrained. Once the delamination progresses, additional bending will impose a mode I contribution.

This difference in fracture surface appearance may seem to indicate that different phenomena occur and hence different delamination resistance characteristics are at stake. However, in discussing the theoretical work of Suiker and Fleck [24, 25], it was demonstrated that the delamination growth tests reported in [4] resulted in similar delamination resistance, despite that theoretically different opening modes can be calculated for both cases illustrated in Fig. 8.2.

In other words, one may have to be careful with calculating SERR values for predicting delamination growth. The primary driving force may be different from what is mathematically calculated using the principles of LEFM.

This aspect of course can be used to our benefit; instead of describing the complexity of the mode—mix, it may suffice to relate a single delamination characteristic, like the ones illustrated in Fig. 8.6, to the driving force. In most fatigue cases, where delaminations occur in conjunction with fatigue cracks in the metal layer, mode II constitutes the major driving force. Any small theoretical contribution of mode I may be neglected [1].

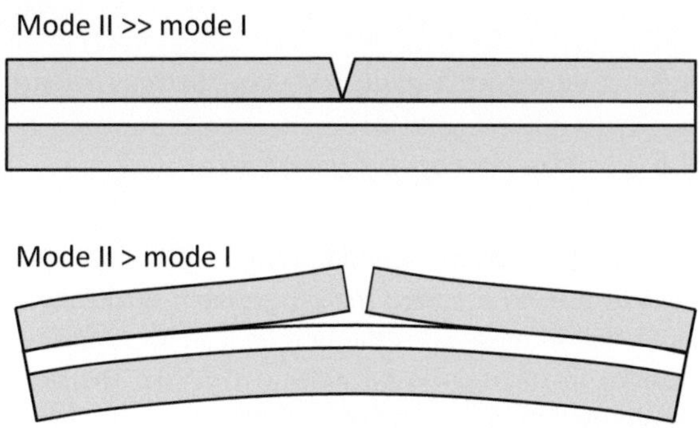

Fig. 8.12 Illustration of the change in opening mode—mix—during delamination growth [23]

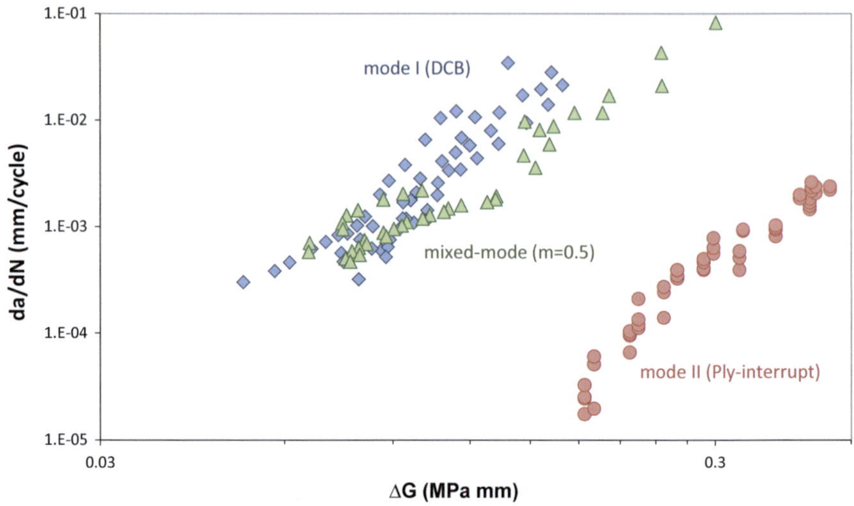

Fig. 8.13 Observed delamination growth characteristics for mode I, mode II and mixed mode with a mode ratio of $m = 0.5$; data from [26]

Despite the above considerations, one is advised to evaluate the delamination resistance in the individual opening modes, mode I and mode II. Take, for example, data obtained by Delgrange [26], illustrated in Fig. 8.13; the difference between mode I and mode II suggests that a significantly higher resistance is present in shear, compared to mode I opening. Figure 8.13 illustrates that a mixed mode of $m = G_{II}/G_{tot} = 0.5$ (equal contribution of both mode I and mode II) results in crack growth data close to that of mode I.

Care has to be taken though, as the exact test conditions of all three opening modes have not been well documented. In addition, as Delgrange reports in [19], the mode II curve obtained with the ENF tests is positioned closer to the mode I curve than the one obtained with ply-interrupt specimens.

8.5 Constant Versus Variable Amplitude Loading

8.5.1 Macroscopic Observations

In case of constant amplitude fatigue, a relationship like Eq. (8.5) may be a convenient way to describe the progression of delaminations for given fatigue loads. However, in most cases the fatigue load spectrum is far from constant, but rather is random in nature. To be able to describe the delamination growth for variable amplitude fatigue, a cycle-by-cycle approach with Eq. (8.5) could be a first start.

The influence of variable amplitude loading on the delamination growth response has been thoroughly investigated by Khan et al. [15, 27]. In [15] a series of delamination tests are reported that vary from constant amplitude, via selective overloads and programmed block loads, to arbitrary variable amplitude load spectra.

The key finding reported is that a sequence effect as generally observed for crack growth in metals is not observed in delamination growth at the interface between metal and fibre layers in FMLs. This seems to suggest that a straightforward cycle-by-cycle analysis suffices, or that linear damage accumulation rules are valid.

Another major observation in [15] is that despite the absence of interaction phenomena, major shear deformations dominate over smaller shear deformations. This means that cycle-by-cycle analysis can only be applied if proper rainflow counting techniques have been adopted. The concept is illustrated in Fig. 8.14; a cycle-by-cycle analysis should not only consider the collection of all small intermediate cycles in a larger cycle. The larger cycle should also be taken into account; Khan et al. [15] illustrate that for a wide-body fuselage spectrum. The cycle-by-cycle analysis predicts delamination growth rates over an order of magnitude lower than the measured rates. If rainflow counting is applied, then the predicted delamination growth rates correlate fairly well with the experimentally measured rates.

This can be understood when looking at the delamination growth curves like the ones illustrated in Fig. 8.6. The linearity of these graphs represents a power law

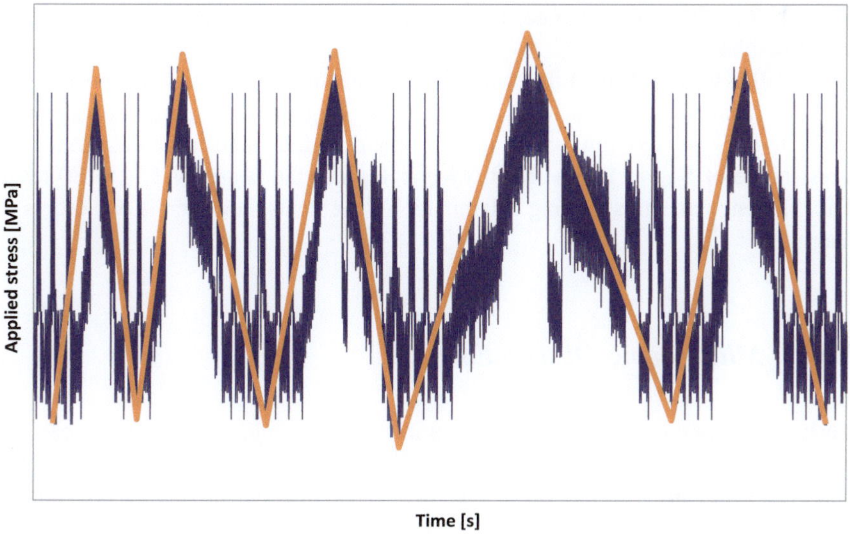

Fig. 8.14 Illustration of major load cycles in a load spectrum that predominantly contribute to delamination growth

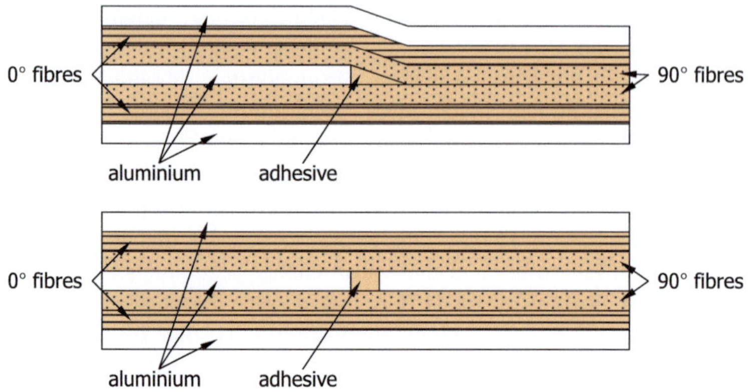

Fig. 8.15 Illustration of the interlaminar doubler run-out configurations tested by Hooijmeijer [10, 28]

relationship. Hence, the accumulated delamination related to the sum of small cycles will be substantially less than the delamination increment related to one large cycle.

One example to further illustrate this is given by the doubler run-out study performed by Hooijmeijer [28], which is further analysed and discussed in [10]. Hooijmeijer tested the interlaminar doubler run-out configurations illustrated in Fig. 8.15. Aside from the quasi-static tests, he tested specimens for 180 kcycles at three different constant amplitude fatigue stress levels; S_{max} = 80, 120 or 180 MPa, all with $R = 0.05$. At that time, Hooijmeijer [28] reported that no delaminations could be observed.

In [10] it was confirmed that with the delamination characteristics illustrated in Fig. 8.6, delamination growth could be expected for these three fatigue stress levels of 2.13, 44.5 and 237 μm, which indeed would be too small to be observed. The stress range of S_{max} = 180 is about twice as high as the range for S_{max} = 80 MPa. The difference in corresponding delamination increments, however, is two orders of magnitude!

8.5.2 *Microscopic Observations*

Various authors have performed scanning electron microscopy to reveal the microscopic features on fracture surfaces. For example, Marissen [6] reported the observations of striation-like surface features at the fracture surface, but was unable to quantitatively relate them to the applied loading. Also Roebroeks [23] illustrated the presence of these features on delamination fracture surfaces obtained from ARALL specimens, stating that they may be present in any composite with low interface strength fibre reinforcements.

More recently, Khan [15] also observed these features in the fibre imprints of delamination fracture surfaces obtained from a GLARE specimen. He, however, was able to quantitatively correlate the spacing of these striation-like features to the stress levels applied in the tests. An example is given in Fig. 8.16, where fracture surfaces were obtained from a specimen that was tested with programmed Hi-Lo block loading. The difference in striation spacing correlated well to the stress amplitudes applied during both blocks of loading.

In addition to the quantitative aspects of fractography, Roebroeks [23] illustrates how the appearance of these striation-like features changes in relation to the major loading or deformation mode of the fibre. The concept is illustrated in Fig. 8.17. When the fibre is pulled out of the matrix in the principal direction of the fibre, the fibre matrix interface is loaded in mode II shear along the entire fibre circumference.

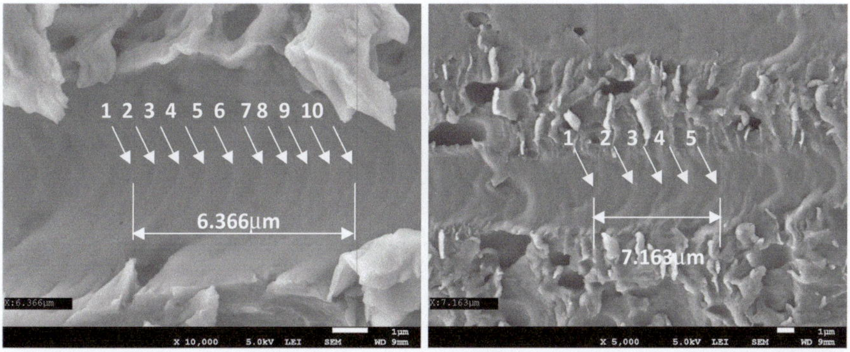

Fig. 8.16 Delamination growth markings (striation) at LO stress amplitude (150 MPa) (**a**) and at HI stress amplitude (169 MPa) (**b**) [15]

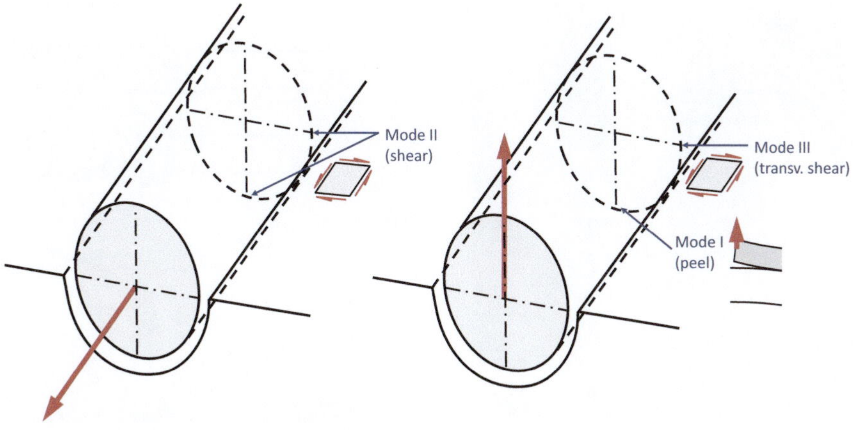

Fig. 8.17 Illustration of the relation between interface loading and opening mode and the major fibre loading/deformation mode [23]

In case the fibre is pulled oud of the matrix perpendicular to its principal direction, the striation-like features change from transverse shear at both sides of the fibre to mode I peeling at the bottom.

Both studies [15, 23] agree with each other in that the striation-like features can be quantitatively related to the applied stress cycles. Later on this was also confirmed by Khan [29] for delamination growth in carbon fibre-reinforced composites.

When fibre stresses are too low, the features become impossible to observe [23], but when fibre stresses are sufficiently high, excellent correlations between stress cycles and striation spacing may be obtained. On the other hand, if fibre stresses are too high, other mechanisms may cause the creation of repeated features at the fibre imprint.

For example, Smulders [30] discusses the so-called ripple marks in the fibre imprint of delamination fractures obtained with ARALL specimen, which he attributes to the high stresses in the fibres.

He explains based on fibre pull-out test results that the region of matrix material adjacent to the fibre, in general referred to as interphase, may explain the severe ripple marks observed when fibre stresses are high, see Fig. 8.18a. The explanation may be considered somewhat specific for the case of ARALL laminates that Smulders studied, because for these laminates the adhesion between fibres and matrix was known to be weak. The adhesion was considered primarily to be provided by van der Waals bonds, rather than chemical bonds. This bond was assisted by the thermal shrinkage after curing which provided clamping forces of the matrix unto the fibres.

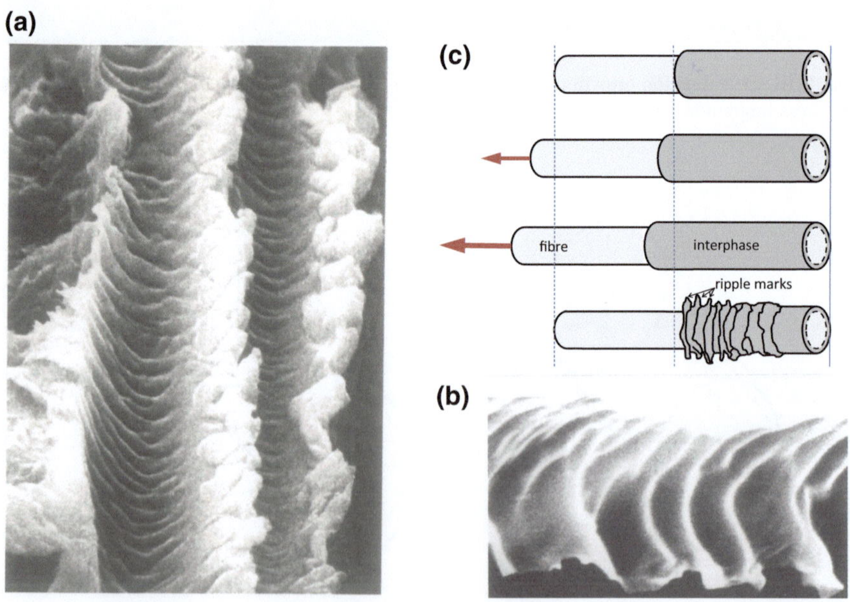

Fig. 8.18 Example of ripple marks in the fibre imprint (**a**, **b**) and illustration of the mechanism of ripple formation indicating high fibre stresses (**c**) [30]

During high stress development in the fibres, the interphase initially deforms along the fibre until the release in clamping pressure at high axial stresses, see Fig. 8.18b, together with the high shear stresses caused the interphase to detach from the fibres allowing sliding of the fibre with respect to the matrix. Upon unloading a slight clamping pressure may then cause the interphase to strip up together with the axial deformation of the fibre. Hence, the ripple features in Fig. 8.18c may not necessarily exhibit a similar appearance as the striation-like features observed at lower fibres stresses and therefore, may not necessarily correlate to the number of load cycles applied.

8.6 Asymptotes in Delamination Characteristics

8.6.1 Static Delamination Growth

It has been explained in Sect. 8.2 that delamination growth can be treated in a similar way as crack propagation in metallic sheet material. The growth rate is related to a stress intensity factor, or in the case of delamination, a SERR using a power law function, such as the Paris relation.

There seems to be an upper bound to such a relation for delamination as indicated in [10], see Fig. 8.19. The position of this upper bound depends on the applied stress ratio.

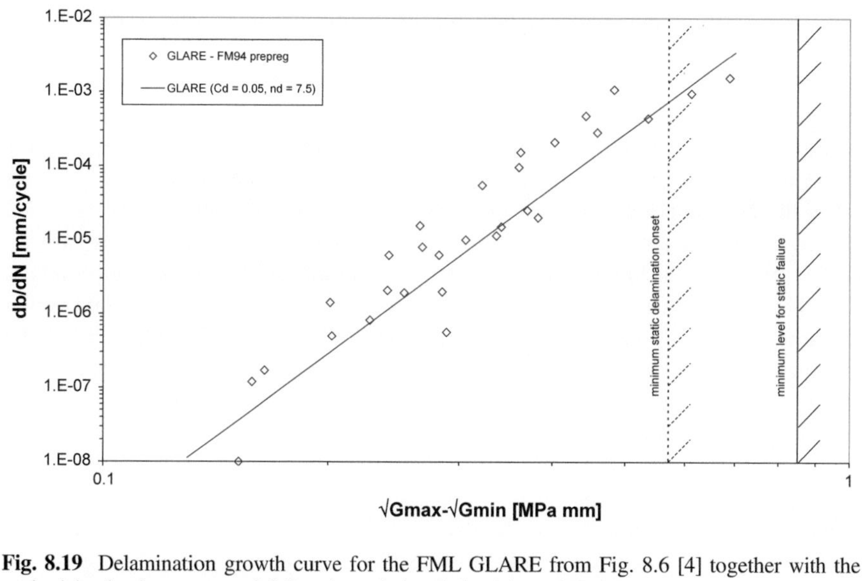

Fig. 8.19 Delamination growth curve for the FML GLARE from Fig. 8.6 [4] together with the static delamination onset and failure boundaries derived from [10, 31]

In fact, similar to the stable tearing region in the diagram of crack growth rate versus stress intensity factor range [32] the upper bound is determined by the critical value of the parameter (G_c for delamination), but the position of the upper bound depends on the applied stress ratio. This is well illustrated and discussed by Rans et al. [33], who demonstrate that with the formulation of the SERR range as in Eq. (8.3), the curves coincide for different stress ratios, but that the position of the upper bound significantly changes for each stress ratio. This change in upper bound is visible in the steep asymptotic increase in delamination growth rate away from the linear trend, which occurs for each stress ratio at another value of the SERR range.

Hence, one may want to normalize each curve by its respective upper G_c, as proposed in the literature [34], but then the position of the upper asymptote still does not coincide with G_{Ic} (in case of mode I data) or G_{IIc}. In the literature several cases are presented where the upper limit coincides with approximately 0.7–0.8 times the G_{Ic} [35, 36].

Khan et al. [37] and Yao et al. [38] discussed the normalization procedure for delamination growth in carbon fibre-reinforced plastics, arguing that this procedure has no physical basis. Indeed, the correlation of upper asymptotes with fatigue delamination growth data, as illustrated in Fig. 8.19, is diluted by the fact that the SERR parameters for the quasi-static and fatigue condition are not consistent with each other.

This has recently been explained in more detail by Amaral et al. [39], who demonstrated how the quasi-static limit correlates to fatigue data based on strain energy dissipation. The SERR is then treated as the average strain energy dissipated per crack increment, which effectively could be treated as the material's resistance to delamination growth. Quasi-static delamination growth represents then the most inefficient form of delamination growth, dissipating the most energy per fracture surface.

The static delamination growth in FMLs is in most cases driven by the load transfer from metal to composite layers. Due to the crack in the metal layers, the deformation of the metal and composite layers is no longer compliant, and load is transferred. The shear stresses at the interface corresponding to the load transfer cause the delamination to extend once the critical value is reached.

However, this absence of compliance in in-plane deformations may also be caused by other mechanisms than cracks, for example severe plastic deformation. This has been illustrated by Rodi [40] who reported quasi-static delamination formation ahead of the crack tip in the metal layers. The development of the monotonic plastic zone in front of the crack tip imposes strain that cannot be followed by the composite layers. This incompliance in deformation causes the delamination to extend ahead of the crack tip, see Fig. 8.20.

Such delamination extension ahead of the crack tip redistributes the high stresses locally, relieving the stress intensity a bit, similar to the mechanisms of crack tip blunting in fatigue crack tips in metals.

For this particular case, Smulders [30] explains that the lateral contraction of the metal layer in the plastic zone induces peel stresses at the location where the

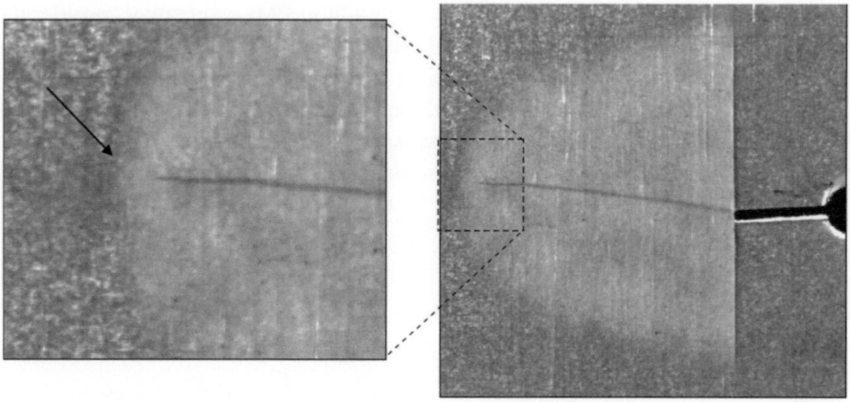

Fig. 8.20 Static delamination ahead of a crack tip in a residual strength test. Image has been obtained with an inverted laminate with only degreasing and sandblasting as surface pre-treatment [40]

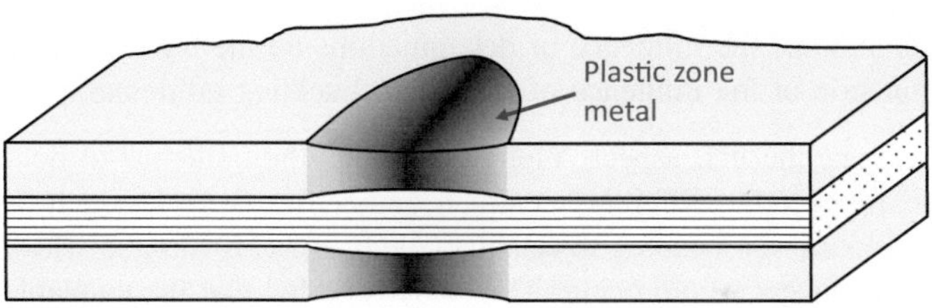

Fig. 8.21 Illustration of the lateral contraction in the plastic zone in front of the crack tip in the metal layer [30]

transition in effective stiffness induces shear stresses, (See Fig. 8.21). This situation implies a combined mode I and mode II load, which may cause formation of delaminations ahead to the physical crack tip in the metal layers. Obviously, the amount of plasticity required to induce this delamination formation ahead of the crack tip corresponds to quasi-static loading of cracked FMLs. In fatigue crack configurations this effect is deemed negligible or not observed.

8.6.2 Delamination Threshold

This automatically brings forward the question whether a lower asymptote could be identified for delamination growth, the so-called delamination threshold. Although some studies seem to suggest delamination thresholds do exist, an equivalent number of studies report delamination growth results without the observation of a lower asymptotic limit.

Either way, for the case of delamination growth in FMLs, one may consider the search for the lower limit in delamination growth unnecessary. Studying the fatigue threshold behaviour of FMLs, Daandels [41] reported a lower limit for crack growth, which after further evaluation [42] appeared to be governed by phenomena at the crack tip in the metal layers, as discussed in more detail in Sect. 9.9. Hence, the easiest (and safest) way is to assume that a lower threshold does not exist by extrapolating the crack resistance curves towards lower delamination growth rates.

8.7 Delamination Buckling

An aspect related to the presence of delaminations at the interfaces between metal and fibre layers is the lack of lateral support in the area of the delaminations. Hence, in case compressive or bending loads are applied, delaminated layers may buckle at substantially lower loads than the entire laminate will.

There are two ways to look at the problem of delaminations in combination with buckling:

- Determination of the influence of delaminations on the buckling load.
- Determination of the influence of (repeated) buckling on delamination growth.

To investigate the first aspect, Vlot [43] studied the influence of delaminations on the buckling loads. Different specimen geometries were tested in axial compression and four point bending to establish the levels of loading at which buckling of delaminated layers would occur. The study revealed that the available buckling theories provide a sound basis for developing prediction models, which could be validated against the experimental results.

Vlot also reported that for all tested conditions, whether quasi-static or fatigue, no growth of delaminations could be observed, induced by the (repetition) of buckling. Hence, the first aspect was satisfactorily addressed, but the influence of buckling on delamination growth could not be tested with the experiments performed.

Bosker [44, 45] studied the growth of delaminations or disbonds without and with the influence of environmental conditions. He tested in both cyclic tension and compression. The case considered was the delamination created during an impact event, i.e. a circular delamination of 60 mm radius. Specimens with both the circular delamination in the specimen centre and at the specimen edges were tested.

Unfortunately, the influence that repeated buckling may have on the growth of the delaminations could not be well addressed. The compressive loading was limited to a level just below Euler buckling of the entire plate, i.e. stress ranged from $S_{min} = -23.4$ MPa up to 0 MPa. In all tests, whether or not specimens were exposed for 1000 h at 70 °C/85% relative humidity, growth of the initial delaminations created with Teflon inserts could not be found.

It seems that in order to obtain growth of delaminations high quasi-static compressive loads must be applied, or a very high number of compression fatigue cycles before detectable delamination growth may be obtained. This was demonstrated by Remmers and De Borst [46, 47] who presented a numerical strategy to assess delamination buckling in FMLs at a mesoscopic level. They calculated that the axial compressive stresses had to reach a value close to 160 MPa, before delamination growth occurs.

This seems in agreement with a recent study on stability aspects of adhesively bonded stiffened aluminium panels [48], which revealed that under quasi-static compression delamination growth may occur before overall panel buckling occurs. Either buckling must be avoided in the design, or the repetition of it must be sufficiently limited in order to avoid disbond growth. The observed influence of the disbond growth on the buckling load indicates that once buckling induces delamination growth, this growth will result in reduction of buckling loads.

For both the case considered by Bosker (impact damage) and manufacturing flaws (disbonds generated during curing), this implies that the effect of a defect is nihil for cases where allowable design stresses are far below the stresses that cause sheet buckling. In fact, these observations and test evidence were reason for the conclusion [9] that: 'A delamination of the size of 150 mm^2 in the far field of GLARE due to a manufacturing error will not show any delamination growth under operational loads, unless the adjacent aluminium layers are also damaged'.

Thus, quality control should mostly focus on damage in the metal layers, because small delaminations visible with ultrasonic C-scanning most likely have no influence at all under operational conditions.

8.8 Effect of Post-stretching

The process of post-stretching is explained in Sect. 3.4.1; the internal residual stress system can be altered or even reversed, depending on the applied strain levels. In general, post-stretching has a positive effect on the fatigue initiation and crack growth behaviour of FMLs because of the favourable internal residual stresses it creates.

Looking solely at delamination growth under cyclic loading, the influence of the internal residual stress may be represented as in Fig. 8.22. The residual stresses after post-stretching tend to close the crack, which means that after application of an external tensile load the stress concentration at the interface will be lower than in the as-cured condition [49].

Smulders [30] also discusses the influence of post-stretching on delamination growth similar to the illustration in Fig. 8.22. He explains that the deformation of the aluminium layers in the as-cured condition adds an additional peel stress to the shear stresses, effectively creating mixed-mode conditions, rather than only shear deformations. Post-stretching limits that deformation, resulting in a more favourable condition with less mode I peeling.

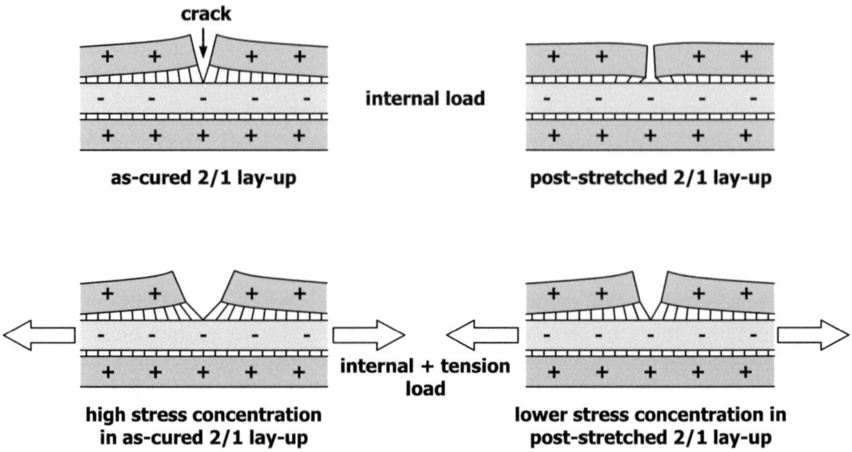

Fig. 8.22 Influence of internal stresses in FMLs on the shear deformation near the metal crack [50]

However, attributing the delamination behaviour of post-stretched laminates to internal residual stresses only may be incomplete and therefore incorrect. When investigating the forming limits of ARALL laminates, Looijenga [51] determined at what post-stretch strain levels permanent damage was induced in the epoxy. In addition, he performed interlaminar shear tests and peel tests to determine the reduction in strength related to this damage in the epoxy. Although the limited data set and scatter in test results do not allow for straightforward conclusions, one may expect that post-stretching has an additional negative effect on delamination growth under fatigue, especially when cracks have been created in the epoxy during the post-stretching process.

The potential negative influence of the post-stretching procedure in addition to the favourable internal stresses created by it has been investigated by Vlot and Van Ingen [49, 50]. They performed mode I (DCB) and mode II (ENF) delamination experiments on ARALL2, ARALL3 and GLARE1 specimens in as-cured condition and in 0.4% post-stretched condition.

They observed no influence of post-stretching on the mode II delamination resistance of ARALL and the mode I and mode II delamination resistance of GLARE, but a reduction in the mode I delamination resistance of ARALL after post-stretching (Fig. 8.23). Their observations are in agreement with the observations reported by Van der Hoeven [52] who investigated the fatigue performance of ARALL3 laminates with bonded doublers.

An aspect not clearly addressed by Vlot and Van Ingen for the case of mode II delamination growth is the contribution of the residual stresses to the delamination resistance. Taking their results and representing them as delamination growth rates versus the SERR according to Eq. (8.5) one may observe that the effect of residual stresses, both as-cured and post-stretched, does not contribute to the delamination

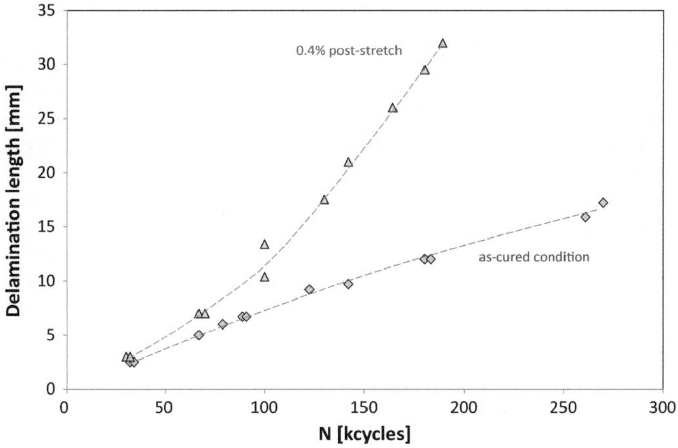

Fig. 8.23 Fatigue delamination growth curves for thin doubler ARALL3 specimens with and without post-stretching, data from [49]

resistance. The effect of residual stresses is cancelled out, as shown by Eq. (8.4). Hence, the negative implication of post-stretching reported by Looijenga [51] may then dominate, reducing the delamination resistance.

References

1. Alderliesten RC (2007) On crack tunneling and plane-strain delamination in laminates. Int J Fract 148:401–414
2. Rans CD, Alderliesten RC (2009) Formulating an effective SERR for linear elastic fracture mechanics description of delamination growth. ICCM, Edinburgh
3. Tada H, Paris PC, Irwin GR (2000) The stress analysis of cracks handbook, 3rd edn. The American Society of Mechanical Engineers, New York
4. Alderliesten RC, Schijve J, van der Zwaag S (2006) Application of the energy release rate approach for delamination growth in GLARE. Eng Fracture Mech 73:697–709
5. Verbruggen MLCE (1986) Aramid reinforced aluminium laminates: ARALL, adhesion problems and environmental effects, Vol. A, Report LR-503. Faculty of Aerospace Engineering, Delft University of Technology
6. Marissen R (1988) Fatigue crack growth in ARALL, a hybrid aluminium-aramid composite material, crack growth mechanisms and quantitative predictions of the crack growth rate, PhD dissertation. Delft University of Technology
7. Roebroeks GHJJ (1991) Towards GLARE—the development of and fatigue insensitive and damage tolerant aircraft material, PhD dissertation. Delft University of Technology
8. Alderliesten RC, Campoli G, Benedictus R (2009) Modelling cyclic shear deformation of fibre/epoxy layers in fibre metal laminates. Compos Sci Technol 67(11–12):2545–2555
9. Alderliesten RC (2005) Fatigue crack propagation and delamination growth in GLARE, PhD dissertation. Delft University of Technology
10. Alderliesten RC (2009) Damage tolerance of bonded aircraft structures. Int J Fatigue 31:1024–1030

11. Wilson G (2013) Fatigue crack growth prediction for generalized fiber metal laminates and hybrid materials, PhD dissertation. Delft University of Technology
12. Roebroeks GHJJ, Hooijmeijer PA, Kroon EJ, Heinimann MB (2007) The development of central. First international conference on damage tolerance of aircraft structures. Delft, The Netherlands
13. Heinimann M, Kulak M, Bucci R, James M, Wilson G, Brockenbrough J, Zonker H, Sklyut H (2007) Validation of advanced metallic hybrid concept with improved damage tolerance capabilities for next generation lower wing and fuselage applications. In: Lazzeri L, Salvetti A (eds) Proceedings of the 24th ICAF symposium. Naples, Italy
14. Vugt AP van (2012) Strain energy release rate concept validation on delamination behaviour of a CFRP wet lay-up scarf repair joint under cyclic loading, -towards a generic approach for assessment of fatigue in composites, MSc thesis. Delft University of Technology
15. Khan SU, Alderliesten RC, Benedictus R (2009) Delamination growth in Fibre Metal Laminates under variable amplitude loading. Compos Sci Technol 69:2604–2615
16. Hoeven W van der, Frijns RHW (1990) Static delamination resistance of ARALL—Interim report. Dutch National Laboratory for Aerospace Research NLR, report NLR-CR-90086C
17. Mangkoesoebroto RH (1987) The effect of fibre volume fraction on the mechanical properties and the fatigue behaviour of ARALL laminates, MSc thesis. Delft University of Technology
18. ASTM D7905/D7905M-14 (2014) Standard test method for determination of the mode II interlaminar fracture toughness of unidirectional fiber-reinforced polymer matrix composites. ASTM International, West Conshohocken, PA
19. Delgrange G, Alderliesten RC, Benedictus R (2009) Delamination growth at interfaces in hybrid materials and structures under various opening modes. In: Bos M (ed) Proceedings of the 25th ICAF Symposium. Rotterdam, The Netherlands
20. Pascoe JA (2012) Delamination of bonded repairs: A damage tolerance approach, Master Thesis. Delft University of Technology, Delft, The Netherlands
21. Benzeggagh ML, Kenane M (1996) Measurement of mixed-mode delamination fracture toughness of unidirectional glass/epoxy composites with mixed-mode bending apparatus. Compos Sci Technol 56:439–449
22. Greenhalgh ES (2009) Failure Analysis and Fractography of Polymer Composites. Woodhead Publishing Limited, Cambridge, UK
23. Roebroeks GHJJ (1986) Observation on cyclic delamination in ARALL under fatigue loading, Report LR-496. Delft University of Technology, Delft
24. Suiker ASJ, Fleck NA (2004) Crack tunnelling and plane-strain delamination in layered solids. Int J Fract 125:1–32
25. Suiker ASJ, Fleck NA (2006) Modelling of fatigue crack tunnelling and delamination in layered composites. Compos Part A 37:1722–1733
26. Delgrange G (2010) Delamination behaviour of bonded structures and hybrid materials, characterization and implementation in a prediction model, Report B2v-10-01. Delft University of Technology, Delft, The Netherlands
27. Khan SU, Alderliesten RC, Benedictus R (2011) Delamination in fiber metal laminates (GLARE) during fatigue crack growth under variable amplitude loading. Int J Fatigue 33:1292–1303
28. Hooijmeijer PA (2002) Doubler run-out fatigue properties, Report B2V-02-16. Delft University of Technology, Delft
29. Khan R, Alderliesten RC, Benedictus R (2014) Two-parameter model for delamination growth under mode I fatigue loading (Part B: Model development). Compos A A65:201–210
30. Smulders EHM (1988) Fibre fracture mechanism in ARALL laminates with Aramid fibres, Master Thesis. Delft University of Technology
31. de Vries TJ, Vlot A, Hashagen F (1999) Delamination behavior of spliced fiber metal laminates, Part 1—experimental RESULTS. Comp Struct 46(2):131–145
32. Schijve J (2001) Fatigue of structures and materials. Kluwer Academic Publishers

33. Rans DC, Atkinson J, Li Ch (2015) On the onset of the asymptotic stable fracture region in the mode II fatigue delamination growth behaviour of composites. J Compos Mater 49 (6):685–697
34. Murri GB (2012) Evaluation of delamination growth characterization methods under mode I fatigue loading, ASC 27th technical conference. Arlington, TX
35. Charalambous G, Allegri G, Hallett SR (2015) Temperature effects on mixed mode I/II delamination under quasi-static and fatigue loading of a carbon/epoxy composite. Compos A 77:75–86
36. Brunner AJ, Stelzer S, Pinter G, Terrasi GP (2016) Cyclic fatigue delamination of carbon fiber-reinforced polymer-matrix composites: data analysis and design considerations. Int J Fatigue 83:293–299
37. Khan R, Alderliesten RC, Yao L, Benedictus R (2014) Crack closure and fibre bridging during delamination growth in carbon fibre/epoxy laminates under mode I fatigue loading. Compos A 67:201–211
38. Yao L, Alderliesten RC, Zhao M, Benedictus R (2014) Bridging effect on mode I fatigue delamination behavior in composite laminates. Compos A 63:103–109
39. Amaral L, Yao L, Alderliesten RC, Benedictus R (2015) The relation between the strain energy release in fatigue and quasi-static crack growth. Eng Fract Mech 145:86–97
40. Rodi R (2010) The residual strength failure sequence in fibre metal Laminates, PhD dissertation. Delft University of Technology, Delft
41. Daandels D (2003) Search for DK threshold of GLARE, Preliminary thesis. Delft University of Technology
42. Alderliesten RC, Rans CD (2008) The meaning of threshold fatigue in fibre metal laminates. Int J Fatigue 31(2):213–222
43. Vlot A (1987) Buckling of delaminations in Z laminates, Report LR-535. Delft University of Technology
44. Bosker OJ (1998) Growth of debonds in GLARE, Part I (no moisture absorption), Report B2V-98-24. Delft University of Technology, Delft
45. Bosker OJ (2000) Growth of debonds in GLARE, Part II (with moisture absorption), Report B2V-00-08. Delft University of Technology, Delft
46. Remmers JJC, de Borst R (2001) Delamination buckling of fibre–metal laminates. Compos Sci Technol 1(15):2207–2213
47. Remmers JJC, de Borst R (2001) Numerical modelling: delamination buckling. In: Vlot A, Gunnink JW (eds) Fibre metal laminates—an introduction. Kluwer Academic Publishers, Dordrecht, The Netherlands
48. Zarouchas D, Alderliesten RC (2015) The effect of disbonds on the stability aspects of adhesively bonded aluminium panels during compression loading. Thin-Walled Struct 96:372–382
49. Vlot A, Van Ingen JW (1998) Delamination resistance of post-stretched fibre metal laminates. J Compos Mater 32:1784–1805
50. Vlot A, Ingen JW (1994) The influence of post-stretching on the delamination resistance of fibre metal laminates. Delft University of Technology, Delft
51. Looijenga KA (1989) Forming limits of ARALL laminates and laminate behaviour at large deformations, Memorandum M-619. Delft University of Technology, Delft (in dutch)
52. Hoeven W van der (1992) Fatigue and residual strength behaviour of ARALL 3 panels with bonded-on doublers. Dutch National Laboratory for Aerospace Research NLR, report 92467 L

Chapter 9
Fatigue Crack Propagation

Abstract This chapter explains the fatigue damage growth phenomena in FMLs, and provides theories to describe the growth of damage based on the individual fracture phenomena. The validity of various geometrical correction factors is discussed. The different crack geometries through-thickness and the crack paths, observed in FMLs, are explained for both in-axis and off-axis loading. Finally, the relation between variable amplitude and constant amplitude loading is covered.

9.1 Introduction

Fatigue crack propagation is an important topic to be considered for FMLs. In an FML, the cracks initiate and propagate in the metal layers, while the fibre layers remain intact and transfer part of the load over these cracks (Fig. 9.1). This leads to low crack propagation rates that often can be approximated by linear growth. This behaviour, which has been observed for various types of FMLs, makes these laminates excellent candidates for areas prone to fatigue, because they give a significant increase of fatigue life and damage tolerance over their monolithic metal counterparts.

Many researchers have attempted to capture the mechanisms occurring during fatigue crack growth in FMLs, and as a result, several prediction models have been developed [1]. Although the phenomena have been approached differently, the mechanisms can be divided primarily into propagation of cracks in the metallic constituents and delamination at the interfaces with fibre layers in the wake of these cracks. These mechanisms have been investigated extensively as separate fracture mechanisms, of which the observations and analyses for the delamination phenomena are summarized in Chap. 8.

Based upon the evaluation of previous studies on fatigue crack growth in FMLs [1] and subsequent detailed investigation of the fatigue fracture mechanisms [2–4], it may be argued that empirical models are often very inaccurate or very limited in their application. In fact, only models that describe the individual fracture phenomena in a balance similar to what is observed in experiments, may lead to generic models that predict the crack growth behaviour accurately.

R. Alderliesten, *Fatigue and Fracture of Fibre Metal Laminates*, Solid Mechanics and Its Applications 236, DOI 10.1007/978-3-319-56227-8_9

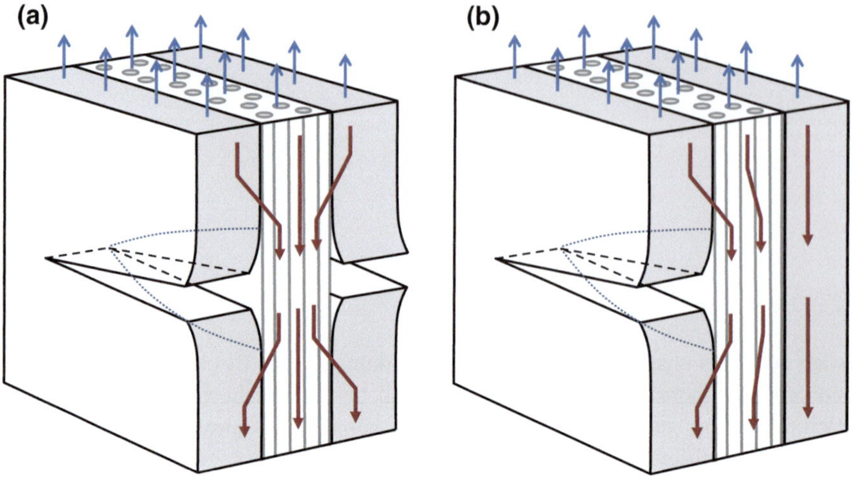

Fig. 9.1 Illustration of fibre bridging: Intact fibres transfer load over the crack; **a** fatigue through crack and **b** surface crack

The theory presented in this section consists of an analytical presentation of the balance between the crack growth mechanisms in the metallic constituents and delamination growth at the metal/fibre interfaces.

9.2 Crack Geometries

Initially, the development of FMLs and the subsequent investigation of fatigue crack growth, focussed on cracks emanating from a starter notch in the metal layers with intact fibres in the crack wakes. These notches were created by creating a saw cut that not only pre-cracks the metal layers, but also the fibre layers.

This may be representative for notches in structures such as fastener holes and cut-outs, but it may not represent fatigue damages initiated as corner cracks or as surface scratches, such as scribe marks. In the latter situation, starter notches may be present as damages in a single metal layer only, without damage to the adjacent fibre layers. Representing such fatigue damages by fatigue cracks with fibres being cut in the starter notch may severely overestimate the crack growth rate and therefore underestimate the crack growth life.

As a consequence, proper analysis of fatigue damage in FMLs needs to identify the relevant fatigue crack scenario before accurate predictions can be performed. The most relevant fatigue damages in FMLs are illustrated in Fig. 9.2 and may be referred to as ‘surface cracks’, ‘part-through cracks’ and ‘through cracks’. Note that ‘through cracks’ in case of fatigue analysis indicates cracks of equal length in all metal layers, while the fibre layers are still intact, whereas in the residual strength

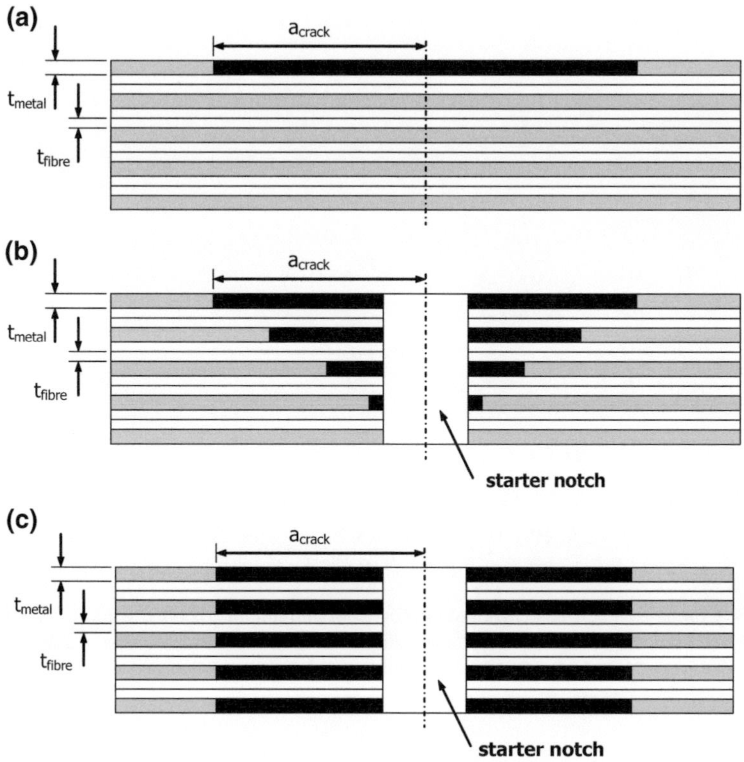

Fig. 9.2 Most significant fatigue crack scenarios in FMLs [3]

analysis, it is often used to describe through-cut cracks, such as saw cuts, see Chap. 10.

9.3 Fatigue Crack Growth Characteristics

The excellent fatigue crack growth characteristics of FMLs have been the main selling point of the concept in the context of damage tolerance certification. The significantly lower crack growth rates compared to monolithic metals, and the approximately linear crack growth curve, see Figs. 9.3 and 9.4, have been the dominant reason to further develop this laminated concept. This crack growth behaviour does not only allow for higher stresses within the structure, but the long crack growth life also allows for large inspection intervals.

There are various ways to look at the growth of fatigue damage in FMLs. The propagation of cracks, illustrated in Fig. 9.4, may be considered the dominant damage mechanism, because it starts with initiation of cracks in the metallic layers of the laminate, as explained in Chap. 7.

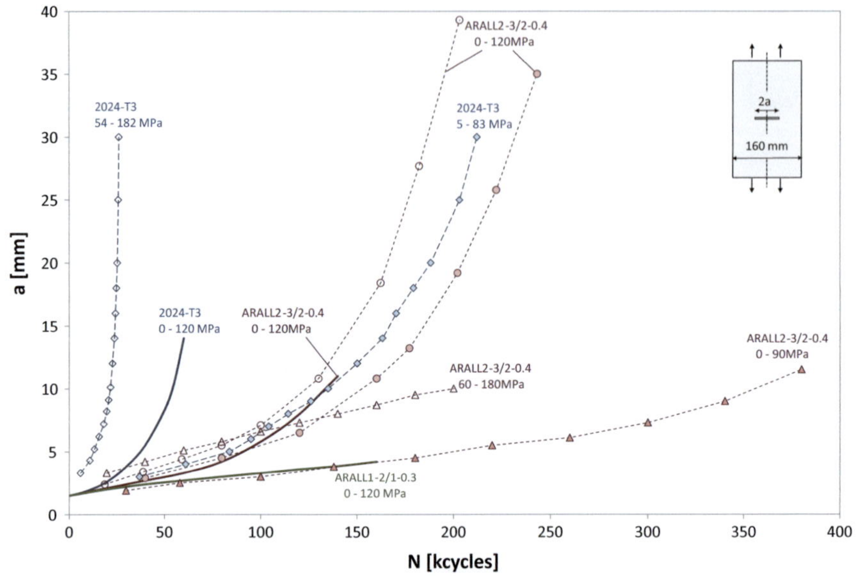

Fig. 9.3 Comparison between crack growth curves of aluminium 2024-T3, ARALL1-2/1-0.3 and ARALL2-3/2-0.4 for constant amplitude fatigue loading at various mean stresses [5, 6]; nonlinearity in the ARALL2-3/2-0.4 curves indicates fibre failure

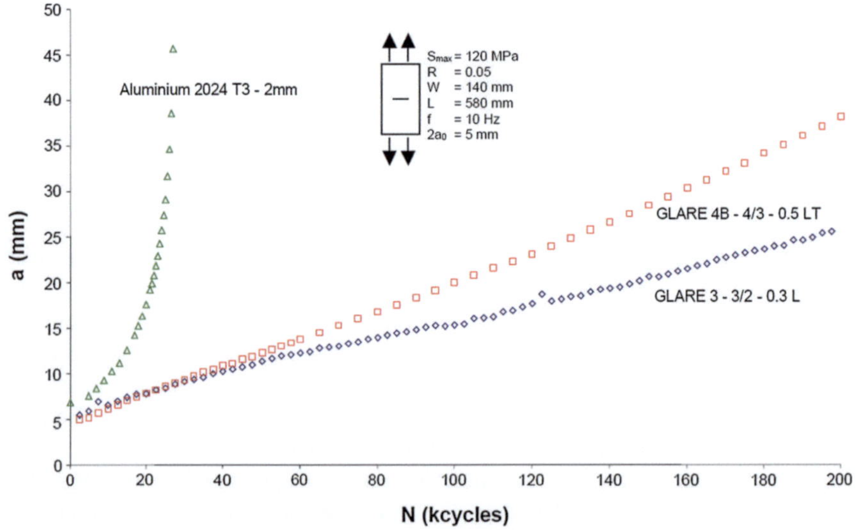

Fig. 9.4 Comparison between crack growth curves of aluminium 2024-T3, GLARE3-3/2-0.3L and GLARE4B-4/3-0.5LT for constant amplitude fatigue loading [7]

However, once a crack progresses, delamination occurs in the crack wake at the interfaces between metallic and composite layers. Here, it must be acknowledged that this delamination to some extent is favourable. Marissen [8] already explained that without delamination, the opening of fatigue cracks in the metal layers would impose severe stresses in the fibre layers bridging these cracks, resulting in failure. Hence, induced by the high strain concentration, delaminations initiate, which subsequently relieve the bridging stresses. This allows the fibres to elongate over a longer length while remaining compliant to deformations in the cracked metal layers.

Within certain limits, delaminations are therefore favourable; too little delamination may cause fibre failure, reducing bridging; too much delamination reduces the fibre stresses, making the bridging mechanisms less effective.

The balance between crack and delamination growth results in an approximately constant crack growth behaviour, as illustrated in Fig. 9.4. As a consequence, people have attempted to develop methods to predict the crack growth rate d*a*/d*N* for given load and geometry conditions, to aid design activities and determination of allowable stresses [9].

For example, Wilson [4] proposed to predict the steady-state crack growth (Fig. 9.5) for FMLs based on the laminate configuration and loading conditions. The method accounted for changes in the residual stress ahead of the crack tip determined with numerical simulations in the pre-cracked conditions. The general observation was that the method did not predict the crack growth entirely correctly, but that the general trend was captured and the error in crack growth prediction was sufficiently small. Hence, this method could serve as a tool for preliminary sizing of FML structures.

Recently, Bied-Charreton [10] investigated the influence of production flaws in friction stir welded aluminium sheets on the structural integrity of the FML GLARE. He concluded that the growth characteristic could be assessed different from Wilson's concept, and that the growth is characterized by exponential growth, as illustrated in Fig. 9.6.

The implication of this observation is that likely a simple method can be developed to predict the crack growth for arbitrary FMLs using:

$$\frac{da}{dN} = C \cdot a \tag{9.1}$$

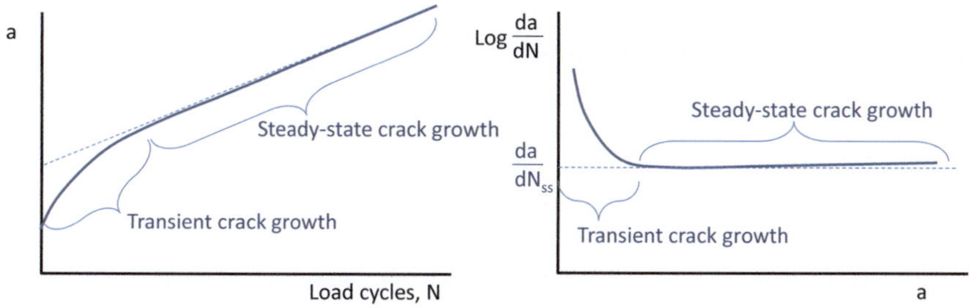

Fig. 9.5 Concept of steady-state crack growth in FMLs [4, 9]

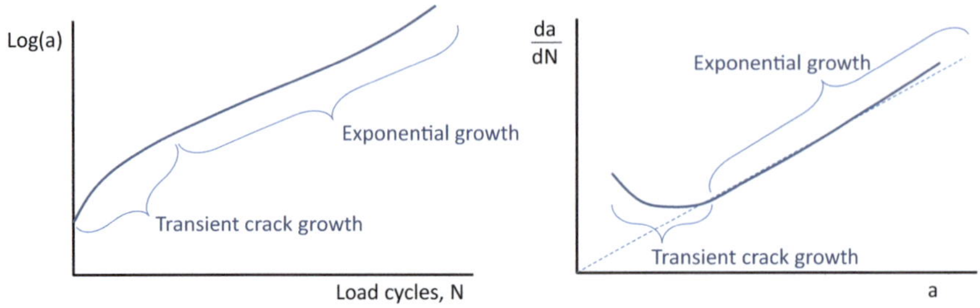

Fig. 9.6 Concept of exponential crack growth in FMLs [10]

where C is the only parameter to be related to laminate lay-up and loading conditions, because the trend seems to go through the origin consistently, as illustrated in Fig. 9.6. The transient regime labelled by Wilson (Fig. 9.5) is also visible in Fig. 9.6 and should be attributed to the notch condition, where fibres are cut. Hence, initially the growth has to find an equilibrium before exponential growth is observed. This could imply that the parameter C implicitly depends on the notch size, as previously a significant dependency on the initial notch condition was observed [11]. This is illustrated in Fig. 9.7.

Aside from a simple tool for preliminary sizing of structures, the exponential growth may also be utilized in a fashion similar to the approach presented by Molent et al. [12] for naturally occurring cracks under operational flight spectrum loading. Rather than the exponents close to 3 in the Paris relationships, Molent et al. observe exponential growth when plotting the crack growth against the number of flights. Plotting the growth curve in a lognormal graphical presentation then allows for easy extrapolation of the initially observed growth up to critical lengths.

With the concept of Bied-Charreton, something similar may be adopted for FMLs, in case of constant amplitude loading. Whether the method can be extrapolated to arbitrary load spectra must still be investigated.

9.4 Superposition of Far-Field Stresses and Fibre Bridging

As mentioned earlier, the fatigue crack initiates and propagates in the metallic layers of the FMLs. To appropriately apply fracture mechanics, the stress intensity factor at the crack tip must be determined. This means that the stress intensity factor should not be treated as a crack tip characteristic assuming a homogeneous material [1], but rather to describe each crack tip in a particular layer individually.

Here, the superposition principle for the stress intensity factor (the factor is linearly dependent on stress, and stresses can be superimposed) enables one to break up the stress intensity factor at the crack tip in multiple components. These components relate to the applied loading, denoted as far field with respect to the

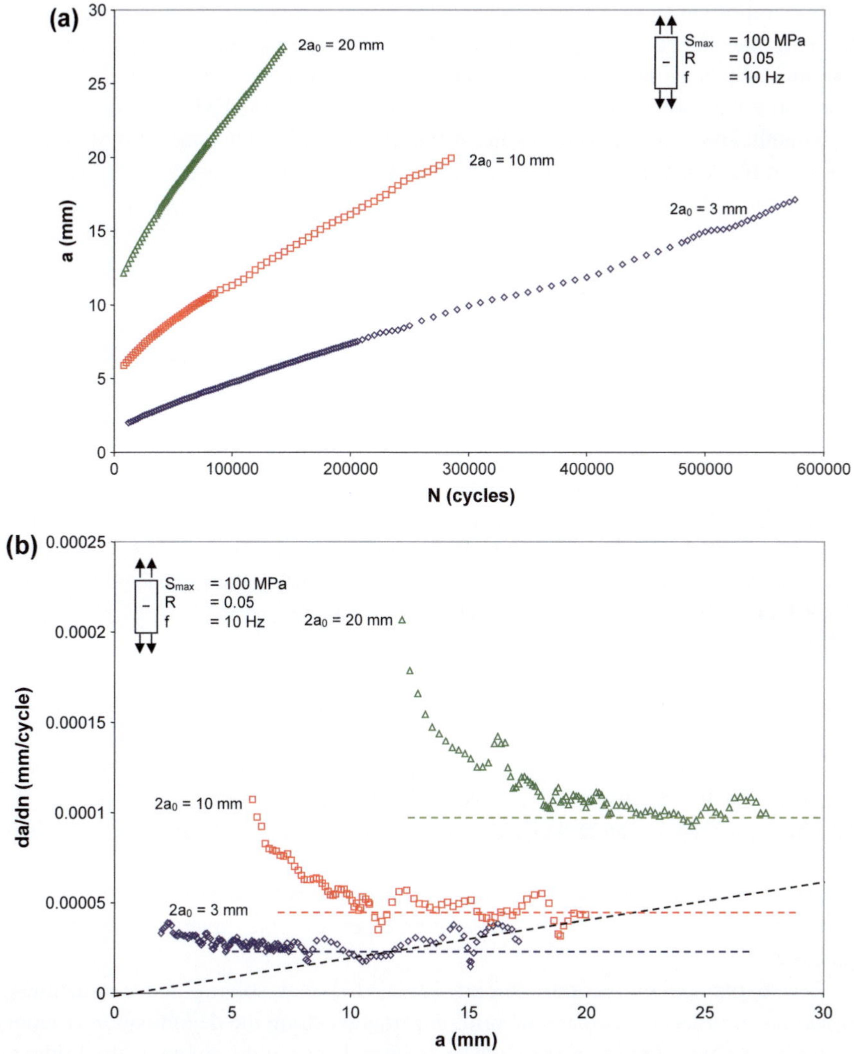

Fig. 9.7 Crack growth curves (**a**) and crack growth rate da/dN versus the crack length a (**b**) for GLARE3-6/5-0.4 for three initial crack lengths $2a_0$ [11]; illustration of difference between exponential growth and steady-state growth

location of the crack, and the reduction of loading, or restraint on crack opening (crack closing contribution) of the bridging of the fibres. In equation form, this may be written as [2, 3]

$$K_{\text{tip}} = K_{\infty} - K_{\text{br}} \tag{9.2}$$

It is important to realize that this superposition principle is generally applicable to stress intensity factors and in case of FMLs provides a sound basis to describe the multiple mechanisms that occur simultaneously. This will become clear when discussing the addition of external stiffening elements to the FML skin in Sect. 9.9.

Nonetheless, one has to be aware of the limitations to this concept of similarity. For example, the superposition principle of Eq. (9.2) allows the calculation of the SIF for the aluminium layer separately for the far-field stresses in that layer, i.e.

$$K_{\text{tip}} = \beta S \sqrt{\pi a} \tag{9.3}$$

These stresses can be calculated using the classical laminated plate theory, as explained in Chap. 4. However, it is discussed in Sect. 9.5 that one has to be aware that the presence of fibres and fibre bridging imposes constraints to parameters like β.

To determine the bridging load itself, which relates to K_{br}, a compatibility equation is needed. This can be provided by the displacements of the metal and fibre layers at the delamination boundary, which must be compatible. Assuming that the deformations in the metal layers within the delamination area are negligible, one may assume that displacements at the crack flanks at a certain location along the crack are equal to displacements at the delamination boundary. Relating the crack flank opening, Fig. 9.8, to the elongation and deformation of the fibre layers within the delaminated area then provides

$$v_{\infty}(x) - v_{\text{br}}(x) = \delta_f(x) + \delta_{\text{pp}}(x) \tag{9.4}$$

The shear deformation component δ_{pp} in this equation depends on the interface geometry itself. As discussed in Sect. 8.3, pre-impregnated layers, such as the S2/FM94 prepreg in standard GLARE, contain fibres equally distributed throughout the thickness, while woven fabric impregnated by two adhesive films, like in ARALL, results in fibres concentrated at the centre with resin-rich layers at both interfaces. The deformation of these two types of fibre layers yields different relations.

It is explained by various authors [2, 4, 13] that solving the compatibility equations in order to calculate the bridging stresses along the delamination contour, requires a rather complex integration procedure, because the shape of the bridging stress distribution along the delamination contour is unknown a priori. As a consequence, the integral of the crack opening displacement as a function of bridging stress

$$v_{\text{br}}(x) = \int_{a_s}^{a} v(x, x_{\text{br}})\, dx_{\text{br}} \tag{9.5}$$

cannot be integrated indefinitely. For this purpose, the problem of Fig. 9.8 has been discretized and solved numerically. Initially, the crack length has been divided into a number of bar elements with identical widths [2, 3], as illustrated in Fig. 9.9.

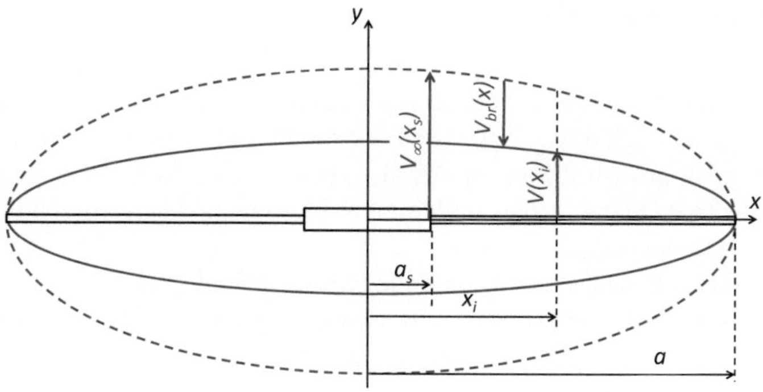

Fig. 9.8 Illustration of the crack opening, obtained by superposition of crack opening under far-field loading $v_\infty(x)$ and a closing contribution due to bridging $v_{br}(x)$ [3]

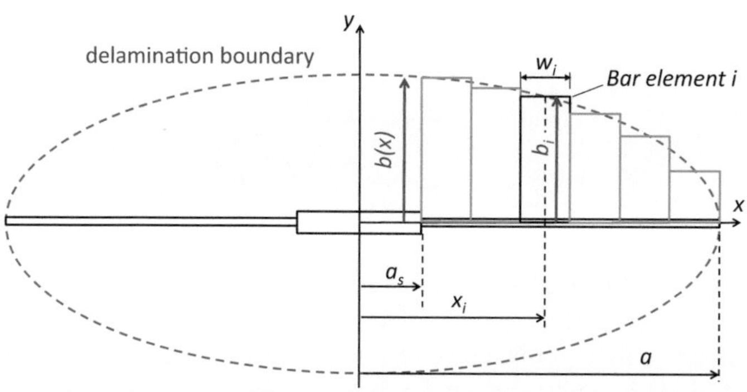

Fig. 9.9 Illustration of the bar element scheme adopted to solve the compatibility equations [4]

The bridging stresses are stable over most of the crack wake length, except for the direct vicinity of the crack tip. At that location significant variations in bridging stresses in relation to the delamination shape have been observed. This caused stability problems in the implementation of the model reported in [2], which at that time were solved with a smoothing operation of the calculated delamination growth near the crack tip [3].

Wilson proposed to bias the discretization of bar elements towards the crack tip, which is a different but more robust solution. He concluded that this bias was only necessary for the crack tip and not for the notch tip. To bias the discretization of the bar elements, Wilson employed the Chebyshev nodes [4], which for a single crack length with n elements yields

$$x_i = a_s - (a - a_s)\sin\left(\frac{\pi}{4}\frac{2i-1}{n}\right), \quad i = 1, 2, \ldots, n \tag{9.6}$$

9.5 Delamination Shapes

The shape of delaminations in fatigue crack configurations has been the subject of many studies, although often implicitly. For example, when Marissen [8] developed his fatigue crack growth theory for ARALL, he considered the delamination to be elliptical of shape. Many studies building on the work of Marissen continued with that shape without proper verification. In fact, despite experimental evidence illustrating different shapes, still often the ellipse was adopted as representing the shape. This can be illustrated with one example, given in Fig. 9.10 reported by Mattheij [5]. Despite the fact that the shape in this figure evidently is not well represented by an ellipse, Mattheij presented the delamination as elliptical of shape.

In subsequent development of bridging theories, Guo and Wu [14] were the first to explicitly consider different shapes as well, analysing both the ellipse and triangle as governing shapes. Reviewing the available evidence revealed that most shapes reported actually did not resemble an ellipse, but rather a shape in between a cosine and a parabolic function [3]

$$\begin{aligned} b(x)_{\cos} &= b(a_s)\cos\left(\frac{\pi}{2}\frac{x-a_s}{a-a_s}\right) \\ b(x)_{\text{par}} &= b(a_s)\sqrt{1-\frac{x-a_s}{a-a_s}} \end{aligned} \tag{9.7}$$

This is illustrated in Fig. 9.11. In this figure, it seems that the cosine shape fits best to the observed and measured delamination shapes. It at least resembles the sharp delamination tip described by either cosine or triangular shapes [3]. However, the analysis of the bridging stresses for these delamination shapes reported in [14] and [3] reveals that the sharp delamination tip corresponds to very high bridging stress peaks at the crack tip in the metal layer. This makes the prediction models based on the cosine and triangular shape numerically very sensitive and easily unstable. It was for that reason that the parabolic shape was used in the method proposed in [2, 3].

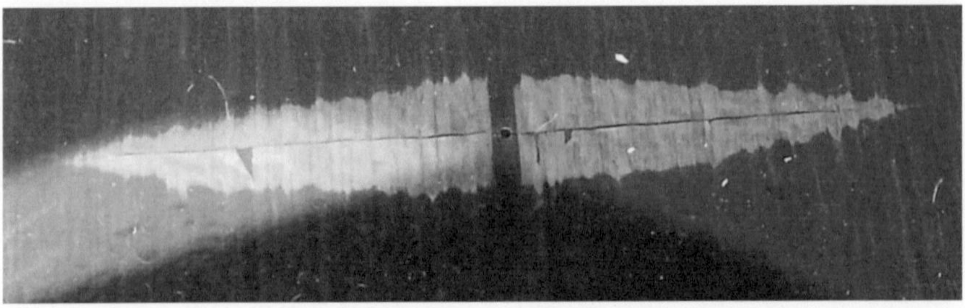

Fig. 9.10 Image of delamination shape labelled by Mattheij as 'elliptical-shaped delamination zone' [5]

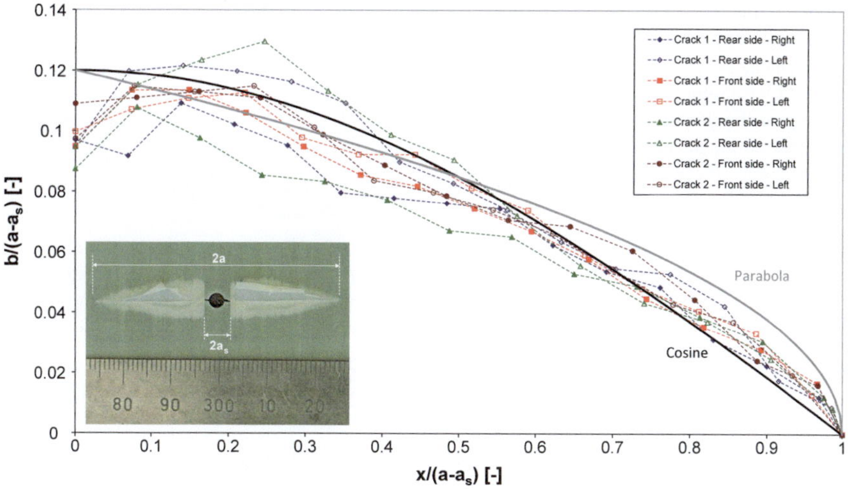

Fig. 9.11 Measured delamination shapes in GLARE3-6/5-0.4 tested in L-T direction with $W = 100$ mm, $a = 17.5$ mm, $a_s = 1.5$ mm and $S_{max} = 100$ MPa with $R = 0.05$ [3]

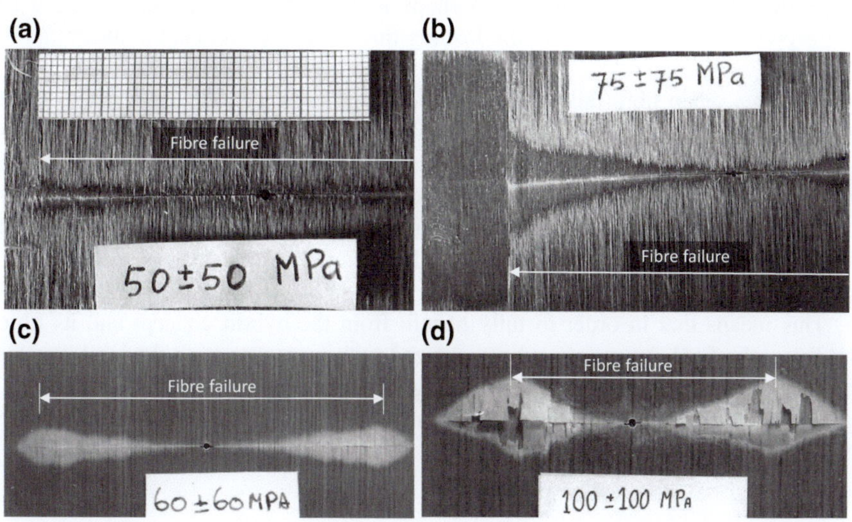

Fig. 9.12 Image of delamination shapes in ARALL containing evidence of fibre failure [5, 16]

Although these delamination shapes, illustrated in Figs. 9.10 and 9.11, are observed in most cases, one has to verify these shapes for each individual case. The shape corresponds to a well-balanced growth of cracks and delaminations. This balance may not always be present.

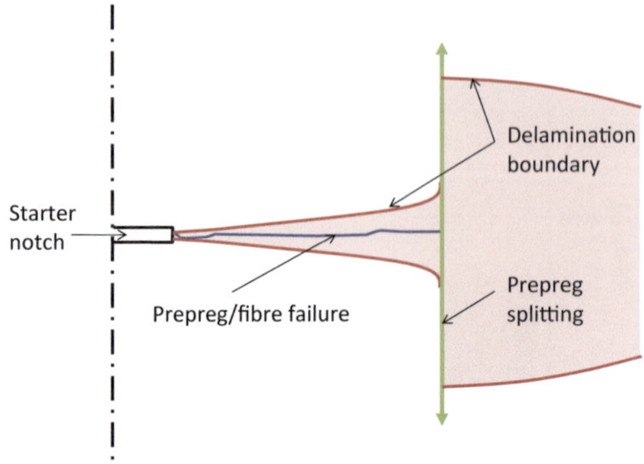

Fig. 9.13 Illustration of the delamination shape development in case of prepreg/fibre failure; after splitting of the prepreg layers the fibres no longer fail and delaminations develop [17]

This can be illustrated with for example delamination shapes observed for ARALL for cases where fibre failure occurred. At the location where fibres fail, load transfer over the interface no longer occurs and hence, delaminations do not grow further. For that reason shapes were often observed in ARALL with very narrow delamination shapes or shapes that seem inverted compared to the shape illustrated in Fig. 9.10. Examples of both observations are given in Fig. 9.12. The large step transition in delamination shape in the right-hand image of Fig. 9.12 corresponds to fibre splitting in the loading direction. This concept is illustrated in Fig. 9.13.

But even when fibre failure does not occur, different delamination shapes have been reported. For example, the delamination shapes between the thick metal layers and the inner FML core of the initial Central concept [15] were much closer to the elliptical shape. These delaminations, however, were excessive in size, significantly reducing the efficiency of fibre bridging. As a result, nonlinear growth was observed similar to that observed for monolithic aluminium, and ARALL in case of fibre failure, see Fig. 9.3.

This means that in order to fully benefit from the hybrid concept and its fibre bridging, the FML lay-up must be optimized with respect to both the load transfer over the interface and the load in the bridging fibres. Once that has been achieved, generally delamination shapes of the form illustrated in Figs. 9.10 and 9.11 are obtained.

9.6 Metal Layer Crack Growth Resistance

The concept of the superposition principle explained in Sect. 9.4 is that the crack tip in the metal layers of an FML experiences effectively a stress intensity that can be described with the stress intensity factor (SIF) K. The similarity principle then is similar to the principle adopted for monolithic metals; similar SIFs result in similar crack growth.

Hence, to describe crack growth in the metallic layer, the SIF from Eq. (9.2) must be correlated with crack growth resistance curves, like the Paris relation. This relationship can be established by performing fatigue crack growth tests on monolithic metals for which the SIF can be determined with the standard methods, like Eq. (9.3).

Here, one has to consider aspects that may influence the similarity principle. It was explained in Chap. 7 that S–N curves obtained from monolithic metals may be used in the methodology for predicting crack initiation in FMLs. The error one makes by using these fatigue lives until failure for initiation until 1 mm has been limited with the definition of that initiation length of 1 mm.

When determining the crack growth resistance or Paris curves for the metallic layers in FMLs, one could similarly rely on the curves obtained for monolithic metals. However, the major difference between the thin metal layers in FMLs and the monolithic specimens tested is the thickness and the applied cure cycle(s). For mode I fatigue cracks in monolithic metal sheet, it is known that shear lips will develop that eventually may slant the entire fatigue crack into a combined mode I and II opening, see Fig. 9.14. This crack slanting has also been observed for thin sheets in FMLs [18, 19], but at different crack lengths. Fatigue cracks tend to slant earlier in the thin sheets than observed for thicker monolithic sheets. The transition of the opening mode on the crack surface has influence on the crack growth resistance, for which reason some authors marked the moment of transition in the crack growth resistance curves.

To address these aspects, Lipzig [20] investigated the fatigue crack growth behaviour of 5-mm-thick and 1-mm-thin Aluminium 2024-T3 (alclad). In addition, he tested bonded laminates of 5 $\times$ 1 mm thin aluminium (Fig. 9.15). The general observation was that the fatigue life of the 5-mm-thick panels was substantially shorter than that of the 1-mm-thin ones, whereas the 1-mm-thin sheets in the laminated composition behaved similarly to a single 1-mm plate.

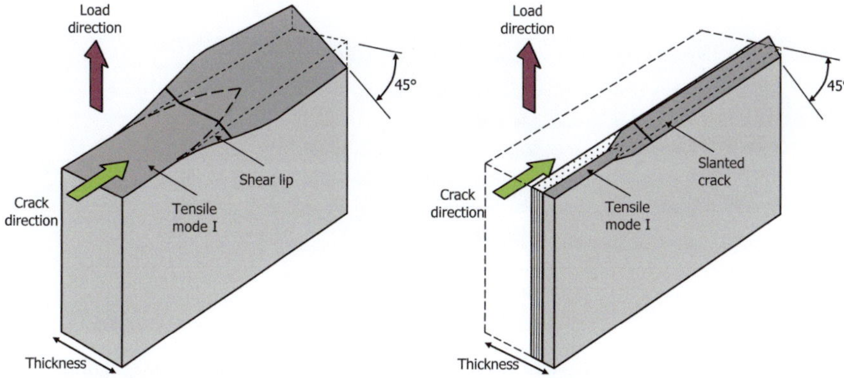

Fig. 9.14 Difference in the development of fatigue fracture surface between monolithic plate and thin sheets of the same material in the FML

Later, Homan [21, 22] investigated the fatigue crack growth behaviour of thinner (0.3–0.4 mm) aluminium sheets of the same alloy and thickness, including the same surface pre-treatment (chromic acid anodizing and BR127 priming) as the aluminium layers in the FML GLARE. To investigate whether the crack growth resistance is affected by the difference in sheet thickness and pre-treatment, Homan compared his results and those of Slagt [23] with data from the Fokker technical handbook [24] and data from the Handbuch Struktur Berechnung [25].

Good correlation was obtained between the experimental results and the handbook solutions, indicating that there is not a significant difference between the crack growth behaviour of the thin sheets in FMLs and the monolithic sheets.

However, Homan did observe a slight difference between the data from Slagt [23] and his data, as illustrated in Fig. 9.16. This could only be explained by the fact that Slagt tested 0.3-mm aluminium adhesively bonded to a laminate with 4 layers and subsequently cured. It is believed that these thin layers in a laminate (or FML) are better supported than the thin sheets tested by Homan, despite the anti-buckling guides used.

A difference between the reports by Homan [21, 22] and later reports is the stress ratio correction adopted. Homan used a correction originally proposed by De Koning [26] defined as

$$\Delta K_{\mathrm{eff}} = \left(1 - R^{1.35}\right) K_{\max} \tag{9.8}$$

while later [4] the R-correction proposed by Schijve [27, 28] was adopted

$$\Delta K_{\mathrm{eff}} = \left(0.55 + 0.33R + 0.12R^2\right) \Delta K \tag{9.9}$$

Recent studies correlating stress-based fracture mechanics approaches with energy principles indicate that most of this R-correction is imposed by the fact that ΔK does not capture well the cyclic work applied to the specimen [29–31].

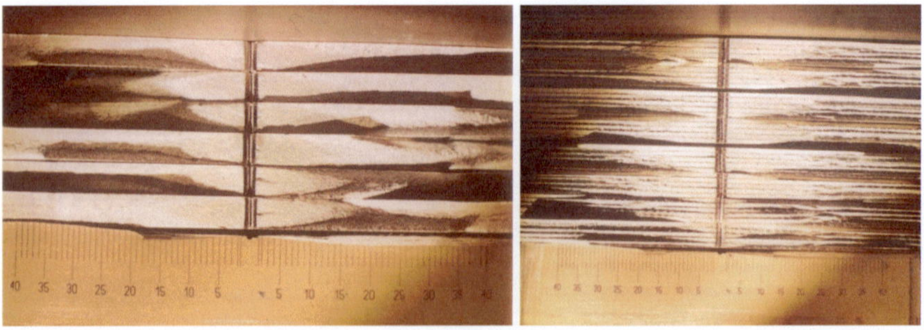

Fig. 9.15 Difference in macroscopic appearance of fatigue fracture surfaces in 5-mm 2024-T3 alclad specimens (*left*) and 5 × 1 mm 2024-T3 alclad laminates (*right*) [20]

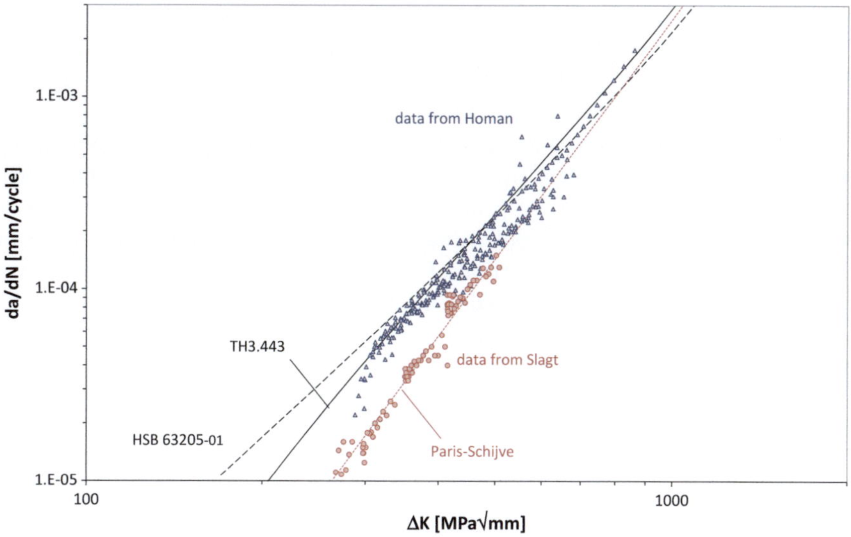

Fig. 9.16 Correlation between crack growth data from Homan [21] and Slagt [23] and the curves from technical handbooks; Paris-Schijve indicates the Paris equation with the R-correction of Eq. (9.9)

9.7 Finite Width Correction Factors

For the case of central fatigue cracks in monolithic metal panels, finite width correction factors have been proposed in the past, with the predominant focus on increasing accuracy, in particular for high values of $2a/W$.

Irwin [32] developed a correction that appeared to be valid at most to $2a/W < 0.5$. Isida [33, 34] developed a correction using series expansion, which appeared to be equally accurate as the equation proposed by Feddersen [35]. The excellent correlation between this secant formulation by Fedderson with Isida's correction, has for a while led to the idea that this Feddersen's equation may be in fact exact. However, Koiter [36] demonstrated the exact limit of $F(a)$ for $2a/W \rightarrow 0$ and proposed a formulation which was slightly more accurate than the formulations of Isida and Feddersen, in particular for $2a/W > 0.8$. In addition, Dixon [37] developed a correction that can be derived straightforwardly from the equilibrium between Westergaard's stress field in the net section and far-field stresses.

Initially, these finite width corrections were considered to be generic geometry corrections that must be applied to consider the change in crack tip stress field induced by the finite dimensions of the panel. For that reason, the methodology explained in previous sections incorporated the Dixon finite width factor for the far-field stress intensity factor in FMLs [3].

However, when validating the methodology with experimental data, the observation was that including such a finite width factor in the model over-predicted the crack growth for long cracks. For that reason, the application of the finite width correction was not mentioned in [2], nor did Wilson incorporate a finite width correction in his generalized form of the methodology [4]. In fact, omitting the finite width correction yielded slightly un-conservative results while including such correction factor gave conservative, but less accurate, results.

To understand the role of these finite width corrections, one may look at the fatigue crack growth problem from an energy perspective [30, 31]. Rather than applying load or stresses to the panel, one has to consider that actually work is applied. Corresponding to the application of load, the panel will deform. In case of force-controlled testing, the reduction of panel stiffness as result of crack growth corresponds to an increase of applied work. One could therefore consider the finite width correction a correction of force or stressed-based models for the fact that they do not consider the energy well.

Hence, the finite width correction could be formulated as the ratio of work applied at any crack length $2a$ compared to the un-cracked panel

$$\beta_{FW} = \frac{W_{2a}}{W_{2a=0}} \tag{9.10}$$

where the stiffness of a monolithic panel theoretically goes to zero for $2a = W$; the corresponding work applied to the panel goes to infinite. Hence, the correction factor β_{FW} asymptotically goes to infinity for $2a/W = 1$.

In case of fatigue cracks in FMLs, the stiffness of the panel at $2a/W = 1$ is defined by the remaining intact fibre layers. Hence, the corresponding work applied to the panel is still finite. One has to keep in mind that since the intact fibre layers define the stiffness at $2a/W = 1$, the actual value of $\beta_{2a=W}$, see Fig. 9.17b, depends on the lay-up of the FML. As a consequence, the application of finite width corrections in describing fatigue damage growth in FMLs requires the determination of this correction as function of the laminate lay-up.

9.8 Other Correction Factors

It was already mentioned in Sect. 9.2 that saw cuts were initially used for FMLs as starter notch or pre-crack similar to the standard practice for monolithic metal specimen. In the latter case, a saw cut may be relevant, because small fatigue cracks in metals are not hindered in propagating through the thickness and have often reached a through-crack configuration at macroscopic lengths used for damage tolerance analyses.[1]

In FMLs, however, the fibre layers act as significant barriers for the crack to extend in thickness direction. This implies for fatigue cracks that have initiated in

[1]Recently, Jones et al. [38] emphasized that most damage tolerance analyses focus on naturally occurring cracks for which the common macroscopic crack growth data for monolithic metals are also deemed inappropriate.

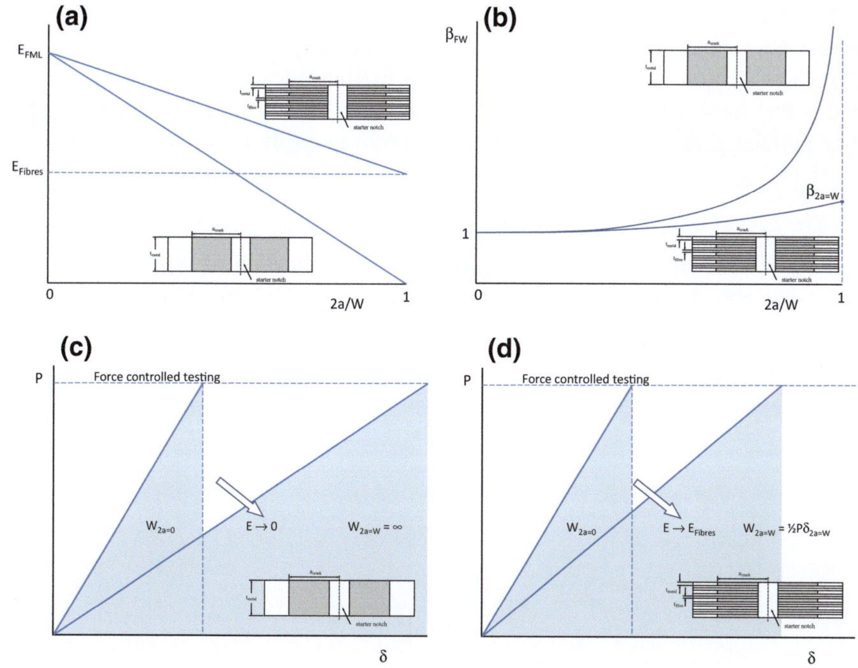

Fig. 9.17 The different stiffness reduction as function of crack length for monolithic metals and FMLs (**a**) leads to different finite width corrections β_{FW} (**b**), which are explained by the corresponding ratio of $W_{2a}/W_{2a=0}$ of monolithic metals (**c**) and FMLs (**d**)

an FML structure as corner cracks or as scratches that saw cuts lead to overestimation of crack growth rates.

This has led to the distinction between the different crack geometries, as explained in Sect. 9.2. But even for the case of through-thickness fatigue cracks, Fig. 9.2c, one has to consider the different notch and crack configurations. Similar to the case of monolithic metals, geometry factors must be considered to account for example for the difference between open hole and pin-loaded holes compared to natural fatigue cracks and to capture the difference between central cracks and edge cracks.

Thus, saw cuts may be considered representative for notches like drilled holes, but the difference between the saw cut and hole notch geometries yields different crack growth behaviour, in particular for small cracks in the notch vicinity. The effect of notch configuration on subsequent growth of fatigue cracks has been investigated for monolithic metallic structures. Often the observed influence has been captured by geometry correction factors that describe the effect of the starter notch geometry on the calculated stress intensity factor [39].

In a similar way, the starter notch effect has been investigated for FMLs. Fatigue tests on specimens containing open holes, saw cuts or pin-loaded holes have been performed and correlated with geometry correction factors developed for monolithic metals [40, 41]. In addition, tests have been performed on double-edge-notched specimens, to correlate the growth of edge cracks to the growth of central cracks [42–44]. Figure 9.18 illustrates the different crack scenarios that have been investigated. What all these studies have in common is that they correlated the observed growth to the stress intensity factor for infinite plates

$$K = S\sqrt{\pi a} \tag{9.11}$$

As a consequence, Homan [45] proposed the application of corrections to that stress intensity factor. For any geometry other than the infinite plate, Eq. (9.11) becomes for FMLs

$$K_{\text{tip}} = f_g f_f S\sqrt{\pi a} \tag{9.12}$$

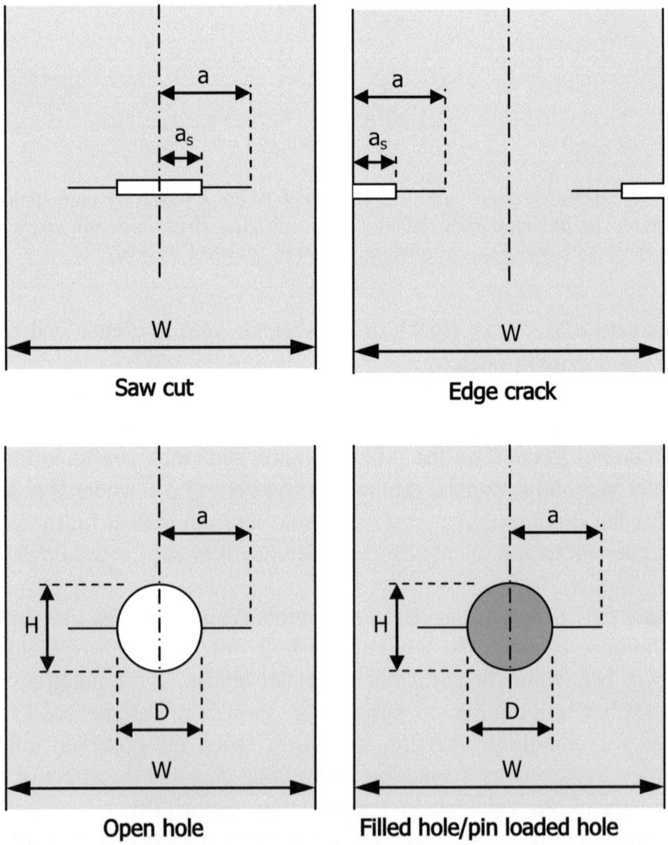

Fig. 9.18 Illustration of the different crack scenarios for which fatigue crack growth experiments have been performed and beta factors have been developed

where f_g is the general geometry correction factor similarly applied to monolithic metals, and f_f represents the contribution of bridging fibres. Given the context of the phenomenological models developed at the time [11, 26, 46], this correction approach can be understood.

One has to be aware that correlating Eqs. (9.2, 9.3, 9.4, 9.5, 9.6, 9.7, 9.8, 9.9, 9.10, 9.11 and 9.12) yields a relation that is empirical and phenomenological

$$f_f = \frac{K_\infty - K_{\text{br}}}{f_g S\sqrt{\pi a}} = 1 - \frac{K_{\text{br}}}{f_g S\sqrt{\pi a}} \tag{9.13}$$

In fact, it is similar to the phenomenological model once proposed by Toi [47] and later adapted by Takamatsu et al. [48], except that the equation for this correction factor was not just a polynomial expression, but rather related to parameters describing the delamination geometry [26, 46].

Homan [45] explains that f_g in Eq. (9.13) is the finite width correction factor commonly calculated with the Feddersen [35] or Dixon [37] relation. The latter is defined as

$$f_{\text{Dixon}} = \frac{1}{\sqrt{1 - (2a/W)^2}} \tag{9.14}$$

This correction was originally applied to the stress intensity factor calculated with Eq. (9.12) based on applied laminate stresses. It implicitly assumed the laminate to constitute a homogeneous plate containing a through crack, which is deemed incorrect [1]. As explained in Sect. 9.5, this correction was later only applied to the far-field stress intensity factor K_∞, see Eq. (9.2). But even then, the correction overestimates the effect of the finite specimen dimensions for the reasons explained in Sect. 9.5.

9.8.1 Open Hole and Pin-Loaded Hole

In line with the concept of Eq. (9.12), Homan [45] proposed a correction for open holes after the correction proposed by Newman [49] for open holes in monolithic panels

$$K = f_{\text{OH}}\sqrt{1 - \frac{D}{2a}}S\sqrt{\pi a} \tag{9.15}$$

in which f_{OH} includes the Bowie correction factor

$$f_{\text{OH}} = \frac{(1.0 - 0.15\lambda_D + 3.46\lambda_D^2 - 4.47\lambda_D^3 + 3.52\lambda_D^4)}{\sqrt{1 - (2a/W)^2}} \cdot \frac{\sqrt{1 - \lambda_D}}{\sqrt{1 - (D/W)^2}} \tag{9.16}$$

with λ_D is defined as $D/2a$. This correction has been developed for through-thickness cracks in monolithic panels, which for FMLs is equivalent to straight cracks through both metal and fibre layers. Obviously, the application of this correction to stress intensity factors for fatigue cracks in FMLs must be inaccurate.

Gonesh [40] investigated the open hole configuration in GLARE and compared it with the known relations for monolithic metals. Figure 9.19 gives the results of Gonesh' comparison in which he calculated the geometry correction factor with

$$f = \frac{\Delta K_{\mathrm{eff}}}{\Delta S\sqrt{\pi\, l}} \tag{9.17}$$

where ΔS is the applied stress range, and l is defined as $l = a - ½D$, see Fig. 9.18, and ΔK_{eff} is taken from the fatigue experiments on GLARE specimen using

$$\Delta K_{\mathrm{eff}} = \left[\frac{1}{C}\frac{\mathrm{d}a}{\mathrm{d}N}\right]^{\frac{1}{n}} \tag{9.18}$$

Note that this approach yields a correction factor that comprises all correction factors, including f_g and f_f, and not only f_{OH}. This makes the correlation with the aluminium correction factors somewhat less straightforward.

The constant C and exponent n were taken from the fatigue experiments on monolithic aluminium sheet, as discussed in Sect. 9.4. Comparison of the

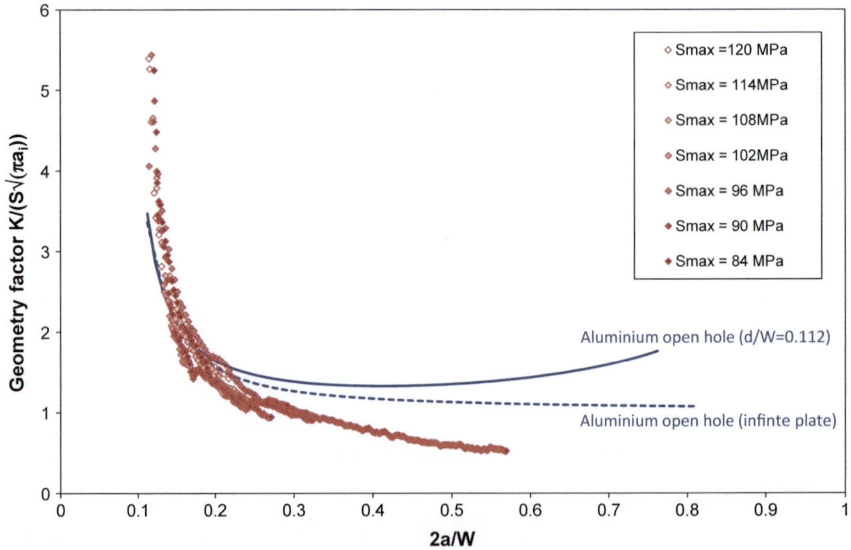

Fig. 9.19 Comparison between geometry correction factors for GLARE3-3/2-0.3 [40] and monolithic aluminium [25], both with $d/W = 0.112$

experimentally determined geometry correction factor according to Eq. (9.17) and the factor established for monolithic aluminium [25] is given in Fig. 9.19.

Evident in this figure is the contribution of the bridging fibres. With increasing crack lengths, the effective stress intensity factor in GLARE is lower than the corresponding stress intensity factor in monolithic aluminium.

Hence, the general observation is that applying the open hole correction factor obtained for monolithic aluminium for GLARE will yield conservative predictions. This observation is in agreement with the finite width correction discussed in the previous section, illustrated in Fig. 9.17b.

Similar to the case of an open hole, Homan related the correction for the pin-loaded hole to Newman's corrections for monolithic metals [49]. According to Newman, the correction for pin-loaded configurations can be derived from the stress intensity factor for open holes with

$$K_{\mathrm{PLH}} = K_{OH}\left(0.5 + \frac{W}{2a\pi}\right) \tag{9.19}$$

Hence, the correction factor for a pin-loaded hole can be presented as

$$f_{\mathrm{PLH}} = \frac{(1.0 - 0.15\lambda_D + 3.46\lambda_D^2 - 4.47\lambda_D^3 + 3.52\lambda_D^4)}{\sqrt{1-(2a/W)^2}} \cdot \frac{\left(\frac{1}{2} + \frac{W}{2a\pi}\right)\cdot\sqrt{1-\lambda_D}}{\sqrt{1-(D/W)^2}} \tag{9.20}$$

Also here, it should be noted that this correction when applied to FMLs assumes the crack to run through both metal and fibre layers, which is evidently incorrect. Because both the correction for open hole and for pin-loaded hole overestimate the stress intensity factor, predictions using these corrections lead to conservative results.

9.8.2 Edge Cracks Versus Central Cracks

For double-edge-notched (DENT) cracks in monolithic panels, illustrated in Fig. 9.18, the correction factor is given by [50]

$$f_{\mathrm{DENT}} = \frac{1.1215 - 1.1215\cdot(a/W) - 0.060\cdot(a/W)^2 + 0.728\cdot(a/W)^3}{\sqrt{1-2(a/W)}} \tag{9.21}$$

To compare the correction factors for double-edge-notched specimens with Eq. (9.21) is not trivial, for the same reasons as explained for centre crack specimens in Sect. 9.5. However, to follow a similar approach as in Eqs. (9.17) and (9.18) one may correlate the obtained curves with the curves for centre crack

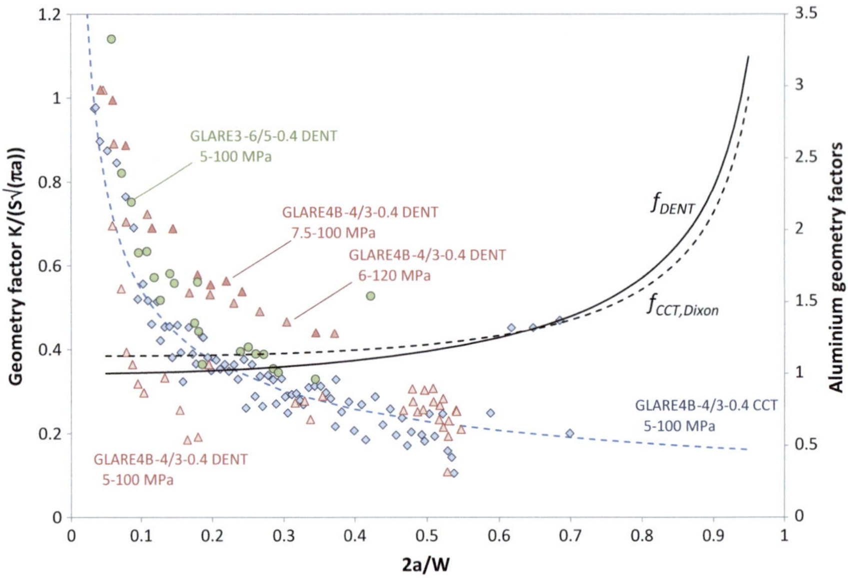

Fig. 9.20 Geometry correction factors for GLARE4B-4/3-0.4 and GLARE3-6/5-0.4 DENT specimens compared to a GLARE4B-4/3-0.4 CCT specimen and aluminium correction factors, data from [42]

tension (CCT) specimens. This is illustrated in Fig. 9.20. The initial difference of factor 1.1215 between the geometry correction factors for DENT and CCT specimen for monolithic metals, is not evident in the GLARE data. In addition, the curves follow different paths obviously related to the different contribution of fibre bridging, which is implicitly present in Eq. (9.17).

Hence, one may apply the correction factors for monolithic aluminium in a similar fashion to FMLs, but evidently these factors are invalid for FMLs, although they yield conservative predictions.

9.9 Fatigue Threshold

The crack growth resistance of the metal layers in FMLs has been discussed in Sect. 9.4. In the Paris representation of this resistance, there is generally a lower asymptote that indicates a threshold ΔK_{th}. To identify whether for FMLs also such lower threshold can be established, Daandels [51] performed an experimental study.

Daandels performed fatigue tests on two GLARE3-3/2-0.4 specimens, one tested in the principal material axis with 0°/90° fibres between the metal layers and one under an off-axis angle with −45°/45° between the metal layers. A procedure was adopted similar to that used for monolithic metals, in which the maximum load is incrementally reduced in small steps to limit the plasticity-induced retardation effects. The procedure is explained in detail in [52]. The obtained crack growth rates were plotted against the effective SIF range ΔK_{eff}, calculated with the phenomenological model of De Koning [26]. Daandels concluded initially that the threshold SIF ΔK_{th} was 83 $MPa\sqrt{mm}$ for the 0°/90° orientation and 115 $MPa\sqrt{mm}$ for the −45°/45° orientation.

In further studying the experimental results and fatigue phenomena, it was concluded that the procedure of stepwise load reduction imposes greater delaminations at the interfaces than would have been obtained for constant amplitude fatigue tests entirely at the corresponding load levels [52]. Hence, the reported threshold values of 83 and 115 $MPa\sqrt{mm}$ are deemed inaccurate.

Probably more important is that the observed threshold in Fig. 9.21 is to be attributed to the threshold in the fatigue crack growth resistance of the aluminium, because, as explained in Chap. 8, no thresholds were observed in similar delamination growth resistance curves.

This means that once the SIF at the crack tip K_{tip} is below the threshold SIF, the fatigue crack in the metal layers retards, but the delamination at the interfaces still grows. Over time, this increase in delamination size decreases fibre bridging and as a result K_{br} in Eq. (9.2) will decay. This increases the K_{tip} until it reaches a level beyond the threshold SIF and the crack may grow again.

In summary, the conclusion of the study by Daandels [51] and the one reported in [52] is that it does not make much sense to establish threshold levels for FMLs by testing FML panels. Rather the crack growth resistance of the metal layers must be characterized including the threshold regime, as explained in Sect. 9.4. Implementation of the crack growth resistance of the metal, including its threshold behaviour, in the methodology explained in Sect. 9.4 will allow evaluation of the fatigue behaviour of the FMLs in the near-threshold regime.

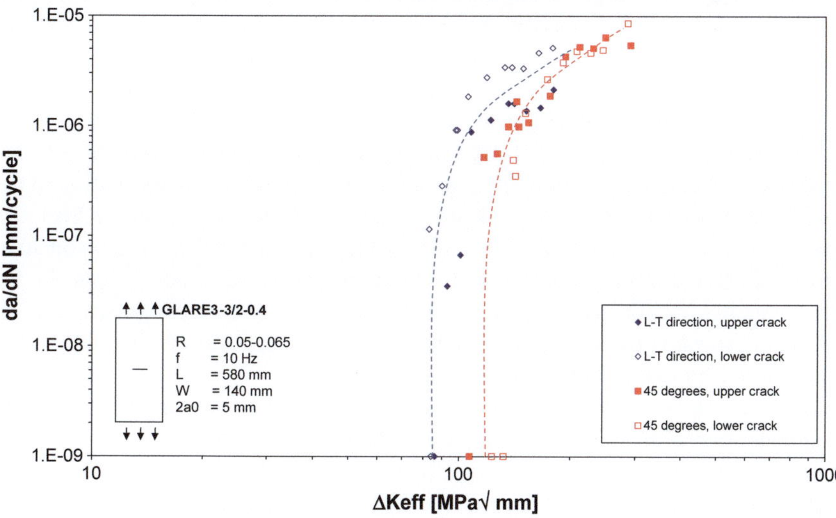

Fig. 9.21 Recorded crack growth rates plotted against the effective stress intensity factor range calculated with the empirical model from [26, 52]

9.10 Surface Cracks

The major difference between surface fatigue cracks and fatigue through cracks is the level of bridging to the cracked layers, see Fig. 9.1. Another difference is the contribution of the starter notch to the initial part of the crack growth curve. In fatigue through cracks, the cracks are initiated in all metal layers by creating a notch through the entire laminate, including fibre layers. Hence, at the start of the test (indicated with ●), the crack growth rate initially is determined by K_{∞} only. Once the crack progresses, fibre bridging will develop and the crack growth rate will decrease, see Fig. 9.7b. For surface cracks, this is different, because the notch is a surface notch. As a consequence, the crack growth rate is determined by both K_{∞} and K_{br} resulting in approximately constant rates after an initial increase at the start. This is clearly visible in Fig. 9.23, where data are plotted for two different surface notch lengths. The difference illustrated in Fig. 9.22 is visible when comparing Fig. 9.23 with the case for through cracks in Fig. 9.7b.

Because the number of intact layers contributing to crack bridging is greater, i.e. both fibre layers and metal layers, the effective SIF at the crack tip of the surface crack is considerably lower than for through cracks.

Depending on the FML lay-up, however, this amount of bridging will be different. The comparison between the 2/1 and 3/2 lay-up in Fig. 9.23 shows that increasing the laminate thickness from 2/1 to 3/2 substantially decreases the crack growth rates.

Aside from the number of intact layers bridging the surface crack, this also is influenced by secondary bending. Secondary bending is the bending of the axially loaded panel, due to a step in the neutral plane of the laminate. For a thin 2/1 lay-up, this step is relatively large, while for a thick lay-up this bending may be marginal.

The consequence of secondary bending is that the delamination at the interface with the intact layers opens not only in a shear mode, but also in a tension mode (mode I). Hence, the delamination propagates slightly faster in a thin laminate containing a surface crack, compared to the thick FML.

The general observation for all the fatigue crack growth tests performed was that the crack growth rate is not only low, but also fairly low for the major portion of the crack growth life. Hence, the crack growth characteristics discussed in Sect. 9.3 follow the steady-state growth illustrated in Fig. 9.5, rather than the exponential growth illustrated in Fig. 9.6. Except for the transient phase, which is opposite for surface cracks as illustrated in Figs. 9.22 and 9.23, the crack growth rate is developing towards the steady state, rather than decreasing from an initially higher value caused by cutting fibres in the notch.

Therefore, to describe the fatigue crack growth behaviour of surface cracks, phenomenological methods were proposed that described a steady state of crack tip stresses independent of the actual crack length [54]. The methods were based on a method initially proposed by Guo and Wu [55], or a variation of the Rose model for patch repairs [56]. The concept for the latter methods was that the intact part of the FML could be described as a patch to a cracked metal layer.

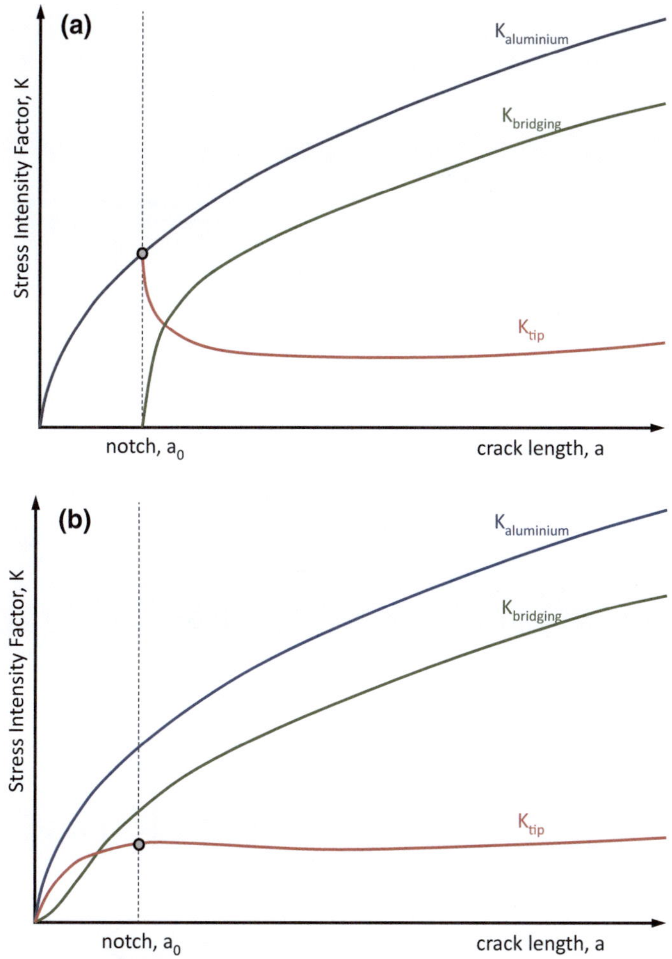

Fig. 9.22 Comparison between the development of K_{tip} with the crack length according to Eq. (9.2) in fatigue through cracks (**a**) and surface cracks (**b**)

Noteworthy is the method proposed by De Koning [57], who derived a SIF for a surface crack based on the superposition of far-field stresses and bridging stresses, similar to Eq. (9.2), with the far-field SIF described by Eq. (9.3). His derivation resulted in cancelling out of the far-field SIF, leaving an expression based on a local crack tip stress field. Hence, the maximum SIF as function of the local stress system (Fig. 9.24) could be described by [54]

$$K_{\max} = \frac{4}{\sqrt{2\pi}} \left(\frac{\gamma \sqrt{\varepsilon}}{\bar{\sigma}} \right) f\left(\frac{a}{W} \right) S_{\max}^2 \sqrt{t} \tag{9.22}$$

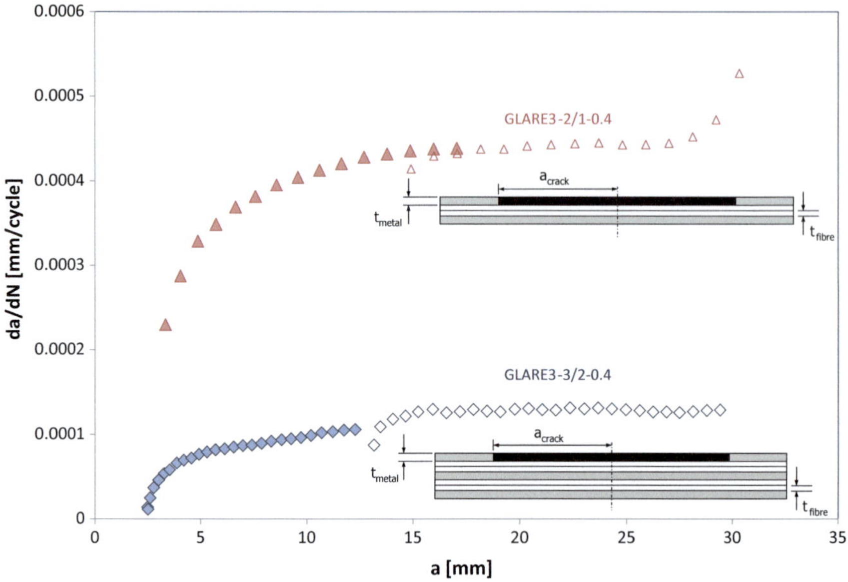

Fig. 9.23 Surface crack growth rates recorded for two surface notch sizes ($2a_s$ = 5 and 25 mm) in GLARE3-3/2-0.4 and GLARE3-2/1-0.4 under $S_{\max}$ = 150 MPa, R = 0.05, data from [53]

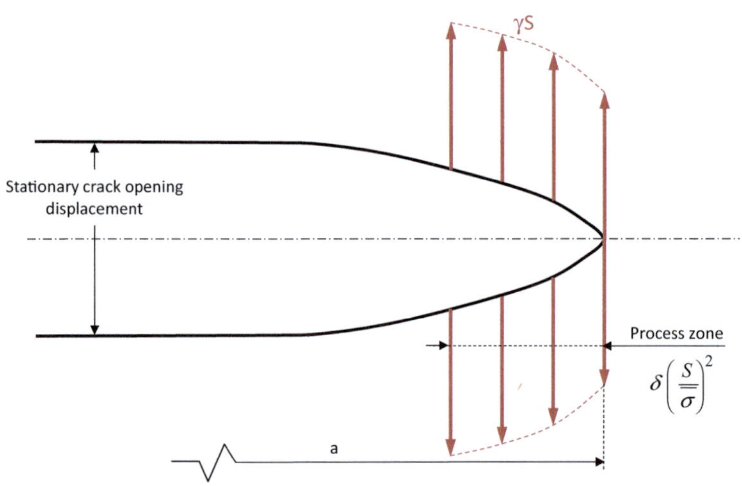

Fig. 9.24 Concept of a local stress system in a process zone near the surface crack tip [54, 57]

The disadvantage of this phenomenological model is the empirical parameters that must be determined based on a number of fatigue crack growth tests on the surface crack configuration. Different to the methodology described in Sect. 9.4, a number of parameters have to be tuned to calibrate predictions.

9.11 Part-Through Cracks

In the case of mechanically fastened joints, secondary bending will impose different initiation and propagation behaviour of fatigue cracks in each individual layer of the FML. In Sect. 7.14, this was discussed with respect to initiation. Similarly, fatigue crack growth is affected by the bending stresses through the thickness of the FML. Hence, to describe the growth of the individual layers in an FML under bending or combined bending and tension, the effective axial stress levels in each layer should be determined.

Initially, prediction methods were developed that predicted the crack growth behaviour of surface cracks (previous section). The assumption was that the tip of the lead crack in the metal layer at the mating surface in the joint (experiencing the highest stresses) will progress in a laminate similarly to a surface crack configuration. The process zone will not be affected by the shorter cracks propagating in subsurface layers.

Hence, the surface crack methodology using Eq. (9.22) was adopted to describe the crack growth in the surface metal layer experiencing the highest bending stress. The crack growth in subsurface layers was calculated with a correction to Eq. (9.22) based on the bending factor and the distance from the subsurface metal layer to the surface metal layer, see Fig. 9.25. In that concept, the subsurface crack growth is assumed to be similar to the surface layer, i.e. not affected by cracks in even deeper layers, but also not by the surface crack [58].

Validation of this phenomenological part-through crack method appeared to be somewhat less straightforward. To this aim combined bending tension tests were performed [59] where throughout the fatigue test the crack growth of the surface crack was monitored, while the crack lengths in the subsurface layers were determined by tear down of the specimens. This tear down approach was also adopted by Randell [60] with the aim to support the development of a better crack growth prediction model for subsurface cracks (Fig. 9.26).

Mortier and Homan [59] report testing five specimens at a gross tensile stress of S_{max} = 80 MPa, with R = 0.1, which each were tested up to a predefined surface crack length. Tear down of the five specimens allowed reconstruction of the crack growth curves for the subsurface layers as illustrated in Fig. 9.27 for GLARE2B-7/6-0.4.

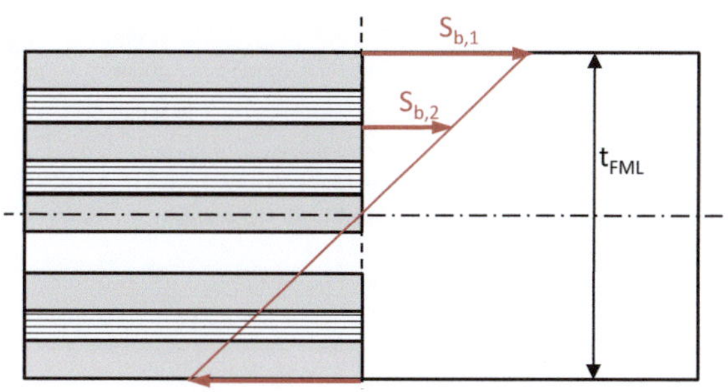

Fig. 9.25 Concept of axial stress due to bending in surface and subsurface layers

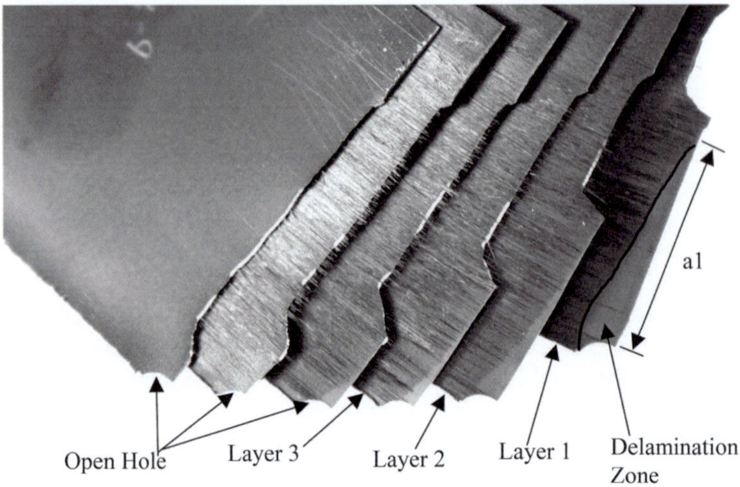

Fig. 9.26 Tear down of four-point bend specimen. One quarter of the specimen is shown with 'Layer 1' labelling the cracked surface layer [60]

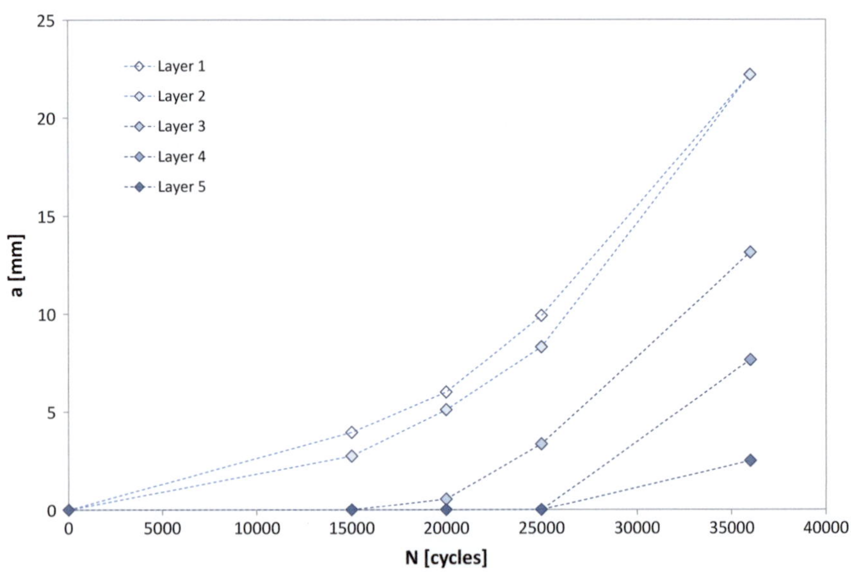

Fig. 9.27 Crack growth curves reconstructed for GLARE2B-7/6-0.4 after tear down of combined bending tension specimens tested under a remote gross stress of S_{max} = 80 MPa, R = 0.1 [59]

In particular, for GLARE2B-7/6-0.4 tested in the LT direction, it was observed that the crack lengths in the surface layer and the first subsurface layer were almost identical, which was not the case for GLARE3-5/4-0.4 and GLARE4B-5/4-0.4 (in LT direction), both tested at the same gross applied stress level of S_{max} = 80 MPa.

According to Mortier and Homan, this difference may be caused by the fact that GLARE2B in the LT direction is significantly more stiff than the other two laminates: having a nominal Young's modulus of E_{lam} = 64.2 GPa versus 57.5 and 56.9 GPa, respectively. The other difference is the nominal laminate thickness, which for GLARE2B was t_{FML} = 4.3 mm versus 3.0 and 3.5 mm for the other two laminates, respectively.

These differences change the bending moment induced by the axial gross stress and, as a result, change the effective stresses in the subsurface layers. If the effective stresses in the surface and the subsurface layer are close, the resulting crack growth will become similar as well.

This was later confirmed by Randell [60] who increased the thickness for the GLARE2 laminate to a lay-up of GLARE2A-11/10-0.4 instead of the 7/6 lay-up. With the increased thickness, the step in neutral line in the combined bending tension specimen was greater, inducing a larger bending factor. Randell observed in his tests a greater difference between the crack growth of the surface and the first subsurface layer.

Later the methodology presented in Sect. 9.4 was further developed by Wilson [4] into a generalized theory that allowed the prediction of the cracks in the individual layers together with corresponding delaminations at the adjacent interfaces. Wilson initially developed the method to enable prediction of fatigue cracks in more complex hybrid lay-ups, like the CentrAl concept, see Chap. 2, which combined thicker outer layers of different aluminium alloys with the thin metal layers of the centre FML. In [4], Wilson demonstrated the wide range of cases and conditions for which the generalized theory applies, including tension and bending, different metal constituents and constituent thicknesses.

Wilson's method was adopted by Spronk [61] to predict the part-through crack propagation in the FML reinforcement of the aluminium frame of the A400 M. The methodology was later validated by Van der Linden [62], who used the Wilson model as adapted by Spronk to predict the required number of fatigue load cycles to generate predefined crack lengths. Tear down after the residual strength tests illustrated that the model was reasonably accurate.

9.12 In-Axis Versus off-Axis Loading

The fatigue crack growth theories presented in [2–4, 8, 13] mostly focussed on fibre bridging in the case that panels were uniaxially loaded in directions of the principal material orientations. The difference between unidirectional laminates like GLARE2 and cross-ply laminates like GLARE3 and GLARE4 was studied because different contribution of fibre bridging could be observed. However, in thin-walled structures such as aircraft fuselage panels or lower wing panels, other load components may be applied leading to situations with the principal loading in directions different from the principal material orientations.

To address this aspect, a number of studies were performed into the crack growth behaviour under off-axis loading conditions. To this purpose, Gonesh [63, 64] performed uniaxial fatigue tests on GLARE2, GLARE3 and GLARE4B centre crack tension specimens with the initial notch perpendicular to the loading direction and the principal material orientation under an angle γ with that loading direction. Similarly, Thibault-Liboiron performed fatigue tests of edge cracks under off-axis loading [43, 44].

Gonesh observed the crack growth rates to increase with increasing off-axis angles, as illustrated in Figs. 9.28 and 9.29. The increase with the off-axis angle is greater for the unidirectional GLARE2 than for the cross-ply laminates like GLARE4.

This can be explained with the effective stiffness of the fibre layers in the direction of loading. The fibre layer stiffness in unidirectional FMLs decreases continuously from 0° to 90°, while for cross-ply laminates that stiffness decreases to lesser extent and increases again towards 90°. As a consequence, the laminate stiffness will vary similarly, as illustrated in Fig. 9.30.

The significant effect of off-axis loading on unidirectional laminates is important, when these laminates are considered for lower wing skin applications. Hence, crack growth prediction models need the ability to predict the higher crack growth rates for principal loading directions misaligned with the principal material direction.

Initially, the off-axis loading was considered in the phenomenological models with another correction factor $\alpha(\theta)$

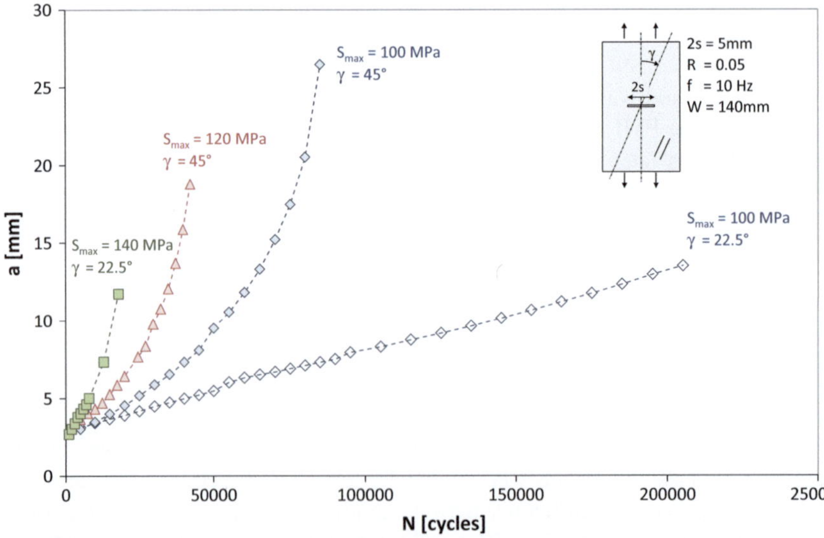

Fig. 9.28 Crack growth curves for various off-axis angles for GLARE2-4/3-0.4 [64]

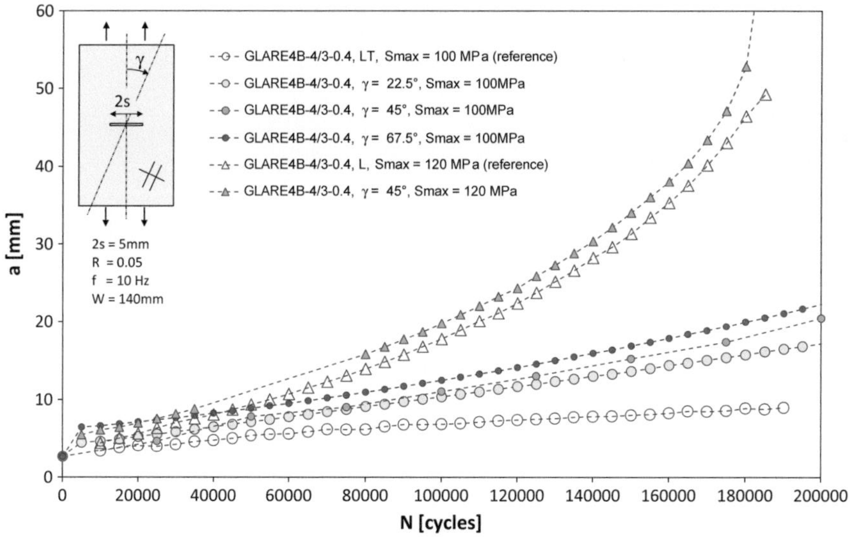

Fig. 9.29 Crack growth curves for various off-axis angles for GLARE4B-4/3-0.4 [63]

$$\alpha(\theta) = \alpha_{L-T}\cos(\theta)^4 + \alpha_{T-L}\sin(\theta)^4 + \alpha_\theta \sin(\theta)^2\cos(\theta)^2 \tag{9.23}$$

where α_{L-T}, α_{T-L} and α_θ were fitted to the experimental fatigue crack growth data obtained for GLARE2, GLARE3 and GLARE4 separately. Note that Eq. (9.23) describes the trend similar to what is illustrated for the laminate stiffness of these three GLARE types in Fig. 9.30.

With the emphasis on providing input for and validation of the phenomenological models [58], less effort was put in developing understanding of all aspects that influence the observed crack growth rate and crack direction. Only recently, Gupta [65] discusses that FMLs tested under an off-axis angle effectively not only experience an axial loading induced by the fatigue test machine, but that the machine effectively also applies a transverse load to the specimen.

This transverse load is obviously not recorded by the machine, but it can be calculated using for example the classical laminate theory. FMLs under an off-axis angle represent orthotropic laminates that are not balanced in the axial loading direction. Hence, to resist the additional in-plane shear deformation of the FML imposed by the axial deformation, the machine has to apply an additional transverse load.

Hence, the crack growth rate and crack direction are governed by both tensile and shear loads, which together create a mixed-mode stress intensity at the crack tip. Gupta [65] explains that this mode-mix at the crack tip constitutes a superposition of the mode-mix induced by the combination of applied axial and transverse load, with the mode-mix created by bridging fibres oriented under the off-axis angle. The concept of this superposition is illustrated in Fig. 9.31.

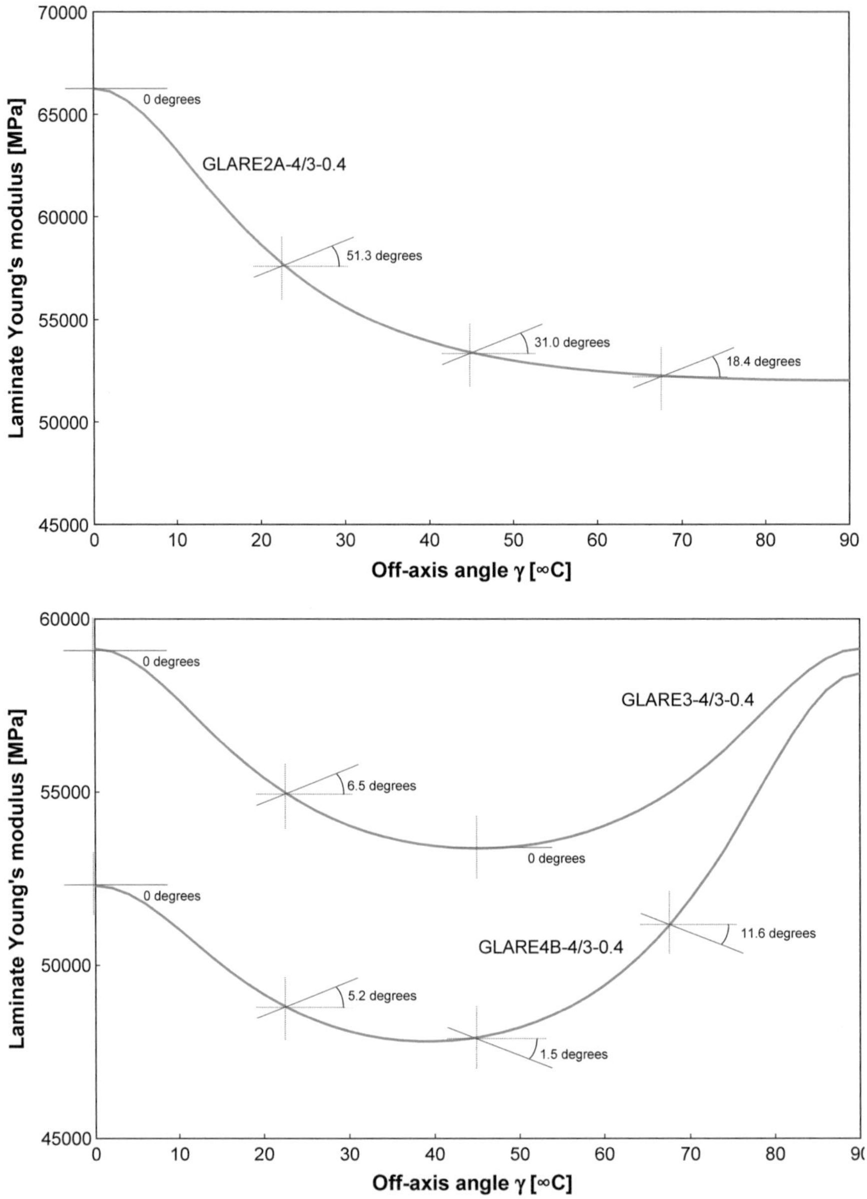

Fig. 9.30 Illustration of the relation between laminate stiffness at given off-axis angle and the observed crack path angle for GLARE2A [64] and GLARE3 and GLARE4B [63]

Following this superposition, one may thus assume that also for the mode II Eq. (9.2) holds, but that K_{∞} in that case is based on the far-field shear stress τ (induced by the transverse load) rather than the far-field stress S given in Eq. (9.3).

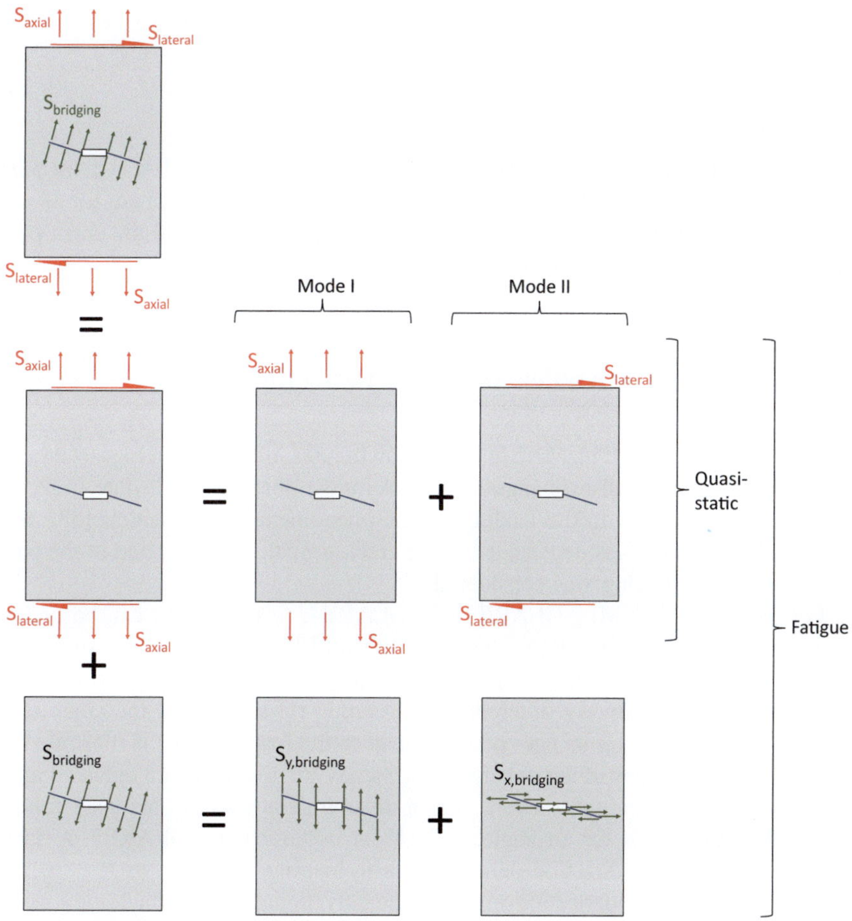

Fig. 9.31 Concept of superposition of mode-mix imposed by far-field axial and transverse loading by the test machine, and the mode-mix imposed by the off-axis fibre bridging [65]

Similarly, the opening of the fatigue cracks in the metal layers comprises two directions equivalent to the opening mode I and shear mode II, which means that Eq. (9.4) also has to be formulated for the tension load and shear load separately.

Gupta [65] illustrates that the methodology presented in [2] can be adopted in a similar fashion for the shear mode to determine the transverse fibre bridging. Hence, the stress intensity factor at the crack tip in the metal layers of the FML is decomposed into two individual opening modes following the concept of Fig. 9.31

$$\begin{aligned} K_{I,\mathrm{FML}} &= K_{I,\infty} - K_{I,\mathrm{br}} \\ K_{II,\mathrm{FML}} &= K_{II,\infty} - K_{II,\mathrm{br}} \end{aligned} \tag{9.24}$$

With the mode-mix defined as

$$M = \frac{K_{II,\text{FML}}}{K_{I,\text{FML}} + K_{II,\text{FML}}} \tag{9.25}$$

One would obtain $M = 0$ for pure tension (in-axis loading) and $M = 1$ for pure shear. The mode difference influences the path the crack follows, where for $M = 0$ the crack propagates straight perpendicular to the applied load, and the crack path tends to deviate when $M > 0$.

9.13 Crack Path Angles and Path Deflections

Aside from the observed increase in crack growth rates, illustrated in Figs. 9.28 and 9.29, with increasing off-axis angle, the cracks were observed to deviate from the direction perpendicular to the loading. In the phenomenological models [58], this crack path deflection was neglected and the crack growth rate projected to the path perpendicular to loading was considered.

For the cross-ply FMLs GLARE3 and GLARE4, this approach can be understood. The results by Gonesh [63, 64] and Thibault-Liboiron [43, 44] illustrate that the cracks deflect under small angles. However, for off-axis loading of unidirectional FMLs this is entirely different. For a certain off-axis angle, the crack may deflect significantly from its path perpendicular to the loading. This is illustrated in Fig. 9.33 where images of the delamination shapes are given that are obtained after etching the outer aluminium layers. These delamination shapes obviously follow the crack path, which for an angle of 22.5° in unidirectional GLARE2 deviates significantly from the direction perpendicular to loading.

Because this crack deflection may significantly impair the methodology crack growth prediction for unidirectional laminates, which are generally considered for lower wing skin structures, a methodology was needed that enabled not only the prediction of the growth, but also the direction of the cracks. To this aim, Gupta [65] recently developed a methodology extending the method in [2], allowing the prediction of the path of the fatigue cracks in off-axis loading conditions. Consistent with the methodologies for describing crack deflections of oblique cracks in monolithic metal, Gupta provided definitions of the relevant angles in the methodology, which are illustrated in Fig. 9.32.

There is a clear difference between the crack path angles in unidirectional FMLs and cross-ply FMLs. This difference is illustrated in Fig. 9.30. The angles of crack paths in unidirectional GLARE2A are significantly larger than the crack paths in cross-ply GLARE3 and GLARE4B, and they are not clearly related to the angular stiffness of the laminate like for the cross-ply laminates. The reason is that the crack initially tends to change path according to what may be expected based on stiffness considerations, but then continues to deflect until it follows a path almost parallel to the fibre orientation.

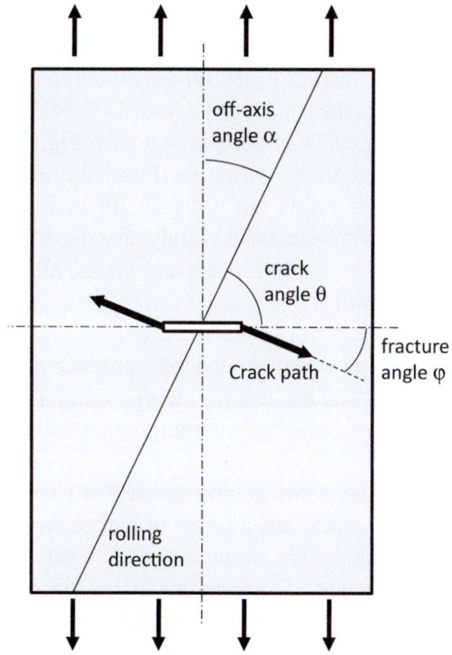

Fig. 9.32 Definitions of the off-axis angle α (between axial loading direction and rolling direction), the crack angle θ and the fracture angle φ [65]

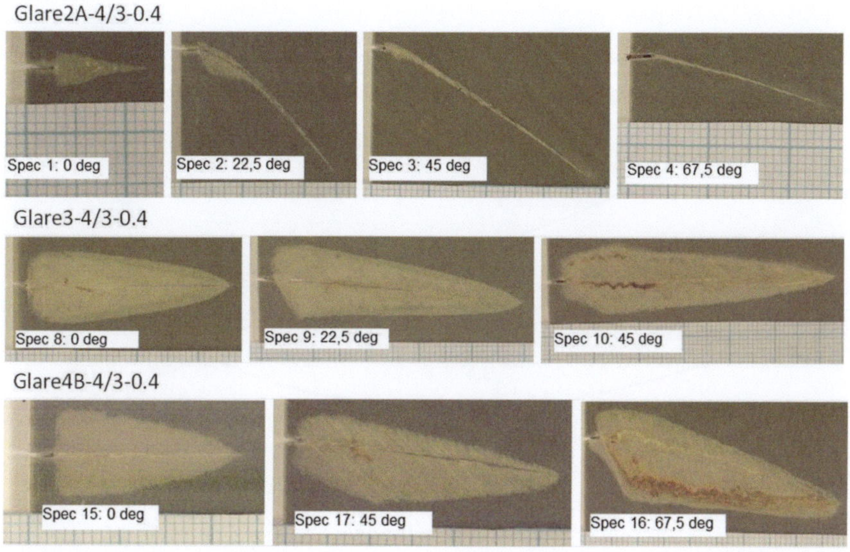

Fig. 9.33 Examples of crack path angles and delamination shapes observed for off-axis tested unidirectional GLARE2A and cross-ply GLARE3 and GLARE4B DENT specimens [44]

Maretti [66] studied the crack path angles of surface cracks in both unidirectional GLARE2 and cross-ply GLARE3 and GLARE4. The interesting observation in his work is that the angles with respect to the off-axis loading angle are different from the through-the-thickness cracks tested by Gonesh [63, 64], Thibault-Liboiron [43, 44] and Gupta [65], see Fig. 9.34 in comparison with Fig. 9.30.

This difference is related to two aspects. First, the effective angularity and directionality of the bridging layers are different for surface cracks compared to through cracks. In case of through cracks, only the intact fibre layers define the mode-mixity at the crack tips, while for surface cracks, all other remaining intact metal layers contribute as well.

However, this is insufficient to explain the observed crack path angles reported by Maretti [66]. This can be demonstrated by applying the theory proposed by Gupta [65] to the case of surface cracks incorporating the intact metal layers' contribution to the intact fibre layers. Different angles will be predicted than presented in Fig. 9.34.

It appears that the fibre layer orientation directly adjacent to the surface metal layer containing the surface crack, has a larger influence on the mode-mixity of the crack tip and then subsequent fibre or intact metal layers. Hence, to be able to describe the crack path direction of surface cracks in case of off-axis loading, one has to account for the through-thickness position of each layer with respect to the layer containing the surface crack.

Another case of off-axis loading was presented by Bradshaw and Gutierrez [67] who tested a hybrid combination of carbon fibre-reinforced polymer composite (CFRP) with thin 0.203-mm aluminium layers at both sides. The lay-up was

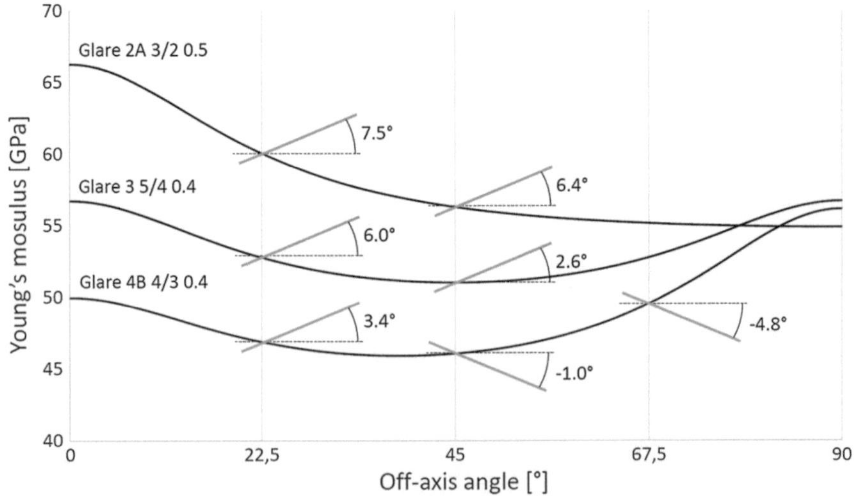

Fig. 9.34 Comparison between the observed crack angles and the FML Young's modulus measured in loading direction [66]

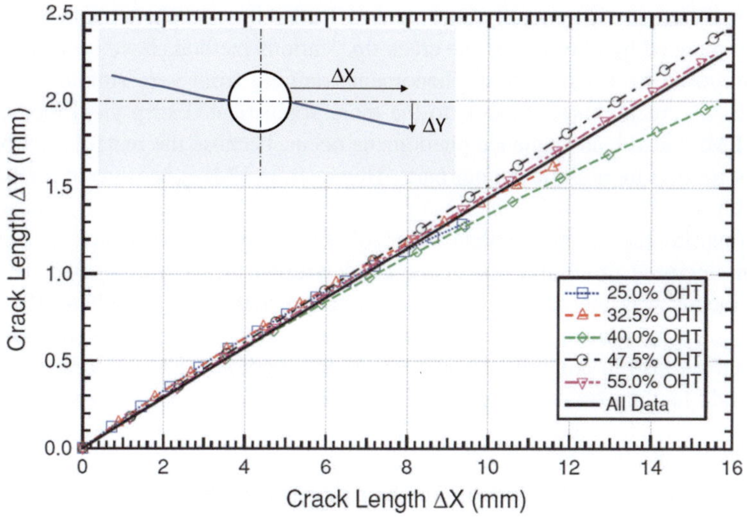

Fig. 9.35 Crack path ΔY versus ΔX for specimens tested at five stress levels along with a linear fit through the entire set [67]

[Al/45/0/−45/90/45/0/−45]$_S$, which basically represented a 2/1 lay-up with the fibre layer consisting of 14 CFRP plies in the 0°, 90° and ±45° directions.

Bradshaw and Gutierrez reported the growth of fatigue cracks in the aluminium outer layers emanating from a hole with similar straight deflected cracks (Fig. 9.35) as reported by Gupta [65] and Maretti [66]. The deflection appears to be independent of the actual stress level, but defined by the orientation of the CFPR layer between the aluminium layers.

Because the fibre ply directly adjacent to the aluminium layer is oriented under an angle of 45°, the aluminium layer experiences a mode II in addition to the dominant mode I of the centre crack configuration.

Thus, despite the fact that the fibre layer constitutes a symmetric and balanced lay-up in itself, it applies a mode-mixity to the crack tip in the metal layer, similar to the observations of Maretti [66].

9.14 Constant Versus Variable Amplitude Loading

Many studies on monolithic aluminium alloys have been reported in the literature that point out the differences between constant and variable amplitude loading. Take for example the typical Paris relationship, like the one illustrated in Fig. 9.16; the exponent of the relationship describing the linear part of the curve typically has a value between 2 and 4, but in most cases close to 3. When testing the same material under various variable amplitude load spectra, the exponent tends to be lower, often close to 2.

This difference is generally attributed to load sequence phenomena that occur, largely induced by plasticity at the crack tip. Various methods have been developed and proposed to describe these phenomena, ranging from very simple yield zone models via crack closure models to the more sophisticated strip yield models.

In FMLs also load sequence phenomena occur, because the metal layers behave similar to their monolithic counterparts. However, in FMLs, because of the balance between crack growth in the metal layers and the delamination mechanisms, additional interaction phenomena occur [68].

This increase in complexity of the crack growth problem compared to monolithic metals was the subject of an investigation performed by Khan [19]. Khan investigated the influence of variable amplitude load variations on the delamination mechanism, also in combination with crack propagation. The influence of variable amplitude loading on delamination growth has been discussed in Sect. 8.5.

Despite the fact that crack growth and delamination growth in FMLs under variable amplitude loading seem more complex than crack growth in monolithic metals, prediction models may in fact be simpler than for monolithic metals while maintaining equivalent accuracy. Take, for example, the simple yield zone models, like the Wheeler and Willenborg models. These models are often inadequate to describe crack growth under distinct load sequences for monolithic metals. However, Khan [69] demonstrated that application of one of these models (Wheeler) in combination with the methods described in Sect. 9.4, yields fairly accurate predictions. The explanation for this observation is that the bridging fibres reduce the crack tip stress intensity in the metal layers of FMLs. When high peak loads are applied in a load spectrum, this bridging also reduces the effect of this peak load in subsequent load cycles. Thus, bridging forms a mechanism that dampens the load sequence effects compared to monolithic metals. Hence, less accurate models like the Wheeler model may describe the effects more accurately in FMLs than they do for monolithic metals. In fact, depending on the load spectrum, predictions based on linear damage accumulation may already be fairly accurate [70].

In [68], the observation is reported that the delamination shape formed under constant amplitude loading changes after the application of an overload. Khan [71] provides a slightly different explanation than in [68], where it was postulated that the overload locally induces a mechanism similar to the stress redistribution obtained with post-stretching. Khan, however, argues that even without such a mechanism the shapes illustrated in Fig. 9.36 may be obtained. The overload induces a larger plastic zone at the crack tip with subsequently reducing the growth, similar to monolithic metals. Because such a retardation effect is not observed in delamination growth at the interfaces, (see Sect. 8.5), the delamination continues to grow, even though the crack in the metal layer is retarded.

After the crack has progressed through the plastic zone created by the overload, the delamination propagates with the same balance as prior to the overload. This results then in the typical kink in the delamination shape, which can be predicted with the method of Khan [19, 71].

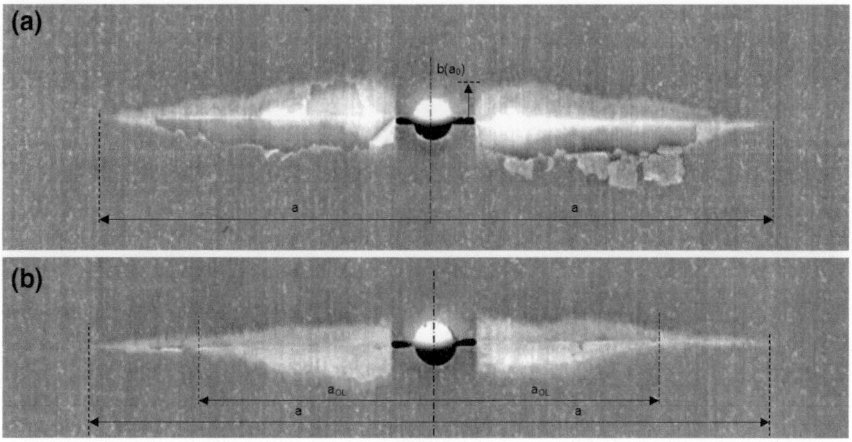

Fig. 9.36 Delamination between outer aluminium layer and adjacent fibre/adhesive layer for GLARE3-3/2-0.3 loaded with S_{applied} = 6–120 MPa, in case no overload is applied (**a**) and in case an overload of $S_{\max}$ = 200 MPa is applied (**b**) [68]

For more realistic load spectra, where high peak loads are arbitrarily positioned between other smaller loads, the effect diminishes and shapes are obtained that remain closer to the ones observed for constant amplitude loading, see Fig. 9.11.

9.15 Post-stretching

The concept of post-stretching consists of changing the residual stress system in the FML, as explained in Sect. 4.8. This reversal of stress primarily changes the initial stress state of the laminate prior to fatigue cycling and can be calculated with the method presented in Sect. 4.8.

This has been illustrated by Khan et al. [72] who extended the theory for crack propagation [2], presented in Sect. 9.4, by implementing the analytical relations for post-stretching into the prediction model. Indeed, the initial stress state calculated with the classical laminate theory (Sects. 4.2–4.4) is changed by implementing the post-stretching procedure, resulting in a reduction of residual stress in the metal layers.

For the propagation of fatigue cracks, this implies that less load needs to be transferred by the fibre layers over the fatigue cracks, and thus less shear load is present at the interface between metal and fibre layers. This reduction of shear stresses gives a reduction of delamination growth, in addition to the smaller stress intensity factor at the crack tip related to the reduced far-field stresses.

Hence, the initial value of the crack tip stress intensity factor, illustrated by the dot at $a = a_0$ in Fig. 9.22, reduces with the reduction of the (initially tensile) residual stresses in the aluminium layers as result of post-stretching. Thereafter, the

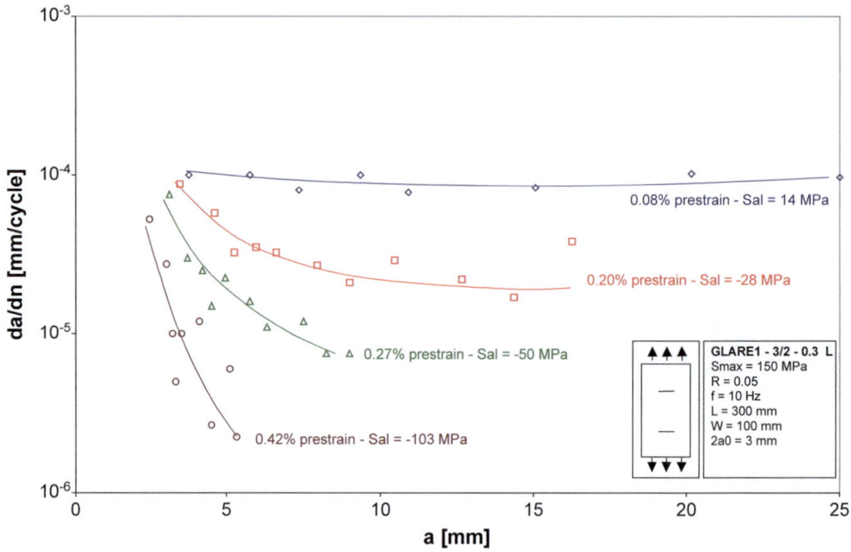

Fig. 9.37 The effect of pre-strain level on the constant amplitude fatigue crack growth rate versus crack lengths for GLARE1-3/2-0.3 tested at 150 MPa, $R = 0.1$ [68, 73]

crack growth rate is reduced in the transient regime of the crack growth phase (Figs. 9.5 or 9.6) as result of fibre bridging. Because less load is transferred over the interface to the bridging fibre layers, the delamination grows significantly more slowly than in the as-cured condition. Smaller delamination areas imply more efficient bridging. As a consequence, the crack growth reduces much more quickly in case of post-stretching as compared to the as-cured condition.

This has been illustrated by the experiments performed by Pegels [73], who tested GLARE1, post-stretched to different levels, in constant amplitude and variable amplitude fatigue crack growth conditions, of which one example is illustrated in Fig. 9.37.

Aside from the variation in residual stresses earlier reported by Verbruggen [74], Pegels did point out that the residual stresses differed for different lay-ups of GLARE1 post-stretched to the same level. With the theory explained in Sect. 4.8, this can be understood. Similarly, the residual stress of as-cured FMLs is dependent on the actual grade and lay-up of the FML.

9.16 Biaxial Fatigue

The discussion of fatigue crack propagation in FMLs in the previous sections has been limited to uniaxially loaded panels. The FMLs considered contain layers with different material properties, which inherently creates a limited biaxial stress state in each individual layer when axially loaded. Although this is inherently the case, all the mechanisms are primarily related to uniaxial fatigue behaviour; the metal layers are cracking under mode I loading similar to uniaxially loaded homogeneous sheets; the fibres elongate and bridge the crack in the loading direction; and the delaminations propagate under mode II in loading direction.

To evaluate fatigue crack growth in FMLs under biaxial fatigue loading in experiments, various options can be considered. The most well-known set-ups are biaxial fatigue machines and barrel tests under internal pressure loading [75]. The first set-up has the complexity of designing an appropriate cruciform specimen geometry where the biaxial stress state must be assured for the entire crack length range of interest. Although the barrel test set-up does not have that complexity, the time required to perform the pressurization results in low test frequencies. This yields long fatigue crack growth tests.

To enable fatigue crack growth testing of FMLs without the disadvantages of the above two experimental set-ups, Heessels [76] tested an experimental set-up that can be applied within a uniaxial fatigue test machine. The concept is illustrated in Fig. 9.38. Two specimens are adhesively bonded at the edges to an internal layer with high stiffness in the transverse direction and little stiffness in the longitudinal loading direction. The prevention of the specimen contraction in the transverse direction effectively imposes a (limited) stress in the direction perpendicular to loading. The loads in the transverse direction reach about 0.3 times the longitudinal load. The Teflon layers between the specimen and the centre layer were added to ensure that the strain fields near the crack tip were not affected by the centre layer.

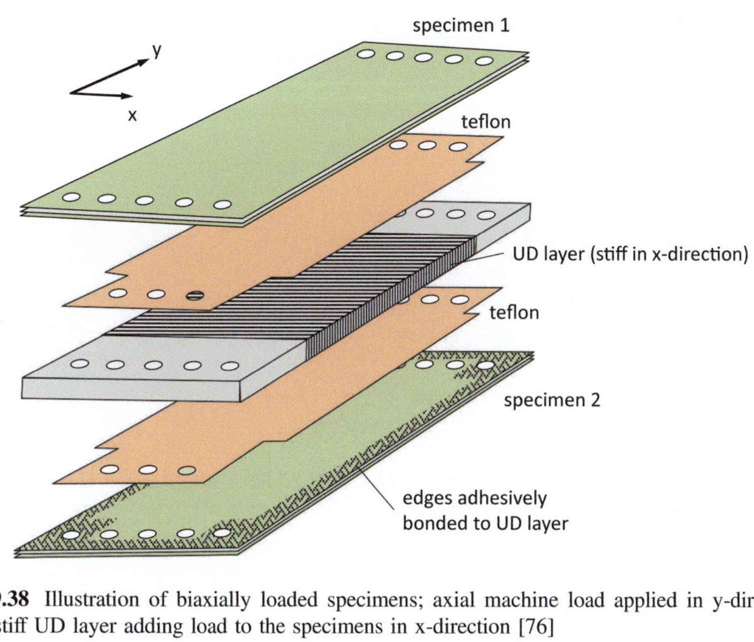

Fig. 9.38 Illustration of biaxially loaded specimens; axial machine load applied in y-direction with stiff UD layer adding load to the specimens in x-direction [76]

The results for ARALL2-2/1-0.3 in comparison with monolithic aluminium 2024-T3 are given in Fig. 9.39. The figure illustrates that for all materials the biaxial loading system has a favourable influence on the crack growth rates. Superposition of both perpendicular uniaxial loads effectively reduces the stress intensity factor at the crack tip and hence the crack growth rate.

The most important observation in this work is that the loads perpendicular to the principal load direction do not negatively affect the fatigue crack growth resistance of unidirectional ARALL, which did not contain reinforcing fibres in the transverse direction.

The main motivation of Heessels' study was the application of FMLs to fuselage structures, which are generally under biaxial loading conditions imposed by pressurization. Hence, although the post-stretched ARALL laminates in Fig. 9.39 performed exceptionally well, as was observed before, they were not considered relevant to the application, because post-stretching entire fuselage panels was not considered practical.

What was interesting to observe, however, is that the load in the transverse direction hardly affected the crack growth of post-stretched ARALL laminates, at least to a significantly lesser extent than for monolithic aluminium and as-cured ARALL.

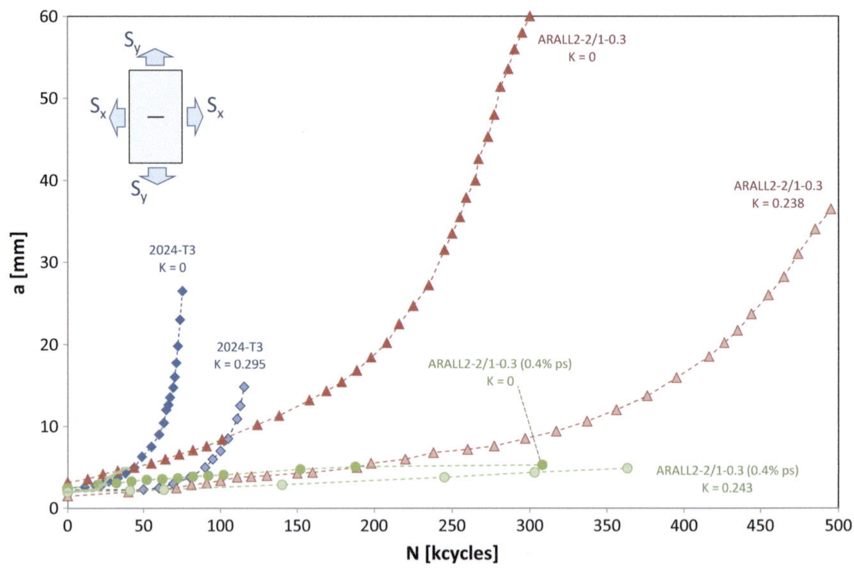

Fig. 9.39 Crack growth curves for ARALL2-2/1-0.3, ARALL2-2/1-0.3 (0.4% post-stretched) and aluminium 2024-T3 under axial ($K = 0$) and biaxial loading ($K = S_x/S_y$), data from [76]

References

1. Alderliesten RC (2007) On the available relevant approaches for fatigue crack propagation prediction in GLARE. Int J Fatigue 29(2):298–304
2. Alderliesten RC (2007) Analytical prediction model for fatigue crack propagation and delamination growth in GLARE. Int J Fatigue 29(4):628–646
3. Alderliesten RC (2005) Fatigue crack propagation and delamination growth in GLARE. PhD dissertation, Delft University of Technology, Delft
4. Wilson G (2013) Fatigue crack growth prediction for generalized fiber metal laminates and hybrid materials. PhD dissertation, Delft University of Technology, Delft
5. Mattheij PC (1986) Constant amplitude fatigue tests on ARALL 2024-T3. Master thesis, Delft University of Technology, Delft
6. Roebroeks GHJJ (1991) The development of a fatigue insensitive and damage tolerant material. PhD dissertation, Delft University of Technology, Delft
7. Alderliesten RC (2001) Fatigue. In: Vlot A, Gunnink JW (eds) Fibre metal laminates, an introduction. Springer Science+Business Media, Dordrecht
8. Marissen R (1988) Fatigue crack growth in ARALL, a hybrid aluminium-aramid composite material, crack growth mechanisms and quantitative predictions of the crack growth rate. PhD dissertation, Delft University of Technology, Delft
9. Wilson GS, Alderliesten RC, Benedictus R (2010) Steady-state crack growth in hybrid fiber metal laminates as a tool for design. In: Proceedings of international SAMPE symposium and exhibition, SAMPE 2010, Seattle, WA, 17–20 May 2010
10. Bied-Charreton AD (2015) Friction stir welding effects of defects in GLARE. MSc thesis, Delft University of Technology, Delft
11. Alderliesten RC (1999) Development of an empirical fatigue crack growth prediction model for the fibre metal laminate GLARE. MSc thesis, Delft University of Technology, Delft
12. Molent L, McDonald M, Barter S, Jones R (2008) Evaluation of spectrum fatigue crack growth using variable amplitude data. Int J Fatigue 30:119–137
13. Guo YJ, Wu XR (1998) A theoretical model for predicting fatigue crack growth in fibre-reinforced metal laminates. Fatigue Fract Eng Mater Struct 21:1133–1145
14. Guo YJ, Wu XR (1999) Bridging stress distribution in center-cracked fiber reinforced metal laminates: modelling and experiment. Eng Fract Mech 63:147–163
15. Roebroeks GHJJ (2007) The development of central, DTAS 2007
16. In 't Velt JC (1987) Fibre failure mechanism in ARALL. MSc dissertation, Delft University of Technology, Delft
17. Smulders EHM (1988) Fibre fracture mechanism in ARALL laminates with aramid fibres. Master Thesis, Delft University of Technology, Delft
18. Rodi R (2012) The residual strength failure sequence in fibre metal laminates. PhD dissertation, Delft University of Technology, Delft
19. Khan S (2013) Fatigue crack & delamination growth in fibre metal laminates under variable amplitude loading. PhD dissertation, Delft University of Technology, Delft
20. van Lipzig HTM (1973) Retarding the growth of fatigue cracks. Graduation report, Technische Hogeschool Delft (in Dutch)
21. Homan JJ (2001) Crack growth properties of thin aluminium sheets. Report B2V-01-16 (issue 2). Delft University of Technology, Delft
22. Homan JJ (2001) Crack growth properties of thin aluminium sheets at various temperatures. Report B2V-02-39. Delft University of Technology, Delft
23. Slagt J (1995) Small crack growth rates in a 2024T3 sheet with a thickness of 0.3 mm. MSc thesis, Delft University of Technology, Delft
24. Fokker Technical Handbook Part 3
25. Handbuch Struktur Berechnung

26. de Koning AU (2000) Analysis of the fatigue crack growth behaviour of "through the thickness" cracks in fibre metal laminates. Report NLR-CR-2000-575, National Aerospace Laboratory, NLR
27. Schijve J (1981) Some formulas for crack opening stress level. Eng Fract Mech 14:461–465
28. Schijve J (1986) Fatigue crack closure, observations and technical significance. Report LR-485, Delft University of Technology, Delft
29. Alderliesten RC (2014) The explanation of stress ratio effect and crack opening corrections for fatigue crack growth in metallic materials. Adv Mater Res 891–892:289
30. Alderliesten RC (2016) How proper similitude can improve our understanding of crack closure and plasticity in fatigue. Int J Fatigue 82:263–273
31. Alderliesten RC (2015) How proper similitude principles could have improved our understanding about fatigue damage growth. In: Proceedings of the 28th ICAF symposium, Helsinki, 3–5 June 2015
32. Irwin GR (1967) Analysis of stresses and strains near the end of a crack traversing a plate. Trans ASME J App Mech 24:361
33. Isida M (1955) On the tension of a strip with a central elliptic hole. Trans Jpn Soc Mech Eng 21(107):507–518
34. Isida M (1966) Stress-intensity factors for the tension of an eccentrically cracked strip. J Appl Mech 33:674
35. Feddersen C (1967) Discussion to: plane strain crack toughness testing. ASTM Spec Tech Publ. No. 410, 77
36. Koiter WT (1959) An infinite row of collinear cracks in an infinite elastic sheet. Ingenieur-Archiv 28(1):168–172
37. Dixon JR (1960) Stress distribution around a central crack in a plate loaded in tension: effect of finite width of plate. J R Aeronaut Soc 64:141–145
38. Jones R, Peng D, Singh Raman RK, et al. (2015) JOM 67: 1385–1391
39. Schijve J (2001) Fatigue of structures and materials. Kluwer Academic Publishers, The Netherlands
40. Gonesh K (2003) Evaluation of correction factors for bore holes and pin loaded holes in GLARE, Report B2V-02-11. Delft University of Technology, Delft
41. Gonesh K (2002) Stress intensity correction factors for pin-loaded GLARE plates, Report B2V-02-59. Delft University of Technology, Delft
42. Quinn BG (2004) Delamination of GLARE. Masters of Engineering Thesis, Queen's University of Belfast, School of Aeronautical Engineering, Belfast
43. Thibault-Liboiron K, Alderliesten RC, Benedictus R, Bocher PH (2007) Off-axis crack propagation and delamination growth in FML's. In: 6th Canadian—international conference on composites CANCOM 2007, Winnipeg, Manitoba, Canada
44. Thibault-Liboiron K (2010) Off-axis edge-crack propagation and delamination in fibre metal laminate GLARE. Université du Québec, École de technologie supérieure, Montreal, Canada, Rapport de projet technique
45. Homan JJ (2004) Stress intensity correction factors for Fibre Metal Laminates. Report B2V-04-07. Delft University of Technology, Delft
46. Bouwman VP, de Koning AU (2000) Fatigue crack growth methodology of 'through the thickness' cracks in fiber metal laminates. Report NLR-CR-2003-032, National Aerospace Laboratory, NLR
47. Toi R (1995) An empirical crack growth model for fiber/metal laminates. In: Proceedings of the 18th symposium of the international committee on aeronautical fatigue, Melbourne, Australia, pp 899–909
48. Takamatsu T, Matsumura T, Ogura N, Shimokawa T, Kakuta Y (1999) Fatigue crack growth properties of a GLARE3-5/4 fiber/metal laminate. Eng Fract Mech 63:253–272
49. Newman JC (1976) Predicting failure of specimens with either surface cracks or corner cracks at holes, NASA TN D-8244
50. Tada H, Paris PC, Irwin GR (1985) The stress analysis handbook, 2nd edn. Paris Productions Inc., St. Louis

51. Daandels D (2003) Search for (DK threshold of GLARE. Preliminary thesis, Delft University of Technology, Delft
52. Alderliesten RC, Rans CD (2008) The meaning of threshold fatigue in fibre metal laminates. Int J Fatigue 31(2):213–222
53. Gonesh KAM (1999) Crack growth prediction of surface cracks in GLARE. MSc thesis, Delft University of Technology, Delft
54. Alderliesten RC, Homan JJ (2003) Fatigue crack growth behaviour of surface cracks in GLARE. In: Varvani-Farahani A, Brebbia CA (eds) Fatigue damage of materials—experiments and analysis. WIT Press, Southampton, UK
55. Guo YJ, Wu XR (1999) A phenomenological model for predicting fatigue crack growth in fiber-reinforced metal laminates under constant-amplitude loading. Compos Sci Tech 59:1825–1831
56. Rose LRF (1981) An application of the inclusion analogy for bonded reinforcements. Int J Solids Struct 13:827–838
57. de Koning AU (2000) Fatigue crack growth of part through the thickness cracks in GLARE 3 and GLARE 4B coupons—final report GTO subproject 2.4.2, Revised edition, NLR-CR-2000-078
58. Homan JJ (2004) Fatigue and damage tolerance methods applied in the FML F&DT Toolbox, Report B2V-04-08. Delft University of Technology, Delft
59. Mortier W, Homan JJ (2004) Crack propagation in surface cracks and part-through cracks, Report B2V-02-31. Delft University of Technology, Delft
60. Randell CE (2005) Subsurface fatigue crack growth in GLARE fibre metal laminates. PhD dissertation, Delft University of Technology, Delft
61. Spronk SWF (2013) Predicting fatigue crack initiation and propagation in GLARE reinforced frames. MSc thesis, Delft University of Technology, Delft
62. van der Linden A (2015) Residual strength of a FML reinforced frame member. MSc thesis, Delft University of Technology, Delft
63. Gonesh KAM (2001) Test results and evaluation of crack propagation in off-axis GLARE, Report B2 V-01-24. Delft University of Technology, Delft
64. Gonesh K (2002) Additional test results and evaluation of crack propagation in off-axis GLARE, Report B2 V-02-02. Delft University of Technology, Delft
65. Gupta M (2015) Directionality of Damage Growth in Fibre Metal Laminates and hybrid Structures. PhD dissertation, Delft University of Technology, Delft
66. Maretti V (2015) Surface crack path in FML: off axis propagation under fatigue loading. Thesis, Politecnico di Milano. MSc dissertation, Italy
67. Bradshaw RD, Gutierrez SE (2007) Characterization of fatigue crack initiation and growth in hybrid aluminium–graphite fibre composite laminates using image analysis. Fatigue Fract Eng Mater Struct 30:766–781
68. Alderliesten RC, Woerden HJM (2003) Loads history effects during fatigue crack propagation in GLARE. In: Proceedings of the 22th symposium of the international committee on aeronautical fatigue (ICAF), Luzern, Switzerland
69. Khan SU, Alderliesten RC, Rans CD, Benedictus R (2010) Application of a modified wheeler model to predict fatigue crack growth in fibre metal laminates under variable amplitude loading. Eng Fract Mech 77(9):1400–1416
70. Khan SU, Alderliesten RC, Benedictus R (2009) Linear damage accumulation for predicting fatigue in fiber metal laminates. J Aircr 46(5):1706–1713
71. Khan SU, Alderliesten RC, Benedictus R (2011) Delamination in fiber metal laminates (GLARE) during fatigue crack growth under variable amplitude loading. Int J Fatigue 33 (9):1292–1303
72. Khan SU, Alderliesten RC, Benedictus R (2009) Post-stretching induced stress redistribution in fibre metal laminates for increased fatigue crack growth resistance. Compos Sci Tech 69(3–4):396–405
73. Pegels CS (1995) A study on the residual stress of GLARE 1. Master's thesis, Delft University of Technology, Delft

74. Verbruggen M (1983) Relaxation due to temperature, moisture and external loading—preliminary results, Memorandum M-491. Delft University of Technology, Delft
75. Müller RPG (1995) An experimental and analytical investigation on the fatigue behaviour of fuselage riveted lap joints, the significance of the rivet squeeze force, and a comparison of 2024-T3 and GLARE 3. PhD dissertation, Delft University of Technology, Delft
76. Heessels F (1987) The fatigue behaviour of ARALL in bi-axial stress state (in Dutch), Internship report, Delft University of Technology, Delft

Chapter 10
Residual Strength

Abstract The residual strength in FMLs can be described differently depending on the geometry of the initial damage. Theories and prediction models are presented for through-cut cracks and fatigue through cracks, based on R-curve and critical crack tip opening angle concepts. For part-through cracks, linear relations based on a damage ratio are presented, while for surface cracks a method is presented based on the through-thickness strain distribution, in combination with the corresponding stress concentration factor. In the end the damage tolerance after impact is discussed, explaining how the residual stress state in the plastically deformed dent improves the residual strength.

10.1 Introduction

Residual strength concerns the remaining strength of a structure or material in presence of damage. In FMLs this damage may have various configurations, as illustrated in Fig. 10.1. Damage may be caused by foreign object damage, like the through-cut crack and impact damage, or it may have been created through usage, induced by design fatigue loading. In the latter case, the location may correspond to stress concentrations of fastener holes, or maintenance-induced scratches.

To assess the residual strength of damaged FML shells, various damage scenarios therefore need to be considered. Ideally, the residual strength of each of these scenarios can be described and assessed with the same methodology. However, ample research has illustrated that specific damage scenarios require specific methodologies. The different damage scenarios that will be discussed in this chapter are illustrated in Fig. 10.1.

The first crack configuration is referred to as through-cut crack and represents the accidental or foreign object damage that creates a through-cut through the panel. The prime difference between this case and the fatigue through crack is that over the entire crack length in the metal layers the fibres are also cut. Hence, the mechanism of fibre bridging, as discussed in the previous chapter, is mostly absent. This case is discussed in Sect. 10.2

R. Alderliesten, *Fatigue and Fracture of Fibre Metal Laminates*, Solid Mechanics and Its Applications 236, DOI 10.1007/978-3-319-56227-8_10

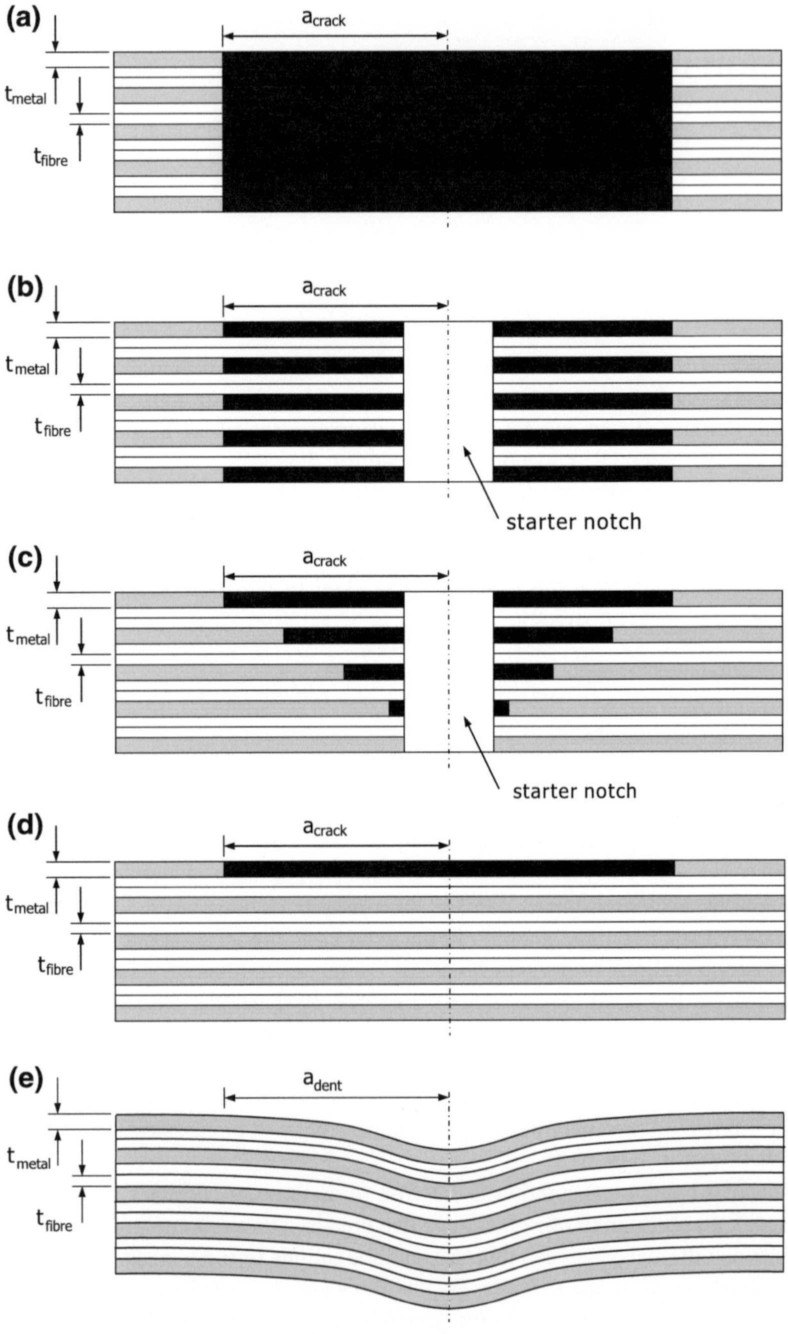

Fig. 10.1 Most significant residual strength scenarios in FMLs: through-cut crack (**a**), fatigue through crack (**b**), part-through crack (**c**), surface crack (**d**), and impact damage (**e**)

The second case is the fatigue through crack, which represents the fatigue crack configuration under in-plane tensile fatigue loading. The methodology for this crack configuration resembles the method discussed in the previous chapter, utilizing superposition principles. This case is discussed in Sect. 10.3.

An alternative methodology that has been proposed for this configuration and the part-through crack configuration is based on the accumulated level of damage. Hence, the third methodology represents the fatigue through crack configuration that may occur under the application of tensile and bending loading. As discussed in Chap. 9, the superposition of tension and bending stresses result in cracks with different lengths in each layer. This case is discussed in Sect. 10.4.

The fourth configuration represents the case of a scratched panel. The damage may remain in the outer metal surface only, or (depending on scratch depth and metal layer thickness) may also occur in the subsurface composite plies. This case is discussed in Sect. 10.5.

10.2 Through-Cut Cracks

In monolithic metals, damage due to fatigue and due to foreign objects is traditionally treated identically. This is possible, because a fatigue crack in a monolithic thin-walled structure often represents a through crack, which resembles a crack created by, for example, a turbine blade cutting through the panel. For FMLs, however, a distinction has to be made, because the fatigue crack configuration contains bridging fibres over most of its crack length, leading to significantly higher strength levels compared to the through-cut crack configuration. One could consider the application of through-cut crack methodology to fatigue cracks conservative, but it turns out to be extremely conservative.

This is illustrated, for example, by De Vries [1], who tested various GLARE2A-2/1-0.3 and GLARE3-3/2-0.3 panels with different initial crack lengths and crack length over width ratios a/W. An example of his results is given in Fig. 10.2. The strength of a panel containing a fatigue crack is often 35–50% higher than the strength of the same panel containing a through-cut crack of the same length. Hence, in order to fully benefit from the superior damage tolerance characteristics of FMLs, one has to treat the fatigue crack configuration separately.

The residual strength of through-cut cracks has been the subject of many studies, of which most present experimental observations. Only a limited number of studies propose methods to predict the residual strength. Among the early studies are Prasad Kadiyali et al. [2], who pointed out that notched ARALL laminates must be evaluated using elastic–plastic rather than linear elastic methods, Macheret and Teply [3], who proposed to combine a flaw criterion with the tensile strength of the composite layers, and Macheret et al. [4], who first proposed to determine the residual strength with a modified net section criterion, but later presented an R-curve approach based on linear elastic fracture mechanics [5]. This latter concept

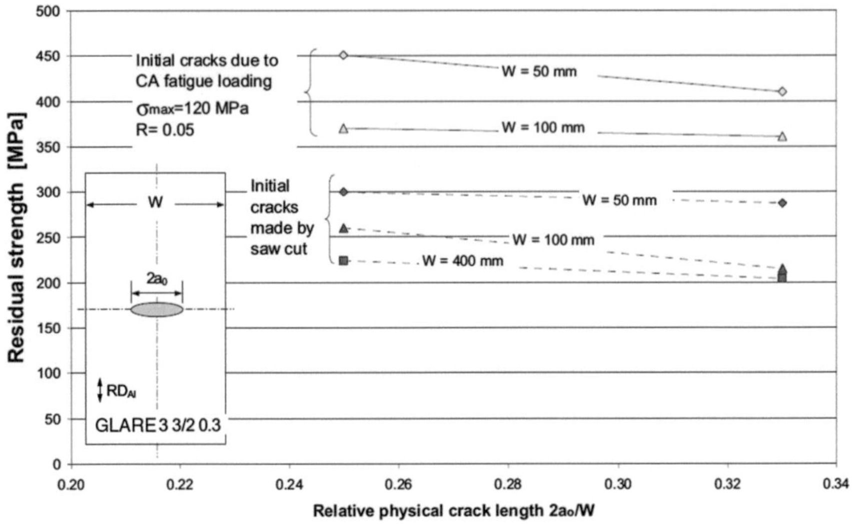

Fig. 10.2 Difference in residual strength of GLARE3-3/2-0.3 containing a through-cut crack or a fatigue through crack [1]

has been exploited further by a number of researchers such as Vermeeren [6, 7] and afterwards De Vries et al. [1, 8–12].

Next to these concepts, Jin and Batra [13] proposed a method based on the Dugdale strip yield zone [14] in combination with a damage zone for the composite layers, and Afaghi-Khatibi et al. [15–18] proposed an effective crack growth model based on a damage zone criterion, which is equivalent to a cohesive zone formulation, while Castrodeza et al. [19–22] proposed to determine the residual strength with the fracture toughness properties determined with traditional compact tension specimens.

Despite validations with experimental data, and claims on accuracy, each of the above methodologies has their drawbacks. Take, for example, the methodology based on the fracture toughness properties. The fracture toughness test for FML is not straightforwardly defined, unless the laminate is considered a homogeneous material with a crack running through the entire thickness, an assumption that has been disputed by most of the other authors. In particular, the phenomenological nature of this concept and its inapplicability to FML fatigue crack configurations, common practice for monolithic metal structures, seem a valid reason not to exploit this direction further [23]. Instead, the individual FML constituents with their respective fracture mechanisms should be acknowledged in the approaches.

This leaves basically two approaches based on linear elastic fracture mechanics that one can adopt to describe and predict the residual strength of FMLs with through-cut cracks. The first method utilizes above-mentioned R-curves or K_R-curves as input material characteristics and predicts the corresponding strength based on the strength versus crack extension curves predicted using stress intensity factors or strain energy release rates.

The other method is based on the relationship between crack (tip) opening and crack tip stress intensity. Here, the crack tip opening angle is plotted against the crack extension, serving as input characteristic in the prediction methods. This method is referred to as the (critical) crack tip opening angle method [1, 23–27].

10.2.1 Fracture Mechanisms

In general, the initiation of the fracture process under quasi-static loading is similar to that in monolithic metals, although the intact fibres ahead of the notch may affect the deformation field. With increasing loads, a large amount of plastic deformation occurs in the notch tip vicinity, which is often accompanied by blunting of the notch tip. These two mechanisms are illustrated in Fig. 10.3.

Initially, the blunting and plasticity delay the initiation of new fracture surfaces, up to the point that from the blunted crack tip a near fracture is created; see Fig. 10.3d. Initially, this fracture surface seems aligned with one of the two branches of plasticity for few tenths of millimetres until it progresses perpendicular to the main loading direction.

The crack propagation phase is characterized by a continuation of the development of plasticity with increasing loading, while the crack tip blunting decreases. In the case of through-cut cracks, the formation of new fracture surfaces is accompanied by fibre failure, because negligible delamination develops at the interface between the metal and fibre layers.

Crack propagation in the thin sheets of FMLs is similar to that in monolithic metals, where rapidly a transition takes place between a flat mode I type of fracture

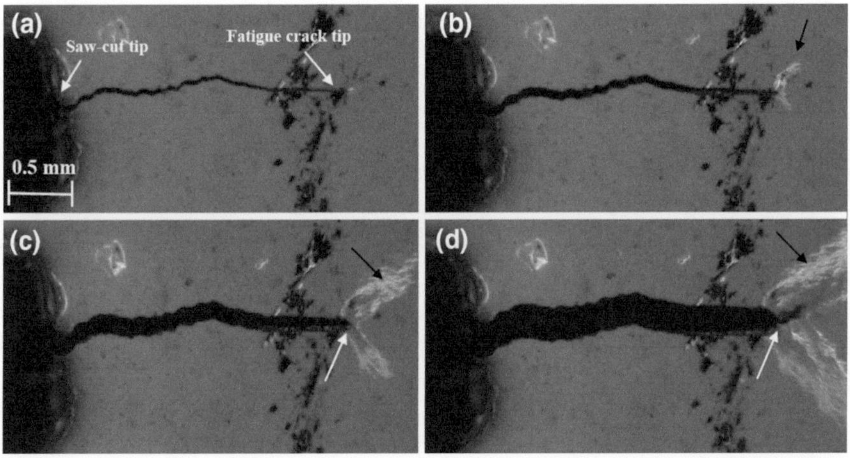

Fig. 10.3 Illustration of crack tip blunting (*white arrow*) and plasticity in front of the crack tip (*black arrows*) in GLARE3-3/2-0.4 [23]

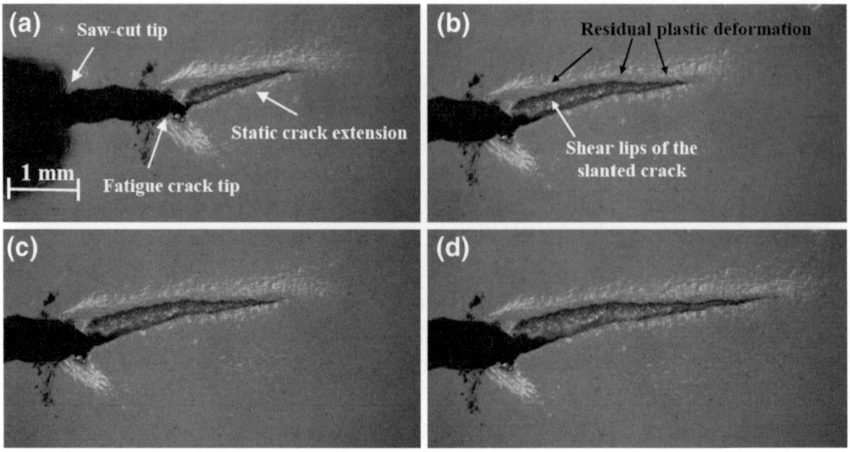

Fig. 10.4 Illustration of crack growth under increasing quasi-static load in GLARE3-3/2-0.4 [23]

surfaces and slanted fracture surfaces that open in a combination of mode I and mode III. Figure 10.4 illustrates this transition in image A and B, where the flat fatigue crack has slanted into a crack with shear lips visible.

Although the FMLs tested with a through-cut configuration hardly show delamination of the plies, resulting in fibre fracture along with the propagating crack in the metal layers, some observations of delamination in front of the crack tip have been reported [1, 6, 23].

The mechanism for this delamination is discussed in more detail in Chap. 8, but here it must be said that similar to fatigue crack growth, delaminations in FMLs are to some extent beneficial. In the small delaminated area in front of the crack tip, deformations in the fibre layers do not need to be compliant to the plastic deformation in the metal layers. They may stretch over slightly longer lengths, delaying fibre fracture.

10.2.2 K_R-Curve or R-Curve Concept

The K_R-curve concept was originally introduced by Irwin based on Griffith's energy criterion for plane stress fracture [28]. The concept states that the crack grows when the driving energy exceeds the crack resistance energy. With the method, stable crack extension and instability can be predicted while accounting for a limited amount of plasticity.

The standard for monolithic metals [25] explains that the unstable crack extension can only be predicted using an effective crack length. To correct the measured physical crack length, either the Irwin correction or the compliance correction can be adopted.

The studies by Vermeeren [6, 7], Zaal [29] and De Vries et al. [1, 8–12] have demonstrated that this R-curve concept can also be adopted for FMLs. Because the method relies on linear elastic fracture mechanics, net section yielding should be avoided, which implies either large widths, or small crack lengths. For example, De Vries and Holleman [9] demonstrated that the effective crack length a_{eff} in case of net section yielding comprises mostly plasticity ahead of a fairly small physical crack length.

Because the ASTM standard [25] prescribes that the K_R-curve should be based on effective crack lengths rather than physical crack lengths for instability predictions, a correction must be applied. Some of the mentioned studies highlight the advantages and disadvantages of the Irwin correction and compliance correction for this purpose.

The Irwin correction is based on the linear elastic fracture mechanics theory, which means that the effective crack length is calculated by adding the theoretical plastic zone size to the physical crack length. Hence, this method becomes invalid once substantial yielding occurs.

This is to a lesser extent the case for the compliance correction, which utilizes the unique relationship between the crack opening displacement COD and the effective crack length a_{eff}. Instead of using theoretical calibration equations, the compliance calibration can be performed experimentally by loading panels with different crack lengths in the elastic range and determining the elastic slopes.

In case of compliance calibration of FML panels, one has to be aware that the standardized compliance calibration using the theoretical Eftis–Liebowitz equation [30] requires a modification to account for the fact that FMLs are not isotropic [31]. This is discussed in more detail in Sect. 10.2.3.

The stress intensity factor calculated in the K_R-curve approach utilizes the finite width correction factor, like the ones from Feddersen [32] or Dixon [33]. Because fibres fracture along with the crack growth in the metal layers, one could assume the entire laminate to consist of a monolithic panel with equivalent properties, containing a crack described by the stress intensity factor. The crack size with respect to panel dimensions then indeed can be captured with well-known finite width corrections. This simplification can only be justified if fibre bridging does not occur. In case fibres bridge the crack, the complexity of the problem increases similar to the fatigue crack configuration, as discussed in Sect. 9.7.

Similar to fatigue crack growth, the crack growth behaviour in case of residual strength is governed by the constituent properties and the lay-up configuration. De Vries [1] studied the relationship between the R-curves and the FML constituents and demonstrated that each GLARE grade and lay-up had its own R-curve.

In order to develop a prediction model, De Vries phenomenologically related each FML R-curve to the R-curve of monolithic aluminium used as constituent in the FML. Aside from the mechanical properties of both metal and composite layers, for which the rolling direction of the metal had to be considered with respect to the loading direction, and the quasi-static delamination resistance of the interface, De Vries identified the influence of the metal volume fraction of the FML together with

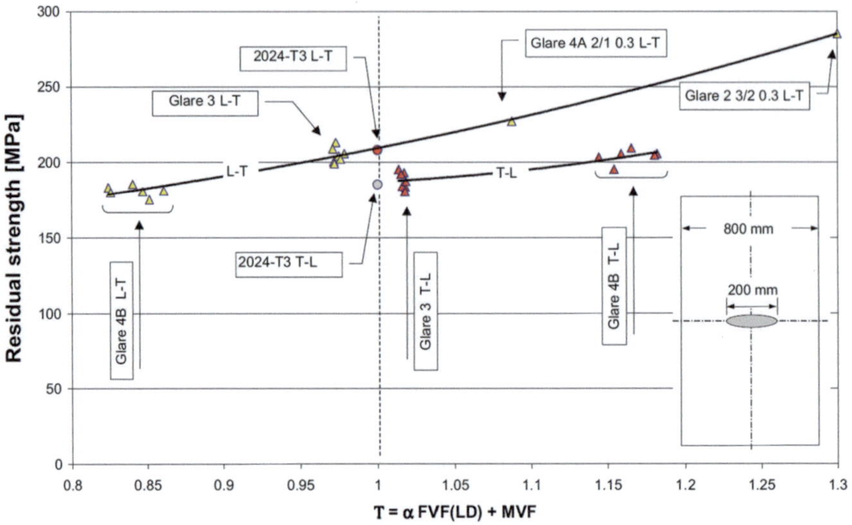

Fig. 10.5 Residual strength as function of T [1]

the fibre volume fraction of the composite layers oriented in loading direction. These parameters he captured with

$$T = \alpha \mathrm{FVF}_{\mathrm{LD}} + \mathrm{MVF} \tag{10.1}$$

This relationship illustrates that the contribution of both metal layers and fibre layers oriented in loading direction could not be captured straightforwardly with a rule of mixtures, like for the mechanical properties (Sect. 4.11) and blunt notch strength (Sect. 5.8). Instead, an empirical parameter α was used to tune the contribution of both metal and fibre layers to the residual strength of the FML.

Figure 10.5 illustrates that the rolling direction of the aluminium constituent in GLARE has to be accounted for in the proposed methodology of De Vries, by testing the monolithic aluminium panels both in rolling direction and transverse to the rolling direction.

The K_R-curve of the FML could then be determined with

$$K_{R_{FML}}(T) = M(T) K_{R_{metal}} \tag{10.2}$$

With

$$M(T) = \frac{aT^2 + bT + c}{a + b + c} \tag{10.3}$$

where for standard GLARE laminates the coefficients α, a, b and c are reported by De Vries [1] for both panels in tested in *L-T* and *T-L* direction separately. These coefficients could only be determined after performing a rather great number of

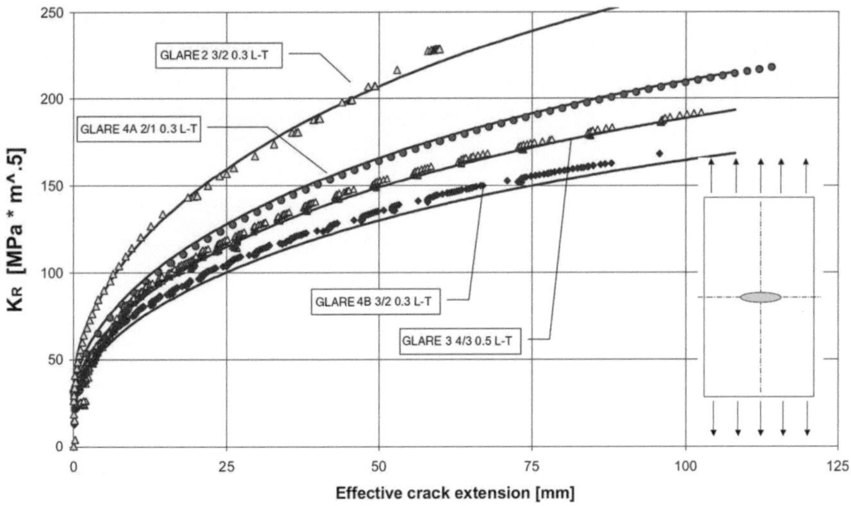

Fig. 10.6 Comparison between predicted K_R-curves and curves determined with experiments for various GLARE grades [1]

large panel residual strength tests, each with its own lay-up configuration, by interpolation between all results. Example curves determined by De Vries are given in Fig. 10.6.

This methodology for predicting the residual strength of GLARE panels was successfully applied for the certification of GLARE application to the fuselage of the Airbus A380. However, the major limitation of this methodology is that it is entirely empirical and phenomenological. In fact, one could argue that the method does not constitute a prediction model, but rather an interpolation method. Hence, in order to apply this methodology to FMLs with different constituents, first a number of residual strength tests on large panels must be executed in order to refit Eqs. (10.1–10.3).

To overcome this shortcoming Rodi [23] decided to approach the residual strength of FMLs from a different angle by relating the residual strength problem to the critical tip opening of the crack, as explained in Sect. 10.2.3.

10.2.3 Compliance Calibration for Orthotropic FML Panels

To determine the experimentally measured physical crack length to the effective crack length a_{eff} adopted in the K_{R}-curve approach, one can apply compliance calibration. The compliance calibration suggested in the ASTM standard [34] and proposed by Eftis and Liebowitz [30] assumes isotopic material properties. Because FMLs are orthotropic in nature, Testi [31] studied the linear elastic fracture mechanics basis for the compliance calibration in order to develop a theoretical orthotropic expression of it that would be valid for FMLs.

Testi using the classical laminate plate theory explained in Chap. 4 for the anisotropic elastic part of the residual strength problem, together with the anisotropic theory of cracks [35, 36]. With this Testi presented a relationship between the mode *I* stress intensity factor K_{I} and the strain energy release rate G_{I}

$$G_I = \alpha \frac{K_I^2}{E} \tag{10.4}$$

with

$$\alpha = \frac{1}{\sqrt{2}} \sqrt{\frac{E_1}{E_2}} \sqrt{\sqrt{\frac{E_1}{E_2}} - v_{12} + \frac{E_1}{2G_{12}}} \tag{10.5}$$

Here, α seems to resemble part of Eqs. (5.4) and (7.10). Taking the local compliance as it relates directly to the measured crack opening displacement, Testi derived the final orthotropic compliance calibration curve as

$$\frac{E_2 \mathrm{COD}}{\sigma W} = 2\alpha \frac{E_2}{E_1} \frac{2a}{W} \left[1 - 0.025 \left({}^{2a}\!/_{W} \right)^3 + 0.06 \left({}^{2a}\!/_{W} \right)^4 \right] \sqrt{\sec\left(\frac{\pi a}{W} \right)} \tag{10.6}$$

in which the latter part comprises Tada's modification to Feddersen's finite width equation. To account for the panel's length, Testi suggested adding the finite panel length correction using the method of Isida [37] in case the panel length is less than three times its width. This correction acknowledges the axial rigidity conditions provided by the significantly higher stiffness of the clamps compared to that of the specimen.

Equation (10.6) is illustrated in Fig. 10.7 for aluminium Alclad 2024-T3, GLARE2, GLARE3, and GLARE4. Unfortunately, Testi does not specify well whether the GLARE4 represents the case of GLARE4A-2/1-0.3 tested in the *L-T* direction, but this seems the most likely case.

It was mentioned before that less attention is paid here to the approach proposed by Castrodeza et al. [19–22]. The prime reason is that this concept treats the FML as a homogeneous material, which could work for the through-cut crack configuration, but not for the fatigue crack configuration [23]. Assessing the residual strength of fatigue cracks based on the through-cut crack configuration underestimates the structural performance and the high crack resistance of FMLs containing fatigue cracks. The level of underestimation for that case can be easily illustrated with Fig. 10.2; one can utilize at most 70% of the full strength, instead of the full strength and damage tolerance.

Nonetheless, it is worthwhile to consider certain aspects related to the fracture toughness approach. In [38] for example, Castrodeza et al. discuss the elastic compliance for the crack resistance curves. Because their concept is based on the standard fracture toughness tested with compact tension specimens, they relate to a different ASTM standard than Testi [31] and the researchers after him.

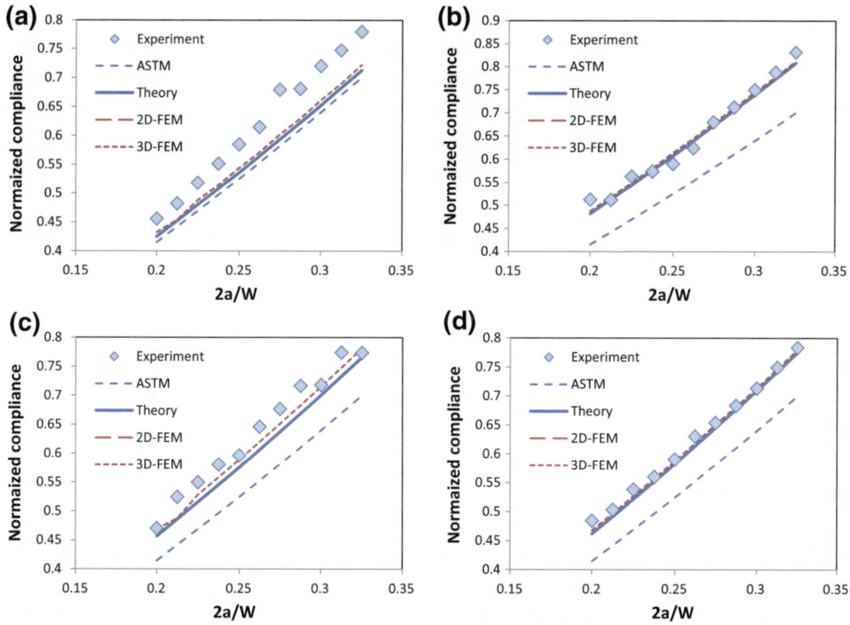

Fig. 10.7 Comparison between the compliance calibration curves from ASTM [34], the orthotropic theory (Eq. 10.3), 2D- and 3D-FEA for Alclad 2024-T3 (**a**), GLARE2-3/2-0.3 (**b**), GLARE3-3/2-0.3 (**c**) and GLARE4-2/1-0.3 (**d**); specimen width is 400 mm and length 800 mm [31]

Evaluating the applicability of ASTM standard E1820 [39], Castrodeza et al. demonstrated that the procedure can similarly be applied to the fracture toughness tests on FML. This observation seems to correspond to the observations of establishing compliance calibration with residual strength tests on centre crack tension panels [1, 23]; the correlation is similar to isotropic materials, but requires a modification for orthotropic conditions as explained by Testi.

10.2.4 Crack Tip Opening Angle or CTOA Concept

More recently, Rodi [23] developed residual strength prediction models entirely based on the crack tip opening angle characteristics. However, he was not the first to adapt the critical CTOA, which was demonstrated by Newman et al. [24] to be suitable for predicting stable crack growth and instability in the fracture process.

A first study into the application of the CTOA method for FMLs was performed by Longhi [26, 27] on 400 mm wide GLARE3-3/2-0.3 panels containing a central notch. Longhi reported an average angle of 7.4° over the constant part of the CTOA data. De Vries [1, 11] continued this research with experiments on a variety of GLARE laminates, ranging from relatively thin GLARE3-3/2-0.3 to fairly thick

GLARE4B-8/7-0.5. The recorded CTOA ranged between 4° and 8°, which De Vries considered too much of a scatter band to exploit this method straightforwardly. Furthermore, by adopting the CTOA concept in a finite element analysis, De Vries [1] illustrated that none of the predicted force versus crack extension curves based on crack tip angles between 6° and 8° correlated well with the experimentally measured curves. Hence, he recommended further investigation before application.

The strength of the critical CTOA method is that the crack tip angle represents a local parameter with a physical meaning. This local parameter may be assumed independent of global geometry, like panel dimensions, or the presence of stringers, frames or doublers, and could be correlated to finite element analyses. The absence of correlation between the predictions and tests reported by De Vries [1] could be explained by the fact that De Vries assumed a constant critical CTOA.

Instead of being constant over the entire range of crack lengths, the CTOA has been reported to initially start at a higher value and decrease until a constant value is reached. Rodi [23] pointed out that incorporating this initially higher value in the predictions could explain the absence in correlation reported by De Vries.

Rodi performed CTOA measurements in both monolithic aluminium 2024-T3 and GLARE2-3/2-0.3 centre crack tension panels. In order to ensure consistency in the measurement, the CTOA was defined with three distances measured from the crack tip visible at the specimen surface [40], as illustrated in Fig. 10.8.

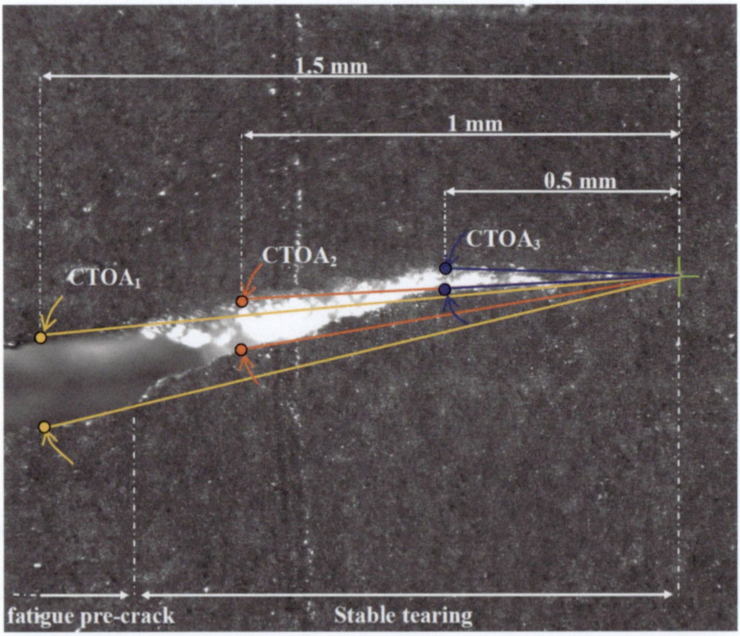

Fig. 10.8 Illustration of the three different CTOA measurement distances adopted in Fig. 10.9 [23]

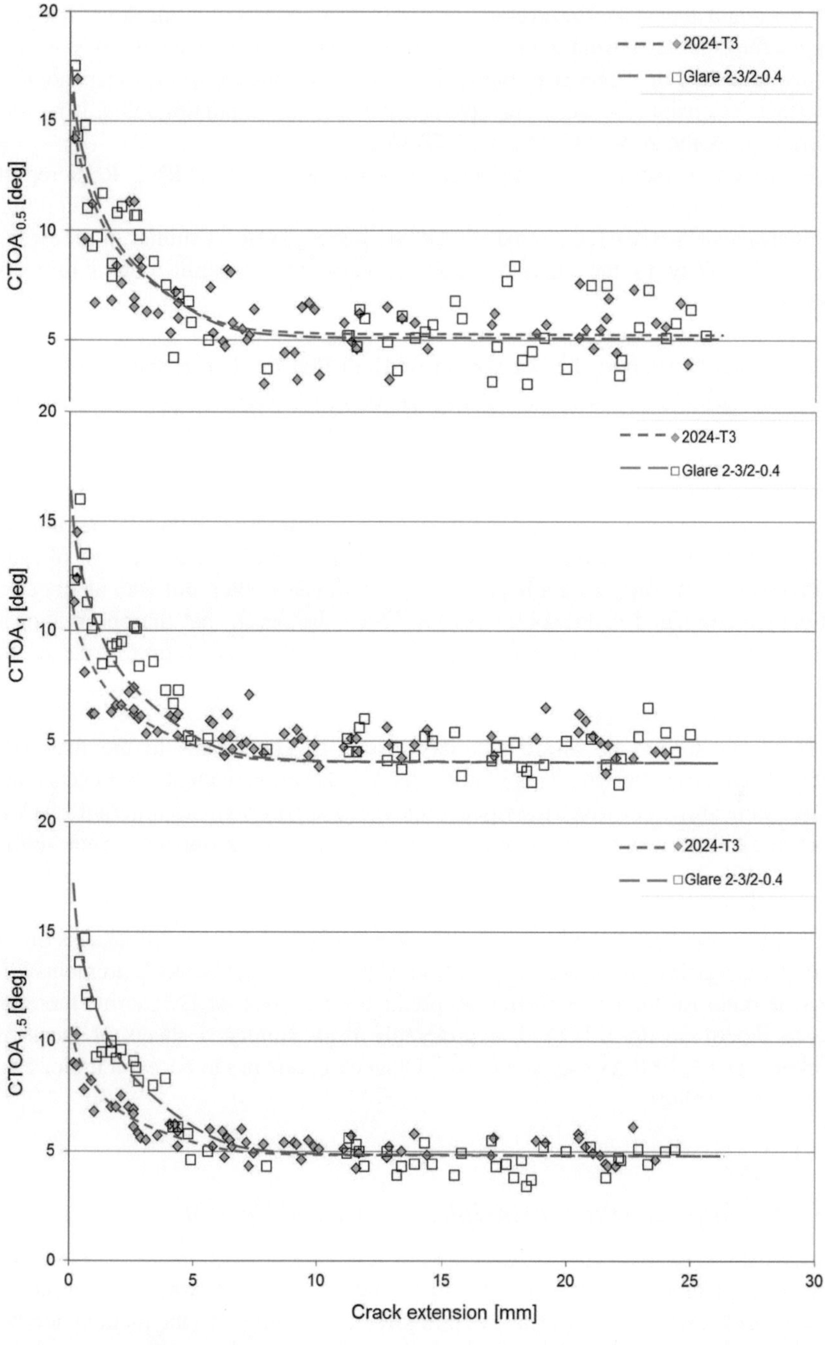

Fig. 10.9 CTOA curves for aluminium 2024-T3 and GLARE2-3/2-0.4 measured with the three distances illustrated in Fig. 10.8 [23]

One could then take the average of the CTOA measured at all three distances against the crack extension to obtain the typical curve. This curve initially starts at a higher value and then decreases until it levels off to a constant value, independent of the crack extension. Because this constant value is the minimum value, it is often referred to as the critical CTOA, i.e. $CTOA_c$.

However, in particular for unidirectional FMLs, like GLARE2, Rodi reports observing differences between the curves obtained from the monolithic aluminium laminates and the GLARE laminates. Although the $CTOA_c$ is similar, it seems that the initial decay in the curve is different between monolithic aluminium and GLARE.

Evaluating the curves for the three measurement distances in Fig. 10.8 separately seems to support the conclusion of Rodi that the longer the length in the CTOA definition, the more the fibre bridging will influence the CTOA measurement in FMLs. The bridging fibres in subsurface layers thus affect the crack opening profile of the aluminium surface layers, influencing the CTOA measurements over larger measurement distances.

In conclusion, one therefore has to make a compromise; either taking a small measurement distance from the crack tip to define the CTOA at the cost of more scatter in the data, or taking a larger distance with less scatter, but with an apparent difference compared to monolithic metals. Note, however, that the observation is that when taking the small distance of 0.5 mm behind the crack tip to measure the CTOA, Fig. 10.8, the differences in CTOA measurements for saw-cut and fatigue crack configurations diminish.

Here one has to be aware that the critical CTOA relates to the material's resistance to crack growth. This implies that CTOA measurements on panels made of the same alloy, but with larger panel thicknesses, will measure different CTOAs. Rodi [23] illustrated this by comparing the measurements obtained from similar panels made of 1 mm thick 2024-T3, or from laminated sheets of either 0.3 or 0.4 mm thin.

Even for the fairly small thickness differences, the effect of the thickness on the fracture toughness becomes apparent in the critical CTOA measurements. To provide input for the residual strength prediction methods for FMLs, one therefore has to determine the CTOA data preferably from laminated sheets of the same thickness as the FML constituent layers. Otherwise, one has to accept a minor error in the input values.

10.2.5 Superposition Principles for Crack Opening

The concept proposed by Rodi [23] to predict the residual strength for both the through-cut crack and fatigue crack configurations is similar to the method adopted to predict fatigue crack growth, discussed in the previous chapter. Hence, assuming linear elastic fracture mechanics, the crack opening contour can be described with Eq. (9.4).

Taking the 0.5 mm distance measured from the crack tip, where $v(x) = 0$, one could theoretically calculate the CTOA. The key aspect that complicates the analysis here, however, is the level of plastic deformation that occurs ahead of the crack tip, left as permanently deformed material behind the crack tip once the crack has incremented further.

Rodi illustrates how an effective crack opening must be calculated based on the linear elastic crack opening profile described with Eq. (9.4), superimposed with the calculated permanent plastic deformation. Neglecting the plastic deformation would overestimate the critical CTOA.

The residual strength test is a quasi-static strength test performed in displacement control. In case the test would be performed in load control, the panel would fail once the maximum strength is reached, resulting in catastrophic energy release. At this point unstable crack growth cannot be established. Performing the test in displacement control, however, allows measurement of the unstable crack growth together with the reduction in load corresponding to a decrease in the residual strength curve.

The method presented by Rodi is based on stresses, which is equivalent to the residual strength test in load control. Hence, in order to capture the unstable growth in the predictions, the method is applied iteratively; the applied load is reduced and then incrementally increased again after each predicted crack increment. This process is illustrated in the form of CTOA against crack extension in Fig. 10.10.

The analytical method presented by Rodi [23] has been validated against a range of data, of which most has been generated by De Vries [1, 11]. The interesting observation here is that with the CTOA curve as input in the method, not only the residual strength curve can be predicted, but also the corresponding K_R-curve. Hence, the empirical method represented by Eqs. (10.1–10.3) is then no longer required.

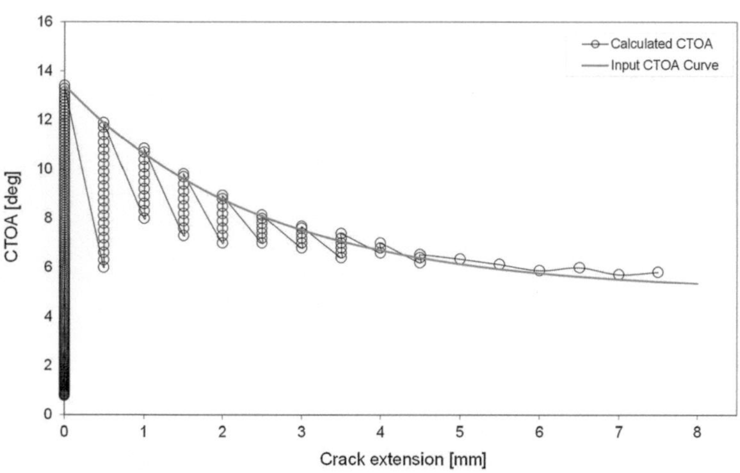

Fig. 10.10 Illustration of the iterative nature of the residual strength prediction method; after predicting crack increment the load is initially reduced to establish the critical stress again [23]

10.2.6 In-Axis Versus Off-Axis Loading

Most research performed on the residual strength of through-cut crack configurations has been performed on centre crack panels under uniaxial loading conditions [1, 6, 23]. Few studies addressed the case of misalignment between the direction of uniaxial loading and the principal material orientations, often referred to as loading under off-axis angles. A small number of studies report the results of uniaxial tensile strength tests on unidirectional and cross-play FMLs under off-axis angles [41–43] (See Fig. 10.11).

There are some important differences between off-axis tensile testing utilizing long and slender specimens and residual strength testing on large panels under off-axis loading. When testing orthotropic laminates under off-axis angles, the applied uniaxial loading will induce a shear deformation in the panel, which is constrained by the test rig. This means that aside from the applied uniaxial load, also a transverse load is applied. For slender specimens, this transverse component is less than one may calculate using, for example, CLT. For large residual strength panels, however, this lateral load implies that the panel in fact is under biaxial loading to certain degree.

For fatigue cracks, this aspect has been addressed by Gupta [45] as explained in Sects. 9.12 and 9.13. However, application of the methodology to cracks under quasi-static off-axis loading illustrates that the larger amount of plastic deformation requires extension of the linear elastic methodology.

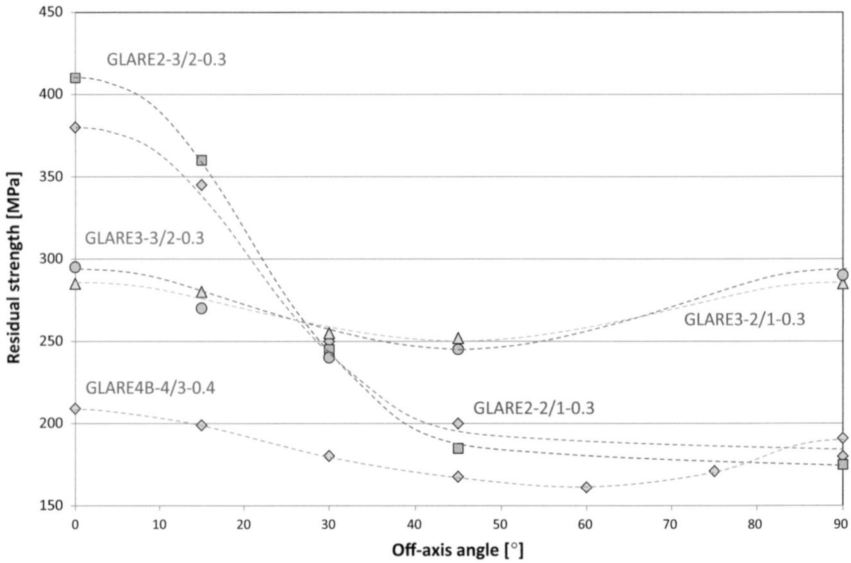

Fig. 10.11 Off-axis residual strength data for various GLARE grades [41, 44]

Without having an analytical methodology to hand, the first studies on residual strength under off-axis loading evaluated the experimental results phenomenologically. Interestingly enough, however, it appears that a modification of the in-axis residual strength methodology using a correction similar to the Tsai–Hill criteria [46] may suffice [47]. One has to keep in mind, however, that such criterion effectively considers the orthotropy of the laminate in a similar fashion as the application of the off-axis fatigue crack methodology of Gupta [45]. Based on that study it appears that the laminate orthotropy of FML panels under quasi-static loading with, relative to fatigue cracks, large amount of plastic deformation requires an elastic–plastic methodology in order to be quantified.

10.3 Fatigue Through Crack

10.3.1 Observations

Several authors have investigated the residual strength of fatigue cracks [1, 6, 23]. The observations reported clearly indicate differences between the through-cut crack configuration discussed in the previous section and the fatigue crack configuration.

For example, Vermeeren [6] illustrates the difference in response to fatigue cracks in unidirectional FMLs and cross-ply FMLs. For this case, he compares GLARE2-2/1-0.3 and GLARE3-3/2-0.3 fatigue tested at two difference fatigue stress levels to identical crack lengths. Where the fatigue crack in GLARE2 generated at a higher fatigue stress results in lower residual strength, he reports the opposite trend for the cross-ply GLARE3.

This difference can be explained with the location of first failure in the residual strength tests. The delaminations in the fatigue crack configuration provide some relaxation in the fibre stresses bridging the cracks, but as a result of the delamination shapes in comparison with the crack opening profile in the metal layers, the corresponding bridging stresses in the fibre layers are not constant over the crack length.

Initially, Marissen [48] assumed these stresses to be constant, but it was later demonstrated by Guo and Wu [49] that these stresses peak towards the fatigue crack tip.

In the methodology of the author, explained in the previous chapter, it appeared that there are actually two peaks in the bridging stress distribution: one near the tip of the original fatigue crack starter notch, and one at the fatigue crack tip.

In the illustration of his methodology, Rodi [23] explains how fibre fracture develops first at the peak near the fatigue starter notch, rather than at the fatigue crack tip; see Fig. 10.12. However, Vermeeren reports the observation of first fibre failure near the fatigue crack tip in case of residual strength tests on unidirectional GLARE2 panels.

Because the fatigue crack growth rates in GLARE2 are substantially lower compared to cross-ply GLARE3, it takes significantly more load cycles in order to

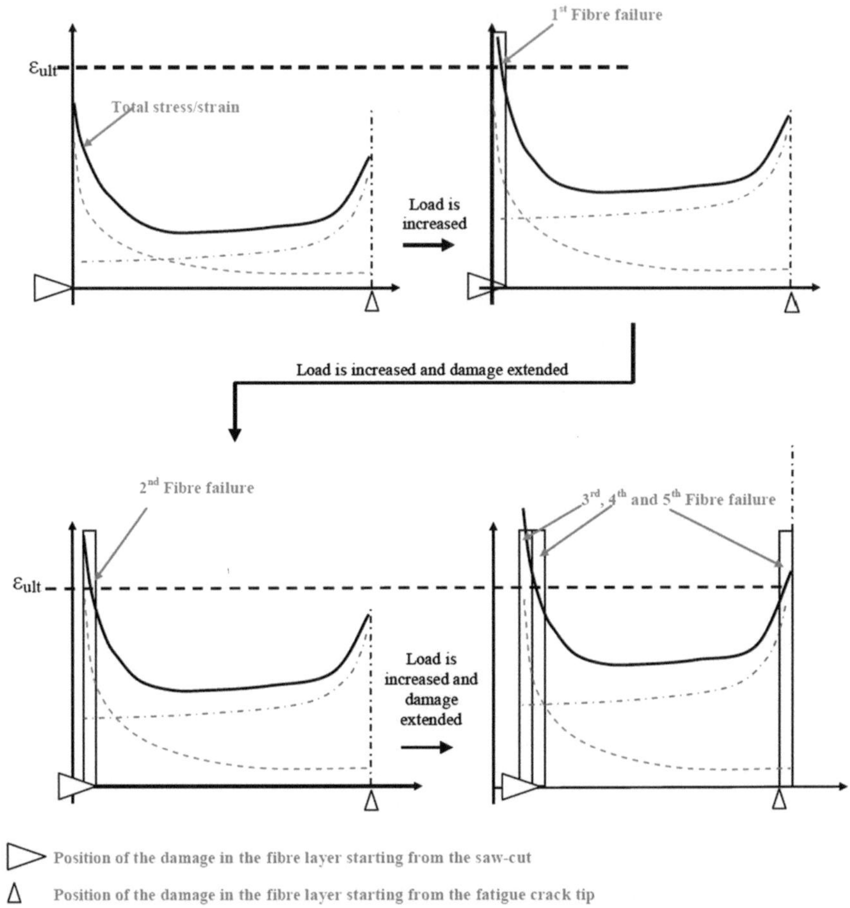

Fig. 10.12 Illustration of the fibre failure sequence in the residual strength prediction model for through-cut cracks in cross-ply FML panels [23]

obtain the same fatigue crack length. This higher number of fatigue load cycles for GLARE2, yields more delamination extension, which in the residual strength tests afterwards imposes quasi-static delamination extension before fibre failure. As a result, the bridging stress distribution for unidirectional GLARE2 may be difference compared to the case illustrated in Fig. 10.12. Instead, a higher peak stress may occur first at the fatigue crack tip, which means that fibres will start to fail at the position of the fatigue crack rather than the fatigue starter notch. The difference between the failure sequences is illustrated in Fig. 10.13.

Although both Vermeeren [6] and De Vries [1] report the failure sequence illustrated in Fig. 10.13, De Vries explains that the relationship between the observed delamination sizes for unidirectional GLARE2 and applied fatigue stress levels, reported by Vermeeren, could not be reproduced.

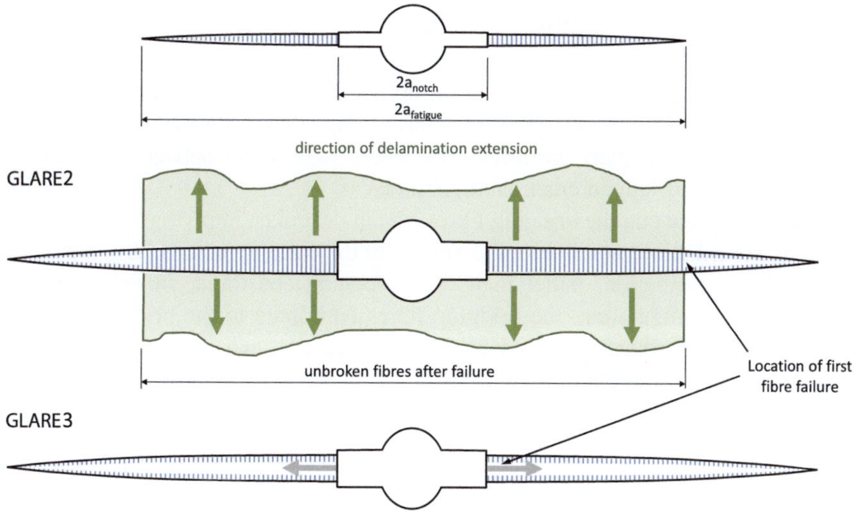

Fig. 10.13 Illustration of the difference in fracture sequence between unidirectional GLARE2 and cross-ply GLARE3 [1]

Hence, the sequence illustrated in Fig. 10.13 may not represent a generic case, but may depend on a combination of parameters. Aspects like the delamination quasi-static and fatigue resistance at the interface, discussed in Chap. 8, and the actual length of the fatigue crack in the residual strength test, may change the observed sequence.

In general, all studies [1, 6, 23] report that the residual strength of fatigue crack configurations are substantially higher than those of through-cut crack configurations, as illustrated before with Fig. 10.2. Another observation reported by Vermeeren [6] is that the increase in fatigue crack length imposes a smaller reduction in residual strength compared to a similar increase in through-cut cracks. This latter observation evidently relates to the higher number of bridging fibres in the longer fatigue crack bearing load during the residual strength test.

What should be considered here though is that the sequence of fracture will differ. Through-cut cracks will develop significantly more plasticity at the crack tip in the metal layers, resulting in a similar ductile fracture sequence as observed for monolithic aluminium, though at higher strength values. The bridging fibre layers in the fatigue crack configuration carry a significant amount of the load prior to fracture, but impose at the time of fracture a much more brittle type of fracture.

This is illustrated in Fig. 10.14 where the typical residual strength for through-cut cracks is compared with a curve for fatigue cracks. The critical crack extension at the point of failure is significantly larger than for fatigue crack configurations. After a little crack extension, bridging fibres will fail, causing complete failure of the panel.

10.3.2 Prediction Methodology

The superposition principle, discussed for fatigue crack growth in Chap. 9, remains similar for residual strength prediction, as discussed for through-cut cracks in Sect. 10.2. However, there are a few differences to consider between the saw-cut configurations and fatigue crack configurations.

In case of a saw cut, the bridging fibres are cut or broken implying that they have either no contribution, or if fibre failure ahead of the crack tip occurs [23], they have an opening contribution, rather than crack opening constraint. In the case of a fatigue crack configuration, the bridging fibres contribute to the bridging, but the interplay between (quasi-static!) delamination extension and possible fibre failure must be evaluated. Fibre failure may either start at the fatigue starter notch tip, or at the crack tip of the extended fatigue crack.

In agreement with the experimental observations, illustrated in Fig. 10.2, the analytical crack growth methodology of Rodi predicts higher residual strengths for fatigue cracks. The intact fibres bridging the fatigue crack bear substantial load until they fracture, which occurs at significantly higher laminate stresses than for the saw-cut configuration.

Based on Fig. 10.14, the difference between the residual strength of through-cut cracks and fatigue cracks can be further illustrated with Fig. 10.15. The through-cut configuration seems to exhibit curves that resemble in shape the crack resistance curves for monolithic metals, although the values in general are higher. The fatigue

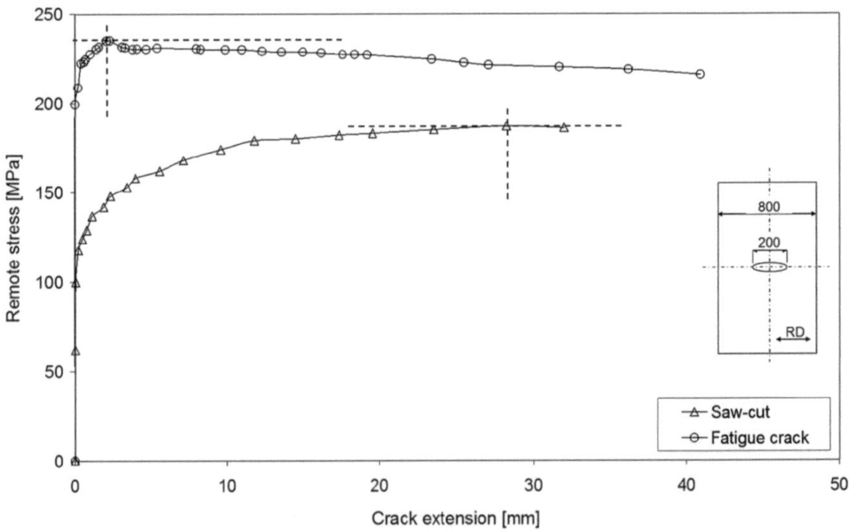

Fig. 10.14 Comparison between residual strength curves for through-cut cracks and fatigue cracks in GLARE4B-3/2-0.4 T-L, dimensions in mm [23]

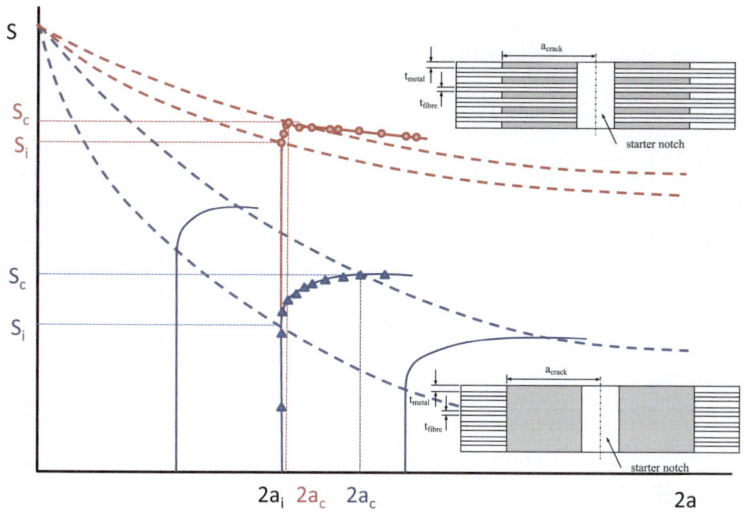

Fig. 10.15 Comparison between the residual strength curves of through-cut cracks and fatigue cracks

crack configuration, however, shows curves that decrease substantially less fast compared to the through-cut crack configuration. This is in agreement with the observation reported by Vermeeren that extending the fatigue crack length has significantly less influence on the residual strength than for the case of through-cut crack configurations. For a given initial crack length for which the residual strength analysis is performed, the strength of panels containing fatigue cracks is significantly higher than for through-cut cracks, although the critical crack length at which this strength is reached is much smaller. This illustrates the more brittle type of fracture of a fatigue crack configuration compared to the more ductile fracture of the through-cut cracks.

10.4 Part-Through Cracks

The case of part-through cracks, Fig. 10.1c, resembles in general the mechanically fastened joint configuration. To assess the residual strength of riveted joints a number of studies have been performed. The focus in this chapter is on the methods proposed for net section failure. Initially, a concept was proposed relating the ultimate strength of the material to the net section stress, empirically corrected with a so called blunt notch factor; see Sect. 5.3.3.

The other approach which is currently adopted for residual strength assessment in riveted joints [50] has been proposed by de Rijck [51] using the work of Müller

[52]. According to De Rijck, the residual strength of a panel containing holes or fasteners can be determined with

$$S_{res} = S_{BN,FML,gross} - MVF \cdot R_D \cdot S_{BN,m,gross} \tag{10.7}$$

Note that the two blunt notch strength values in Eq. (10.7) are gross stress, different from the net section blunt notch strength discussed in Chap. 5. The damage ratio R_D is defined as the cross-sectional area of cracked metal over the total cross-sectional area of metal in pristine condition

$$R_D = \frac{A_{m,cracked}}{A_{m,pristine}} \tag{10.8}$$

Here, one has to keep in mind that both the area of cracked metal and the area of pristine metal are net section parameters. Figure 10.16 illustrates how the crack lengths are measured, excluding the fastener hole. Similarly, the area of pristine material for tested specimen geometries is determined with

$$A_{m,pristine} = (W - nD)n_m t_m \tag{10.9}$$

and for arbitrary mechanically fastened joints with

$$A_{m,pristine} = (S - D)n_m t_m \tag{10.10}$$

The parameters W, D and S are defined in Fig. 10.16. An illustration of the correlation between residual strength and the damage ratio is given in Fig. 10.17. This figure illustrates how the linear trend for monolithic aluminium approaches zero at 100% damage, while the FMLs still have a significant strength left, which is due to the intact fibre layers.

The linear trend lines in Fig. 10.17 illustrate how the strength reduction against the damage ratio for FML is significantly better than for monolithic aluminium. The intact fibre layers in the part-through-the-thickness crack configuration contribute to the strength, which ideally reduces at most to the strength of the intact fibre layers in case all metal layers are cracked. The monolithic aluminium configuration will reduce to zero strength for that case.

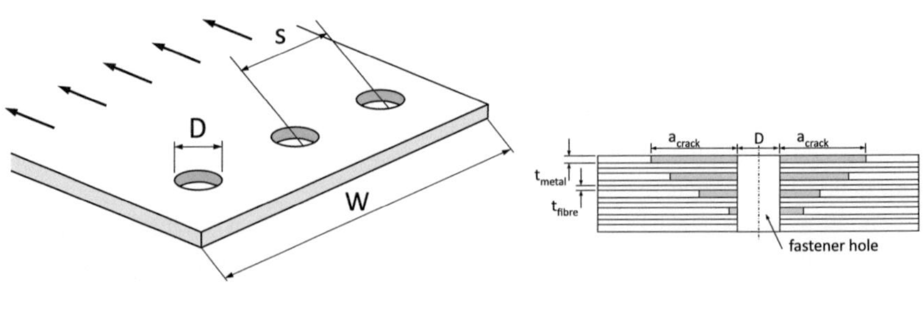

Fig. 10.16 Definition of diameter D, specimen width W, and the corresponding crack lengths

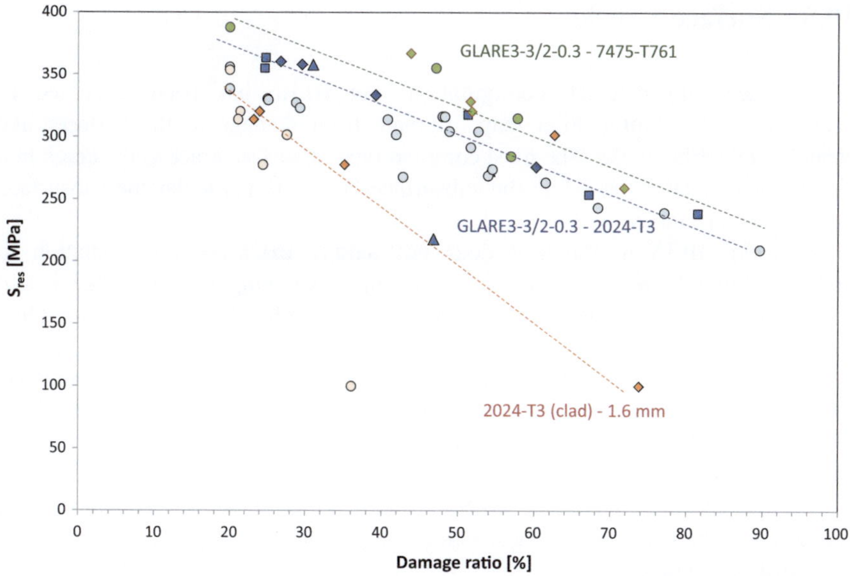

Fig. 10.17 Correlation between residual strength of mechanically fastened GLARE and aluminium joints and the damage ratio, data from [52, 53]

Here an analogy can be made between the part through crack case, discussed here, and the residual strength case for the through-cut and fatigue cracks discussed in Sects. 10.2 and 10.3. This correlation is illustrated in Fig. 10.18. Both the fatigue crack configuration and the part-through-the-thickness crack configuration reduce the strength of the intact fibre layers for a damage ratio of 100%, whereas the monolithic aluminium and through-cut crack configuration would reduce to zero strength.

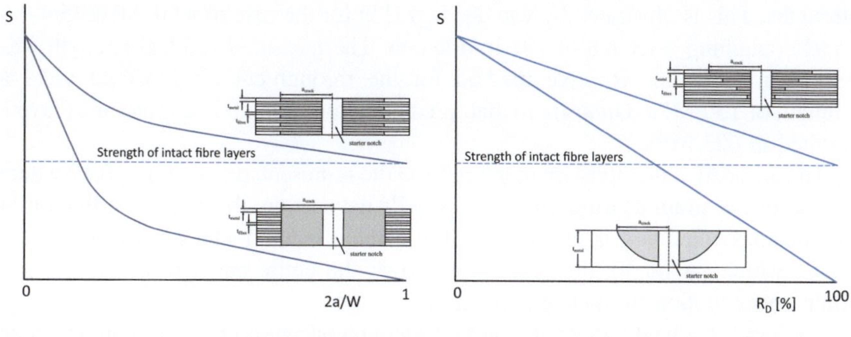

Fig. 10.18 Illustration of the similarity between residual strength of fatigue and through-cut cracks (*left*) and the part-through fatigue cracks in FMLs and monolithic metals (*right*)

10.5 Surface Cracks

The surface fatigue crack configuration, Fig. 10.1d, has been discussed in Sect. 9.10. This configuration may originate from damage to the surface, like scratches or scribe marks. The most common type of surface crack is the crack in a single outer metal layer, while the subsurface layers, in particular the subsurface fibre layers, are intact.

With Fig. 10.15 in mind, it does not require extensive argumentation to understand that the remaining strength of a panel containing a surface crack is still very high. In fact, extending the surface crack length will have only marginal influence on the residual strength of the panel, in particular for thicker FMLs.

This has been illustrated by Hagenbeek [54] who performed experiments on FML panels containing scratches. The panels were made from GLARE3-5gg/4-0.4 L-T and GLARE4B-5/4-0.4 T-L laminates and contained an artificial notch of 50 mm length, which was subsequently extended to about 80–100 mm under a fatigue load spectrum of S_{max} = 140 MPa, and R = 0.05. In the subsequent residual strength test, the panels failed mostly in the net section of the panel containing the longest fatigue crack.

The GLARE3-5/4-0.4 panel, for example, failed at 430 MPa, which is well above the laminate's yield strength. For any FML structure sized to avoid yielding, these scratches would not have imposed a strength reduction below ultimate load. Hence, they would not constitute a damage tolerance problem.

There are different ways to look at the problem of FML panels containing scratches. One may, for example, assume that the scratched layer does not contribute to the panel strength, and calculate the strength of the remainder of the panel thickness. However, this severely overestimates the panel strength, because it ignores the contribution of the stress concentration the scratch imposes in thickness direction.

On the other hand, one could consider similarity between the scratch and a through-cut crack for which the R-curve method has been developed, or the fatigue through crack assuming similar dimensions. However, both these cases would penalise the design, because the predicted strengths will be lower than the actual strengths. This is illustrated by van der Jagt [55] for the case of a GLARE3-5/4-0.4 panel containing a scratch of 101 mm length. The measured residual strength was 430 MPa, while the R-curve method for the through-cut crack would yield a strength of 130 MPa. Opposite to that, predicting the strength based on intact layers resulted in 623 MPa.

The alternative methods approach reflects the representation of Fig. 10.19 where instead of the strain distribution in the panel's net section, the strain distribution in the thickness direction is considered. The scratch in the metal layer represents a notch, which (assuming the scratch runs over the entire panel width) imposes a strain concentration in thickness direction.

The strain gradient corresponding to the strain concentration factor could then be related to the ultimate strain of the constituent layers, in particular the adjacent

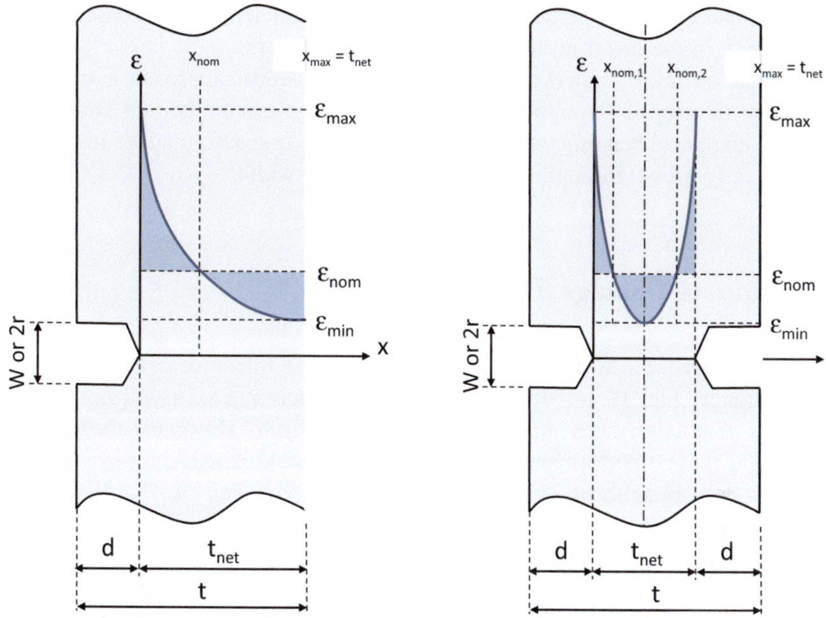

Fig. 10.19 Illustration of the strain distribution through the thickness in presence of a surface crack [55]

intact fibre layer. Initially the idea was to take the configuration illustrated in Fig. 10.19, and take stress concentration factors from handbook solutions like Peterson's stress concentration factors [56]. However, taking the stress concentrations factors for notch geometries corresponding to the cross section of the panels containing scratches overestimated the contribution of the notch. As result, residual strength values were underestimated compared to the tests.

Calibrating the magnitude of the strain concentration factor to the experiment, however, revealed that a fairly low strain concentration factor had to be adopted in order to make reasonable predictions. Hence, the concept illustrated in Fig. 10.19 may be worthwhile to explore further, and the effective geometry of the notch may not relate straightforwardly to handbook solutions.

Comparison between the two cases shown in Fig. 10.19, either having an elliptical or parabolic strain distribution through the thickness, revealed that the predicted residual strengths were similar for the same strain concentration factor applied. It appeared to make no difference which of the two configurations was adopted for the predictions.

van der Jagt [55] did advise, however, to investigate the influence of other, probably more steeper, strain gradients, because such strain distributions may be more applicable to certain cases tested. In addition, he advised to explore the strain concentration factor further, because the values taken from the handbook assume isotropic linear elastic properties, which may not be sufficient for the condition of a surface crack in the metal layer, with adjacent fibre layers.

It should be noted, however, that the two-dimensional approach assuming a crack fully developed through the panel width, may assume certain strain concentration and panel bending, which overestimates the true deformations in case the crack has only grown through a portion of the panel width.

10.6 Impact Damage Tolerance

The last damage type to be considered in the damage tolerance evaluation is the impact damage, Fig. 10.1e. Here the impact resistance of FMLs will not be discussed, as that constitutes an entirely different subject. However, the residual strength and fatigue performance after impact must be assessed.

To this end, Hagenbeek [57, 58] impacted unidirectional GLARE2B-3/2-0.3, GLARE2B-4/3-0.4 and cross-ply GLARE4A-4/3-0.5 and subsequently tested the panels under fatigue loading. He then compared the results to the post-impact performance of 3.8 mm aluminium 2024-T3. Different impact energies were considered, among which a very low impact of 6 J, and impacts of 80–85 J. The low energy impact created at most a dent in the FML, whereas the 80–85 J impacts imposed either only a crack in the outer aluminium layers at the non-impacted side, as reported by Moriniere et al. [59, 60] and various other researchers in literature, or through-thickness cracks. Delaminations were observed to remain within the dented area of the impact.

The fatigue tests consisted of constant amplitude load cycles at S_{max} = 120 MPa and R = 0.05. All tests ran for 200 kcycles without failure, but in case of cracks after impact, the cracks propagated to longer lengths. A comparison between the crack growth in the outer aluminium layer of GLARE4A-4/3-0.5 at the impacted side and the 3.8 mm aluminium 2024-T3 is given in Fig. 10.20.

The reason why only cracks in the outer aluminium layers in GLARE at the impacted side propagate during the fatigue tests is explained by the secondary bending imposed by the geometry of the dent. The neutral line of the laminate is not aligned with the fatigue load application axis, which causes additional tensile bending stresses at the impacted side, and compressive stresses at the rear, non-impacted side.

This immediately also explains the difference with the monolithic aluminium panel, where the fatigue cracks propagate through the entire thickness.

After the fatigue tests, a residual strength test was performed. The GLARE panels containing through-thickness cracks after the impact, and subsequently fatigue tested at S_{max} = 120 MPa, R = 0.05 for 200 kcycles, exhibited the greatest strength reduction.

Hagenbeek [57] reports residual strength values for GLARE4-4/3-0.5 of 404 MPa in *L* direction and 198 MPa in *LT* direction. Compared to a gross blunt

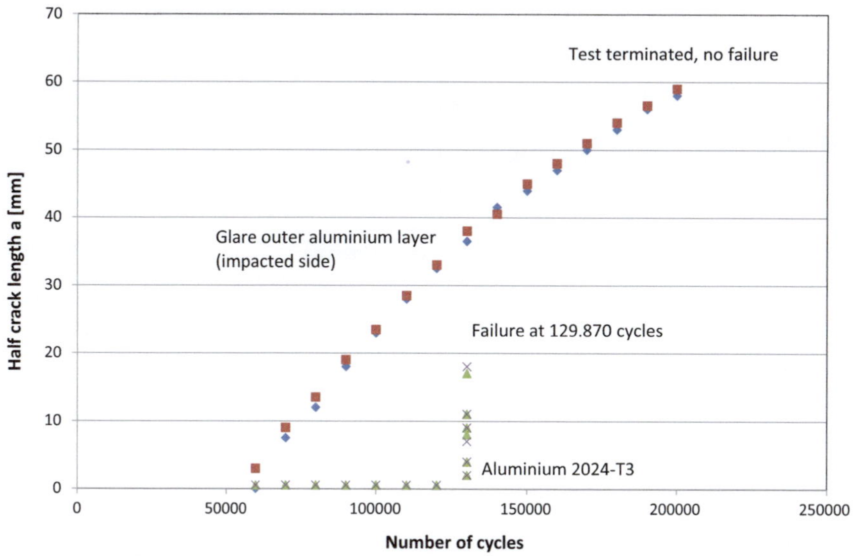

Fig. 10.20 Fatigue crack growth curves of GLARE4A-4/3-0.5 and 2024-T3 both tested in the L direction after low velocity impacts of, respectively, 78.5 and 87.8 J, which created (first) cracking [57]

notch strength for this laminate of about 500 MPa in *L* direction and 350 MPa in *LT* direction, this implies a reduction of about 20% in *L* direction and 40% in *LT* direction.

In [57] Hagenbeek already made the recommendation to perform the fatigue and residual strength tests after impact on panels with larger widths. In [58], he therefore reports similar experiments performed on unidirectional GLARE2B-3/2-0.3, GLARE2B-4/3-0.4. Again panels were impacted to create only a dent, or first cracking at the non-impacted side, or through-thickness cracks. For GLARE2B-3/2-0.3, this required 8, 16 and 25 J impacts, respectively, while for GLARE2B-4/3-0.4 this required 30, 45 and 58 J impacts.

Subsequent fatigue testing in the *LT* direction at S_{max} = 120 MPa with R = 0.05 for 180 kcycles caused fatigue cracking at the impacted side only, for reasons illustrated in Fig. 10.21, with the maximum crack length of a = 44 mm.

Interestingly enough, all specimens reached very high strength values during the residual strength tests. For example, all GLARE2B-3/2-0.3 panels reached a residual strength of at least 900 MPa, while GLARE2B-4/3-0.4 panels reached at least 736 MPa.

These results seem in line with observations reported before by Bosma [61] who compared the residual strength of panels containing an open hole or with a saw-cut with those containing impact damage. The residual strength of both the open hole and saw-cut are distinctively different due to the difference in stress concentration and stress intensity.

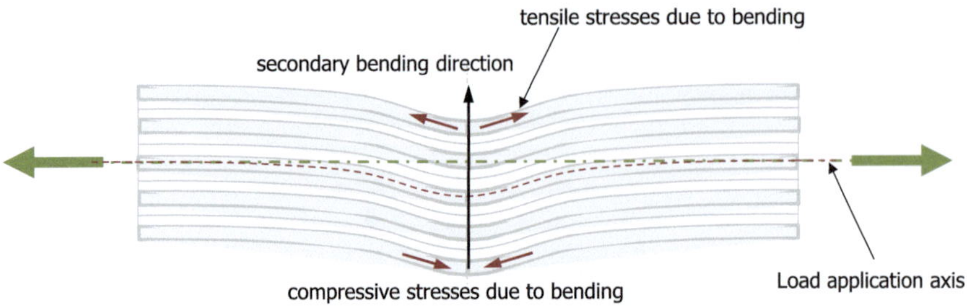

Fig. 10.21 Secondary bending induced by the misalignment of the neutral line with the load axis, causing tensile stresses at the impacted side and compressive stresses at the rear side [57]

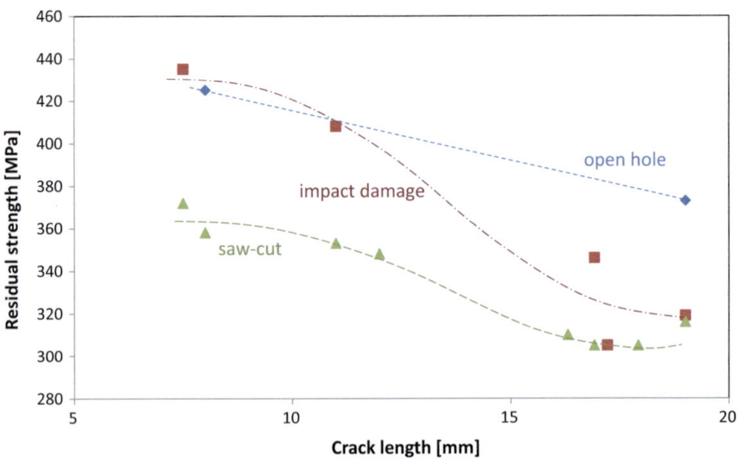

Fig. 10.22 Comparison between residual strength of 2H32 specimens containing either an impact damage (with cracks), or a saw-cut, or an open hole [61]

However, comparing impact damages with different crack lengths with equivalent open hole diameters or saw-cut lengths revealed a transition, which is illustrated in Fig. 10.22. This transition Bosma attributed to the fact that for smaller crack lengths, the impact had only created cracks at the non-impacted side, while for the larger crack lengths, cracks were present at both sides having closer resemblance with the saw-cut configuration.

Although the values seem somewhat different, the results reported by De Vries [1] on residual strength after impacts that created full penetration, illustrate a similar trend between unidirectional and cross-ply FMLs. Comparing the low velocity impacts with energy levels creating full penetration, the residual strength always exceeds a through-cut crack configuration with length $2a$ = 30 mm.

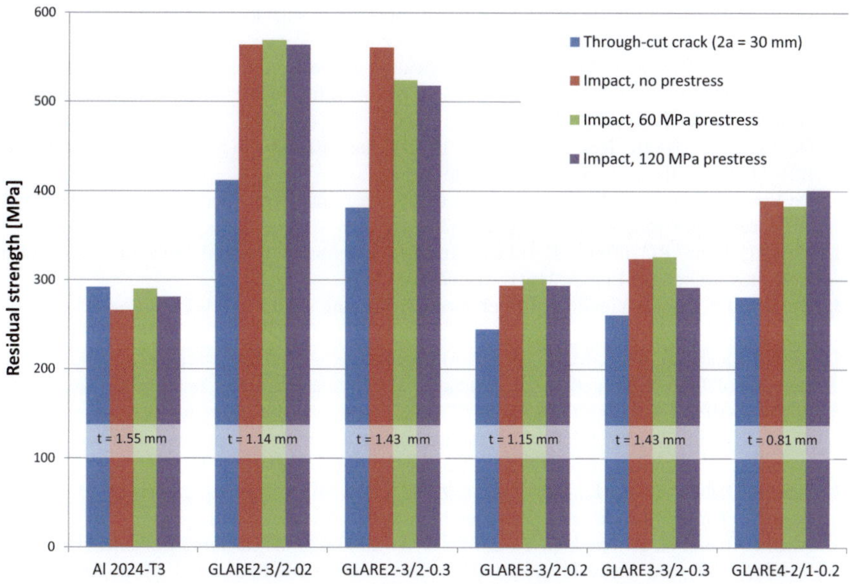

Fig. 10.23 Residual strength of impacted aluminium and GLARE specimens [1]

Similarly, the residual strength of the unidirectional FMLs is substantially higher compared to the cross-ply laminates (Fig. 10.23).

References

1. de Vries T (2001) Blunt and sharp notch behaviour of Glare Laminates. PhD dissertation, Delft University of Technology, Delft
2. Prasad Kadiyala S, Gurdal Z, Herakovic CT (1988) An experimental and numerical study of notch sensitivity of ARALL laminates with holes. In: 4th Japan-US conference on composite materials, pp 179–188
3. Macheret Y, Teply JL (1988), Residual strength of ARALL laminates. AMD symposia series, ASME, pp 53–61
4. Macheret Y, Bucci RJ, Kulak M (1990) Metal plasticity and specimen size effects in evaluation of ARALL laminates notched panel residual strength, fracture behaviour and design of materials and structures. In: Proceedings of the 8th European conference on fracture, pp 288–295
5. Macheret J, Bucci RJ (1993) A crack growth resistance curve approach to fiber/metal laminate fracture toughness evaluation. Eng Fract Mech 45(6):729–739
6. Vermeeren CAJR (1995) The residual strength of fibre metal laminates. PhD dissertation, Delft University of Technology, Delft
7. Vermeeren CAJR (1998) The residual strength of fibre metal laminates: GLARE 2 and GLARE 3. In: International SAMPE technical conference series, pp 471–483
8. De Vries TJ, Vermeeren CAJR (1994) R-curve testdata: 2024-T3, 7075-T6, Glare 2 and Glare 3, Memorandum M-705. Delft University of Technology

9. De Vries TJ, Holleman E (1995) An R-curve approach to fracture toughness analysis of some Glare grades, Memorandum M-687, Delft University of Technology
10. De Vries TJ, Pacchione M (1996) The application of the R-concept for the prediction of the residual strength of fibre metal laminates. In: Proceeding of the USAF structural integrity program conference, USA
11. De Vries TJ (2000), Residual strength of flat glare panels—experimental results. Report B2V-00-15, Delft University of Technology
12. De Vries TJ, Vlot A (2001) The influence of the constituent properties on the residual strength of Glare. Appl Compos Mater 8(4):263–277
13. Jin ZH, Batra RC (1996) Residual strength of centrally cracked metal/fiber composite laminates. Mater Sci Eng A 216(1–2):117–124
14. Dugdale DS (1960) Yielding of steel sheets containing slits. J Mech Phys Solids 8(2):100–104
15. Afaghi-Khatibi A, Ye L, Mai Y-W (1996) Evaluations of effective crack growth and residual strength of fibre-reinforced metal laminates with a sharp notch. Compos Sci Technol 56 (9):1079–1088
16. Afaghi-Khatibi A, Ye L (1997) Residual strength simulation of fibre reinforced metal laminates containing a circular hole. J Compos Mater 31(19):1884–1904
17. Afaghi-Khatibi A, Ye L, Mai Y-W (1997) Effects of constituent properties on residual strength of notched fibre reinforced metal laminates. Adv Fract Res 1–6:643–650
18. Afaghi-Khatibi A, Lawcock G, Ye L, Mai Y-W (2000) On the fracture mechanical behaviour of fibre reinforced metal laminates (FRMLs). Comput Methods Appl Mech Eng 185(2–4):173–190
19. Castrodeza EM, Ipina JEP, Bastian FL (2002) Experimental techniques for fracture instability toughness determination of unidirectional fibre metal laminates. Fatigue Fract Eng Mater Struct 25(11):999–1008
20. Castrodeza EM, Bastian FL, Ipina JEP (2003) Critical fracture toughness, J(C) and delta(5C), of unidirectional fibre-metal laminates. Thin-Walled Struct 41(12):1089–1101
21. Castrodeza EM, Bastian FL, Ipina JEP (2004) Residual strength of unidirectional fibre-metal laminates based on J(c) toughness of C(T) and SE(B) specimens: comparison with M(T) test results. Fatigue Fract Eng Mater Struct 27(10):923–929
22. Castrodeza EM, Ipina JEP, Bastian FL (2004) Fracture toughness evaluation of unidirectional fibre metal laminates using traditional CTOD (delta) and Schwalbe (delta(5)) methodologies. Eng Fract Mech 71(7–8):1107–1118
23. Rodi R (2010) The residual strength failure sequence in fibre metal laminates. PhD dissertation, Delft University of Technology, Delft
24. Newman JC, James MA, Zerbst U (2002) A review of the CTOA/CTOD fracture criterion. Eng Fract Mech 70(3–4):371–385
25. ASTM-E2472-06 (2006) Standard test method for determination of resistance to stable crack extension under low-constrain conditions. Annual Book of ASTM Standards, American Society for Testing and Materials
26. Longhi A (1998) On some preliminary measurements of the crack tip opening angle in glare sheets. Report B2-98-04, Delft University of Technology
27. Longhi A (1998) An innovative approach to the evaluation of the residual strength of fibre metal laminates. Report B2-98-08, Delft University of Technology
28. Broek D (1988) The practical use of fracture mechanics. Kluwer Academic Publisher, Dordrecht, The Netherlands
29. Zaal KJJM (1995) Residual strength analysis of structures in aluminium alloy and fiber metal laminates. PhD dissertation, Delft University of Technology
30. Eftis J, Liebowitz H (1972) On the modified westergaard equations for certain plane crack problems. Int J Fract Mech 8(4):383–392
31. Testi S (1995) A study on the compliance calibration for GLARE®, Master Thesis. Politecnico di Milano & Delft University of Technology, Delft

32. Feddersen C (1967) Discussion to: plane strain crack toughness testing. A.S.T.M. Spec. Tech. Publ. No. 410, p 77
33. Dixon JR (1960) Stress distribution around a central crack in a plate loaded in tension: effect of finite width of plate. J Roy Aeronaut Soc 64:141–145
34. ASTM-E561-81 (1982) Standard practice for R-curve determination. Annual Book of ASTM Standards, Part 10, American Society for Testing and Materials
35. Wu EM (1967) Application of fracture mechanics to anisotropic plates. J Appl Mech 34 (4):967–974
36. Sih GC, Liebowitz H (1968) Mathematical theories of brittle fracture. In: Liebowitz H (ed) Fracture, vol 2. Academic Press, NY
37. Isida M (1971) Effect of width and length on stress intensity factors of internally cracked plates under various boundary conditions. Int J Fract Mech 7(3):301–316
38. Castrodeza EM, Schneider Abdala MRW, Bastian FL (2006) Crack resistance curves of GLARE laminates by elastic compliance. Eng Fract Mech 73:2292–2303
39. ASTM-E1820 (1999) Standard test methods for measurement of fracture toughness. Annual Book of ASTM Standards, American Society for Testing and Materials
40. Rodi R, Alderliesten RC, Benedictus R (2010) Experimental characterization of the crack tip opening angle in fibre metal laminates. Eng Fract Mech 77(6):1012–1024
41. Teigen R (1991) Determination of off-axis properties of Glare. MSc thesis, Delft University of Technology
42. Kawai M, Morishita M, Tomura S, Takumida K (1998) Inelastic behaviour and strength of fiber-metal hybrid composite: GLARE. Int J Mech Sci 40(23), 183–198
43. Kawai M, Hachinohe A, Takumida K, Kawase Y (2001) Off-axis fatigue behaviour and its damage mechanics modelling for unidirectional fibre–metal hybrid composite: GLARE 2. Compos A 32:13–23
44. Tinga T (2003) Off-axis residual strength of Glare laminates, part 1; experimental results Glare4B-4/3-0.4. Report NLR-CR-2003-065-PT-1, National Aerospace laboratory NLR, The Netherlands
45. Gupta M (2017) Directionality of damage growth in fibre metal laminates and hybrid structures. PhD dissertation, Delft University of Technology
46. Sun CT (1996) Comparative evaluation of failure analysis methods for composite laminates. Report DOT/FAA/AR-95/109, U.S. Department of Transportation, Federal Aviation Administration, Office of Aviation Research, Washington, D.C
47. Tinga T (2003) Off-axis residual strength of Glare laminates, part 2; prediction method. Report NLR-CR-2003-065-PT-2, National Aerospace laboratory NLR, The Netherlands
48. Marissen R (1988) Fatigue crack growth in ARALL, a hybrid aluminium-aramid composite material, crack growth mechanisms and quantitative predictions of the crack growth rate. PhD dissertation, Delft University of Technology
49. Guo YJ, Wu XR (1999) Bridging stress distribution in center-cracked fiber reinforced metal laminates: modelling and experiment. Eng Fract Mech 63:147–163
50. Beumler T (2004) Flying GLARE®, a contribution to aircraft certification issues on strength prediction in non-damaged and fatigue damaged GLARE® structures. PhD dissertation, Delft University of Technology, Delft
51. de Rijck JJM (2015) Stress analysis of fatigue cracks in mechanically fastened joints, an analytical and experimental investigation. PhD dissertation, Delft University of Technology, Delft
52. Müller RPG (1995) An experimental and analytical investigation on the fatigue behaviour of fuselage riveted lap joints, the significance of the rivet squeeze force, and a comparison of 2024-T3 and Glare 3. PhD dissertation, Delft University of Technology, Delft
53. Soetikno TP (1992) Residual strength of the fatigued 3 rows riveted Glare3 longitudinal joint. MSc thesis, Delft University of Technology, Delft
54. Hagenbeek M (2000) Fatigue and residual strength effect of scratches. Report B2v-00-14, Delft University of Technology, Delft

55. van der Jagt OC (2002) Residual strength of damaged glare shells. Report B2v-02-56, Delft University of Technology, Delft
56. Pilkey WD (1997) Peterson's stress concentration factors, 2nd edn. John Wiley & Sons, Inc. New York
57. Hagenbeek M (1999) Fatigue and residual strength after impact. Impact properties of GLARE. Report B2v-98-16, Delft University of Technology, Delft
58. Hagenbeek M (2000) Fatigue and residual strength of GLARE2 after impact. Impact properties of GLARE. Report B2v-00-21, Delft University of Technology, Delft
59. Moriniere et al (2013) Low-velocity energy partitioning in GLARE. Mech Mater 66:59–68
60. Moriniere et al (2013) An integrated study on the low-velocity impact response of the GLARE fire-metal laminate. Compos Struct 100:89–103
61. Bosma H (1991) Gevolgen van lage snelheden impact op aramide ARALL en GLARE. MSc thesis, Delft University of Technology (in Dutch)

Chapter 11
Effect of Temperature

Abstract The influence of the environmental temperature is first discussed with respect to the residual stresses and thermal properties. The second part of the chapter explains how the temperature influences the mechanical and fatigue initiation properties, the delamination and crack growth resistance. It is illustrated how the relation between ambient temperature and constituent properties explains the FML behaviour.

11.1 Introduction

For certification of primary aeronautical structures made of fibre metal laminates (FMLs), static strength, fatigue and damage tolerance must not only be evaluated in laboratory air at room temperature. The effect of the environment on the formation and progression of damage must also be investigated.

It is important to realize that treating fatigue and damage tolerance separately from environmental aspects (durability) may have implications for the justification of composite structures in general, but for these hybrid laminates in particular. The fatigue performance and damage growth resistance may be influenced by the characteristics of the dominant environment the FML is operated in.

This chapter describes the effect of temperature as a specific environmental aspect for structural application of FMLs. The next chapter will address the different environmental media in which the laminates may operate, including the effect of moisture on the composite constituents of the FMLs.

11.2 Temperature-Induced Residual Stresses

The effects of temperature on the performance of FML structures may be diverse. It should therefore be considered in its broadest sense, which may require different studies in the initial phase of the FML development process. However, because the effects may be attributed either to one of the constituents or to the combination of constituents, the effect should be understood on the constituent level, before conclusions can be drawn on the FML performance.

R. Alderliesten, *Fatigue and Fracture of Fibre Metal Laminates*, Solid Mechanics and Its Applications 236, DOI 10.1007/978-3-319-56227-8_11

Chapter 4 explains the presence of residual stresses within FMLs due to the difference in thermal expansion coefficients of the metallic and composite constituents. After curing, the metallic layers want to shrink more than the composite layers, in particular if glass fibre layers are involved. This yields an equilibrium in which the metallic layers face tensile stresses and the composite plies compressive stresses.

Once the FML operates at a different ambient temperature, the residual stresses due to curing will differ. An illustration of this relationship between curing stresses and ambient temperature is given in Fig. 11.1. This also means that when the environment applies a temperature gradient to the laminated material, either by cyclically heating and cooling, or by exposing a portion of the structure to heat, this may induce either thermal fatigue or asymmetric residual stress systems within these laminates.

Depending on the curing cycle, and in particular the temperature cycle, the residual stresses may be higher or lower. An illustration is provided by Lalberté et al. [1] who reported a study in which they adopted different curing processes for the same epoxy system. Tuning the consolidation temperature in the cure process allows for modification and decrease of the residual stress system.

11.3 Thermal Properties of FMLs

11.3.1 Thermal Conductivity

Thermal conductivity refers to heat diffusion through solid bodies or fluids. Especially for spacecraft the thermal conductivity of a material is of importance, because in the absence of convection, conduction and radiation are basically the only mechanisms to control temperature. Whereas radiation relates to material surfaces and coatings, conduction relates to the material itself.

In addition, this aspect might be of interest for thick FML structures exposed to either high or low temperatures. The stress analysis, including residual stresses related to thermal coefficients of expansion, might be affected by the different temperature levels in each layer through the laminate thickness. An example application where this is a subject of interest is for example the anti-icing and de-icing systems. These are discussed further in Sect. 11.6 with respect to thermal fatigue.

Mensink [2] proposed an approach to calculate the transverse thermal conductivity of a laminate material built up from different materials. According to him, the energy flux or power Q, per unit area A, is given by

$$q = \frac{Q}{A} = -k\frac{\mathrm{d}T}{\mathrm{d}x} \tag{11.1}$$

where k represents the thermal conductivity. Integrating this equation yields

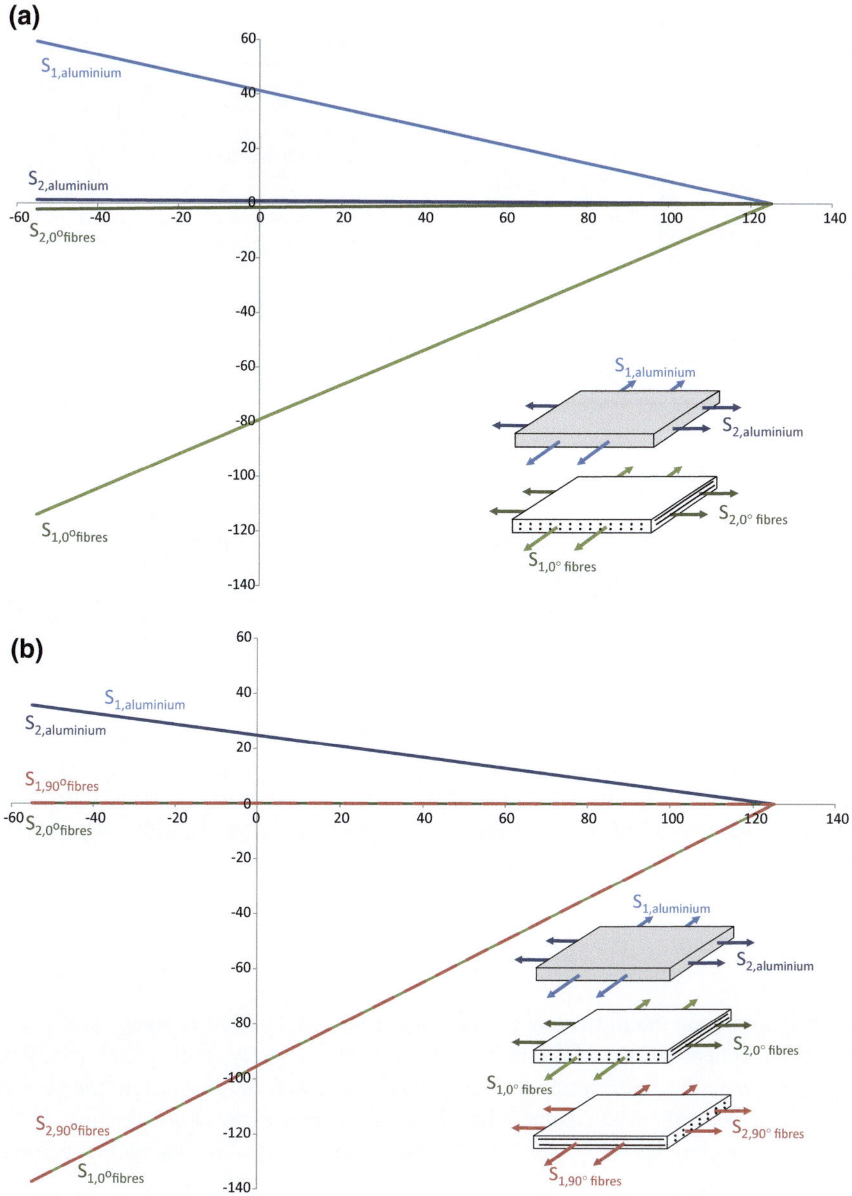

Fig. 11.1 Correlation between the principal stresses S_1 and S_2 in the aluminium layers and 0° and 90° fibre layers of unidirectional GLARE2-5/4-0.4 (**a**) and cross-ply GLARE3-5/4-0.4 (**b**) calculated with the classical laminate theory explained in Chap. 4

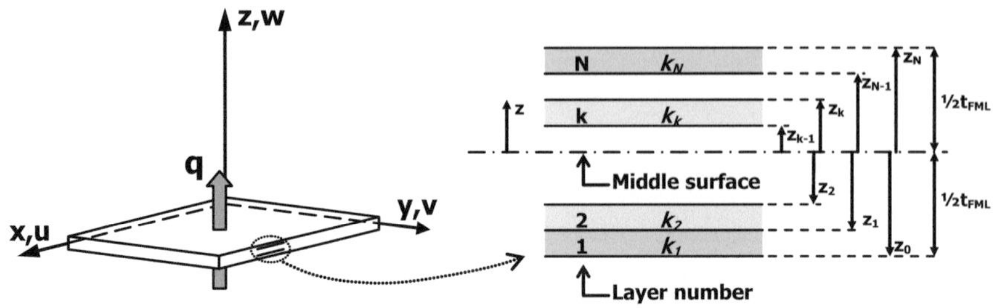

Fig. 11.2 Definition of the heat conduction through an n-layered laminate containing dissimilar materials

$$q = -k_k \frac{T_k - T_{k-1}}{z_k - z_{k-1}} \tag{11.2}$$

where z_k represents the position of the kth layer in thickness direction, see Fig. 11.2. Analogously to a system of electrical resistances in series, the energy flux over the total thickness of that laminate can then be described as

$$q = -\frac{T_N - T_1}{\sum_1^N \frac{t_i}{k_i}} = -k_{\text{transv}} \frac{T_N - T_1}{t_{\text{lam}}} \tag{11.3}$$

where T_N and T_1 are, respectively, the cold and hot side of the laminate, t_i the thickness of each individual layer, and k_i and k_{transv} the transverse thermal conductivity of, respectively, each individual layer and the laminate. This transverse thermal conductivity of the laminate can be approximated with a rule of mixtures [2–4]

$$k_{\text{transv,FML}} = \frac{t_{\text{FML}}}{\sum_1^{n_m} \frac{t_i}{k_i} + \sum_1^{n_f} \frac{t_j}{k_j}} \tag{11.4}$$

where t_i and t_j are the individual thicknesses of all metal and fibre layers, and k_i and k_j are the transverse thermal conductivities of, respectively, each metal and fibre layer. Because the contribution of highly conducting metals (like aluminium alloys) to the transverse thermal conductivity of the laminate is fairly small, this parameter can be estimated based on the fibre/epoxy conductivity only using the volume fraction approach

$$k_{\text{transv,FML}} = \frac{k_f}{\text{FVF}} \tag{11.5}$$

One should be aware, however, that this relation will only provide sufficiently accurate results if the difference in conductivity of the metal and composite components in the laminate is very high.

The in-plane conductivity can be treated analogously to a system of parallel electrical resistances.

$$k_{\text{plane,FML}} = \frac{k_m}{\text{MVF}} + \frac{k_{\text{plane},f}}{\text{FVF}} \tag{11.6}$$

where $k_{\text{plane},f}$ is the plane thermal conductivity of the fibre/epoxy layer and a function of the fibre orientation in that layer. In case of high metal conductivity and very small fibre/epoxy conductivity (so not for CFRP!) this relation can be simplified to the first component of Eq. (11.6).

The work of Mensink [2] has been further evaluated and validated by Graafsma [3] and Verolme [4], who all report that the through-thickness conductivity of GLARE is low in comparison with monolithic aluminium alloys, while the in-plane conductivity remains similar. In the latter case, the conductivity of the aluminium layers dominates the in-plane conductivity of the laminate.

11.3.2 Specific Heat

The specific heat, often also denoted as heat capacity, is the quantity of heat required to increase the temperature of a unit of mass of a solid by one degree under specified conditions. This property is particularly important for thermal analyses of spacecraft, and for assessing fire resistivity of aircraft structures. Up until today, the specific heat has received only little attention for FMLs. For example, without detailed experimental assessment, Mensink [2] proposes to estimate the specific heat of an FML in a similar way as the conductivity, using a simple rule of mixtures

$$C_{\text{FML}} = \frac{t_m \rho_m C_m + t_f \rho_f C_f}{t_m \rho_m + t_f \rho_f} \tag{11.7}$$

with C being the specific heat, t the thickness, and ρ the density of the material. The specific heat properties are known for the constituents, which are for the GLARE constituents C_m = 0.9 kJ/kg °C and C_f = 0.88 kJ/kg °C [2].

According to Mensink the specific heat of GLARE is close to 1 kJ/kg °C, which has recently been confirmed in [5], where a slight increase of the heat capacity is reported when measured against higher ambient temperature.

11.4 Fatigue Initiation

The fatigue initiation performance of FMLs is discussed in detail in Chap. 7. The methodology to assess fatigue initiation, defined as the life until a crack of 1 mm has been reached, is based on constituent layer stresses. The fatigue initiation life can be determined using S–N curves obtained for the metallic constituents of FMLs based on fatigue failure life data. The error induced by assuming the application of failure life data on initiation life data is fairly small, as discussed in Chap. 7.

The fatigue initiation behaviour of FMLs is thus directly related to the cyclic stresses in the individual metallic constituents. With the previous section one may conclude that the fatigue initiation behaviour is therefore dependent on temperature, or even thermal cycles.

Three aspects can be identified here that influence fatigue initiation. First, the difference in thermal expansion coefficients induces residual stresses in the constituent layers after curing. The magnitude of the tensile residual stress in the metallic layer is linearly dependent on the ambient temperature, as illustrated in Fig. 11.1. Hence, without any other aspect involved, the temperature dictates the mean stress of the stress cycle in the metallic layers. High temperatures imply low mean stresses, while low temperatures induce high mean stresses. Here one would expect that decreasing ambient temperatures would have a negative effect on the fatigue initiation performance. However, that appears not to be the case.

11.4.1 Temperature Effect on Mechanical Properties

There are two more aspects to be considered: the second aspect is the influence of temperature on the constituent stiffnesses, and the third aspect is the influence of temperature on the fatigue performance of the metal itself. The influence of temperature on the constituent stiffnesses is relatively small over the temperature ranges often considered for primary aircraft structures. Various authors have addressed the influence of temperature on the mechanical properties; among them were Mensink [2], Horst [6], Boertien [7], van der Hoeven and Schra [8, 9] and Hagenbeek [10].

As for the metallic constituents, for example aluminium alloys used in GLARE, the dependency has been described in, for example, MMPDS [11]. The relation from this handbook has been used by Kieboom [12] and Beumler [13] to determine the Young's modulus for GLARE. Figure 11.3 illustrates the relation between temperature and constituent stiffnesses and temperature and GLARE stiffnesses as reported by some of the authors.

Note that for the Young's modulus of S2-glass/FM94 prepreg often E_0 = 54 GPa and E_{90} = 9.4 GPa [12] have been used. These values have been reported to be slightly too high for the S2-glass/FM94 prepreg used in GLARE. Hence, several sources present the values E_0 = 48.9 GPa and E_{90} = 5.5 GPa [10, 14, 15].

The theoretical curves for the GLARE laminates in Fig. 11.3 are based on these lower values, in agreement with the curves from [10]. This provides good correlations with the measurements by Mensink [2] and van der Hoeven and Schra [8, 9].

In addition to the variation in constituent stiffnesses, Hagenbeek [10] investigated the dependency of the coefficient of thermal expansion on the temperature. Based on the results, one may conclude that the above-mentioned dependency of mean stress on the temperature is not exactly linear. However, as a first-order approximation the error remains fairly small.

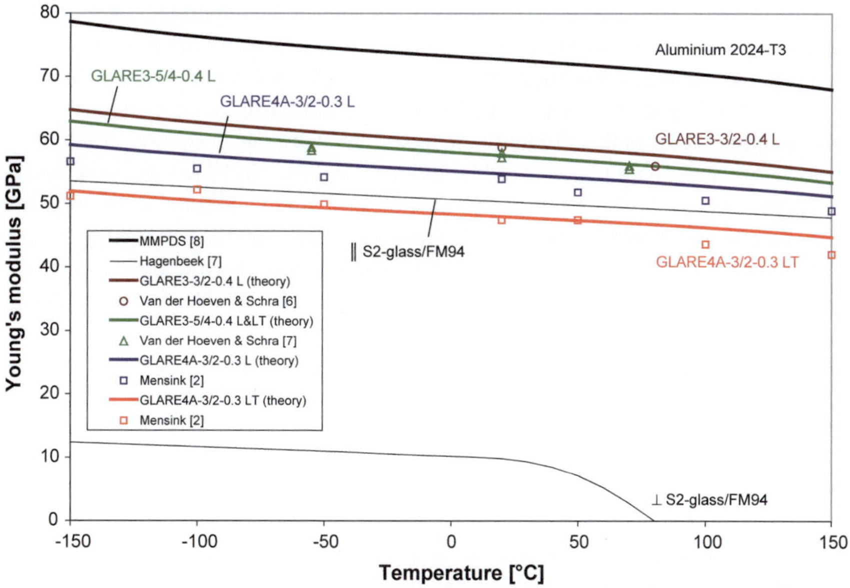

Fig. 11.3 Temperature dependency of the metallic and composite constituents of GLARE and comparison between theoretical stiffness relations and measurements [2, 8–11]

11.4.2 Temperature Effect on Fatigue Properties

The third effect that can be observed when addressing the influence of temperature on fatigue properties of FMLs is the influence of temperature on the initiation and propagation in the metallic constituents. As mentioned before, based upon the mean stress dependency, one may assume that decreasing the temperature would negatively influence the fatigue initiation and propagation behaviour.

However, this effect is completely reversed by the temperature effect on the fatigue properties of the metallic constituent. Schijve [16] explains that not only the mechanical properties change with reduction of ambient temperature, but also the effects of environment are reduced because reaction rates of chemical processes and diffusion are lower at low ambient temperatures. Water vapour in air is known to affect the fatigue properties of metals. In this context, the illustration provided by Schijve is particularly of interest, in which he illustrates the similarity between crack growth curves obtained with fatigue tests in ambient temperatures ranging between −25 and −75 °C and one obtained at room temperature in very dry air.

This increased crack growth resistance dominates the fatigue resistance of FMLs at low temperatures. Despite the high residual stresses induced by the combination of curing and ambient temperature, the resistance to initiation and crack propagation increases.

Kieboom [12] investigated the influence of the ambient temperature on the fatigue initiation properties of GLARE3-3/2-0.3. The fatigue initiation life N_i was recorded for open hole tension specimens with a theoretical stress concentration factor of $K_t = 2.7$. In reference to Chap. 7, the reader should realize that the effective stress concentration factor in the aluminium layers of this laminate is slightly different.

Kieboom's results illustrate how the fatigue properties of FMLs are governed by the fatigue properties of the constituent materials. Figure 11.4 shows the comparison of the fatigue initiation life curves for 1.6-mm monolithic aluminium and the 0.3-mm aluminium layers in GLARE3-3/2-0.3. One can observe that the curves for room temperature and 70 °C are similar, but that the curves for −55 °C are clearly higher.

The consequence of this observation is that once ambient temperatures are considered in the fatigue initiation assessment with methods as described in Chap. 7, then different S–N curves should be used for the constituent material. Because the stress amplitudes for GLARE3-3/2-0.3 in Fig. 11.4 are determined using the method described in Chap. 7, the strong correlation between both sets of curves supports this concept.

Another observation one may make based on Figs. 11.1 and 11.4 is that at elevated temperatures, the residual stresses due to curing are smaller (reducing the mean stresses of the stress cycles), but that the initiation behaviour at 70 °C is similar to the one at room temperature. Hence, the initiation life at elevated temperatures may not really be affected by the elevated ambient temperatures [13]. This, however, does not hold for fatigue crack propagation, see Figs. 11.5 and 11.6.

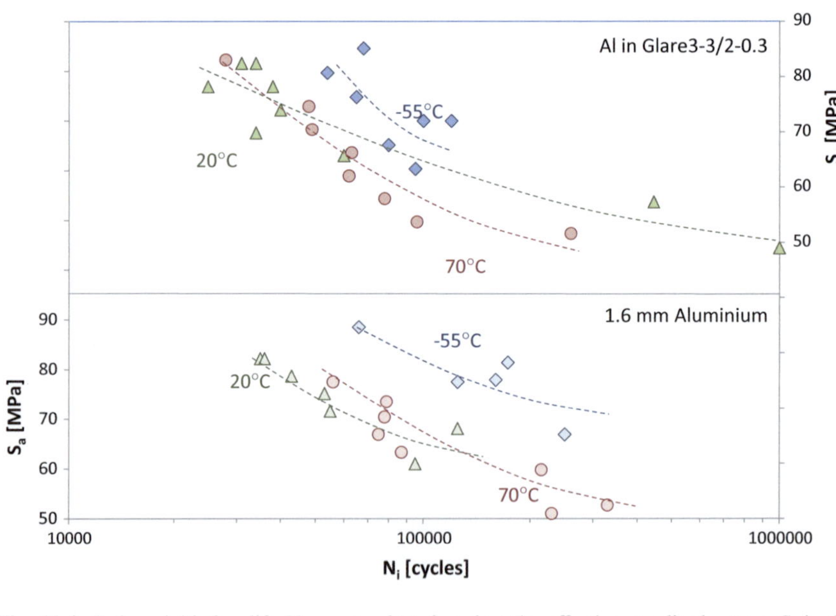

Fig. 11.4 Fatigue initiation life N_i curves plotted against the effective amplitude stress S_a in the metal sheet, data from [12]

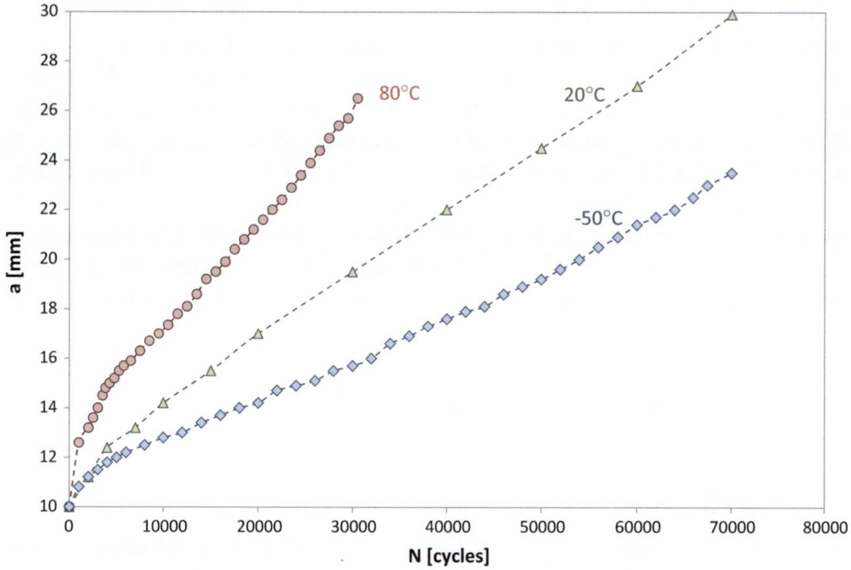

Fig. 11.5 Fatigue crack growth curves of GLARE3-3/2-0.3 (R-glass/AF163-2) tested at S_{max} = 140 MPa, R = 0.1. Initial notch length $2a_o$ = 20 mm [17]

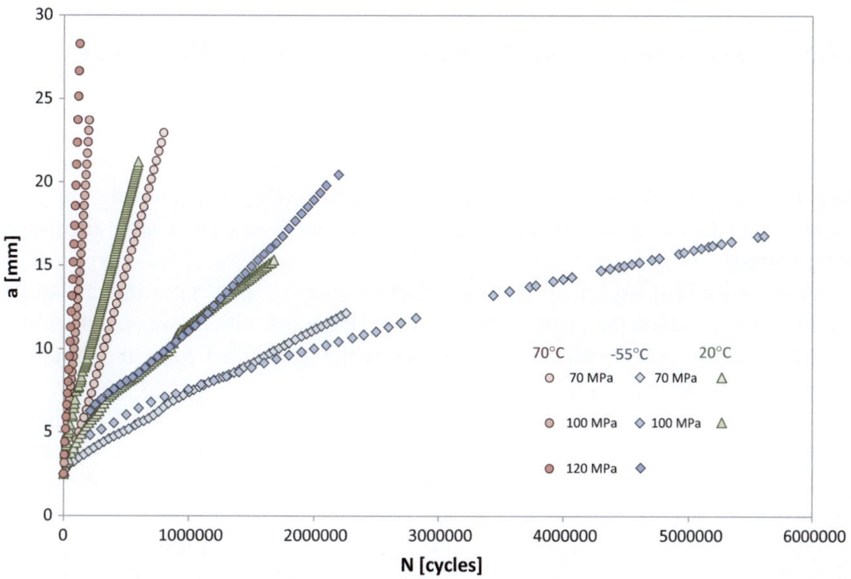

Fig. 11.6 Fatigue crack growth curves of GLARE3-6/5-0.4 tested at three different maximum stress levels (R = 0.05) at three different ambient temperatures [19]

Here, one should note that the above discussion assumes constant ambient temperature. The ultimate proof whether the fatigue crack initiation life of riveted FML joints can be predicted, however, requires the results from variable temperature fatigue tests as reference. Extensive experimental investigations of the fatigue performance under variable laminate temperature conditions are reported by Beumler [13, 18]. Heating and cooling specimens, e.g. between −30 and +70 °C, demand considerable time depending on the specimen volume. Consequently, the frequency dependency on crack initiation in the metal layers must be considered as additional variable if fatigue crack initiation under variable temperature condition is compared to constant laminate temperature tests results, often generated at 10 Hz.

11.5 Fatigue Damage Growth

Once the fatigue crack has initiated in the metallic layers of FMLs, subsequent damage growth is no longer attributed to the metallic layer properties alone. With increasing crack lengths the adjacent fibre layers contribute to the crack opening restraint by bridging the crack. The damage mechanisms that play a major role in the damage growth phase are discussed in Chaps. 8 and 9. Each of these mechanisms is affected by the ambient temperature in its own way. In the following sections the individual phenomena observed and reported in the literature are discussed in more detail.

11.5.1 Temperature and Fatigue Crack Growth Resistance of Metals

Several authors have investigated the influence of environment and temperature on fatigue crack growth resistance of aluminium alloys. Depending on the environment, either faster or slower crack growth rates were observed with increasing temperatures.

Vogelesang [20] explained that the effect of temperature relates directly to the environment in which the crack growth occurs. He reported increasing crack growth rates for increasing temperature in salt water, while in distilled water the trend was observed to be the opposite.

Although early studies on crack propagation in aluminium alloys are highly informative with respect to the micromechanical aspects influencing the crack growth, basic crack growth tests are sufficient to feed the damage growth models discussed in Chap. 9.

For that reason Homan [21] performed fatigue crack growth tests on aluminium 2024-T3 sheets with thicknesses of 0.4 mm, as typically applied in GLARE fuselage shell structures. The observed trend corresponds with the fatigue crack growth tests in salt water reported by Vogelesang [20]; the crack growth rates were higher

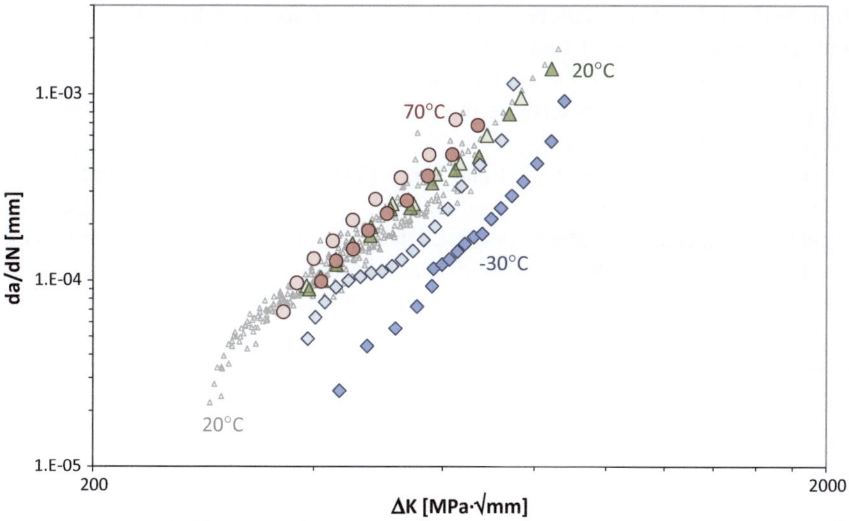

Fig. 11.7 Fatigue crack growth resistance of thin aluminium 2024-T3 sheets tested at three levels of ambient temperature, data from [21, 22]

at 70 °C compared to room temperature and lower at −30 °C. A correlation between the data and data obtained earlier by Homan [22], see Fig. 9.15, is presented in Fig. 11.7.

Aside from the effect temperature may have on the crack growth rate, the transition between the tensile mode and the shear mode was reported to be affected as well [20]. Because the transition is preceded by the formation of shear lips, Schijve [16] stated that the transition is more evident in thin sheets compared to thick sheet material, as discussed before in Sect. 9.6.

11.5.2 Temperature and Fatigue Delamination Resistance

The decohesion of the individual lamina in an FML, known as delamination, often occurs as a cohesive failure mechanism rather than an adhesive failure mechanism between the metal layers and the composite plies. Hence, the delamination growth characteristics are greatly determined by the matrix material, as discussed previously in Chap. 8.

This implies that when ambient temperatures are concerned, the delamination resistance in FMLs is governed by how well the matrix material responds to different temperatures. Most epoxies are known to be more brittle at low temperatures, while they tend to be more ductile at elevated temperatures.

Few studies address the influence of the ambient temperature on the delamination resistance of FMLs. Most studies addressing the influence of temperature limit their work to the overall fatigue crack growth performance of the FML. However, with the demonstration of how that fatigue performance relates to the individual fatigue characteristics of the metal constituent and the metal/fibre interface in Chap. 9, it is necessary to separate the influence of ambient temperature on these two constituents.

To that end, van Kesteren [23] and Schut [24] performed ply-interrupt delamination growth experiments at low and elevated temperatures. The advantage of the ply-interrupt specimen configuration over the end-notched flexure test geometry is that the strain energy density at the delamination tips remains independent of the delamination length. Hence, one could estimate the delamination length throughout the test, simply based on the specimen compliance. After a first calibration there is no need to visually measure the delamination lengths from the specimen's side [25]. In particular when investigating the influence of temperature and environment on the delamination growth resistance this is helpful, because it does not require opening the climate chamber for each measurement.

Obviously, one has to calibrate the compliance with the delamination growth first, which requires paying attention to the thermal expansion of the constituents in the delaminated and non-delaminated regions of the specimens. In addition, the ply-interrupt specimen has the disadvantage that not all delaminations propagate at the same rate. Compliance calibration then averages the response over all delaminations. A more precise method of evaluating the ply-interrupt delamination test configuration for that purpose has been discussed by Pascoe et al. [26].

In the delamination growth experiments performed by van Kesteren [23] and Schut [24] the four delaminations were measured individually, after which they were averaged in order to obtain a single delamination growth rate. This procedure has been explained in more detail in [27].

The results of the delamination growth experiments at elevated temperatures are illustrated in Fig. 11.8. From room temperature up to 40 °C, only a marginal decrease in delamination growth resistance was observed, while the tests at 70 °C revealed a significant reduction.

Interesting in particular in the study by Schut [24] is the observation that the delamination growth resistance curves for temperatures below 0 °C all seem to coincide with the room temperature curve in Fig. 11.8. Hence, one would assume that the delamination resistance for low temperatures is not different from room temperature. However, when testing centre crack tension fatigue crack growth specimens made of GLARE, the delamination shapes for low temperatures are substantially smaller than for room temperature. When assessing the results with the crack growth prediction methodology explained in Chap. 9, it appears that the smaller delamination shapes could only be explained when assuming delamination resistance curves well below the curve for room temperature. This discrepancy has not yet been well explained.

The observations by van Kesteren [23] and Schut [24] correspond, however, with the observations by Deutekom [28] who performed fatigue crack growth tests on GLARE in both laboratory air and water environments at two ambient temperatures. The measured delamination areas at the end of the tests are plotted in

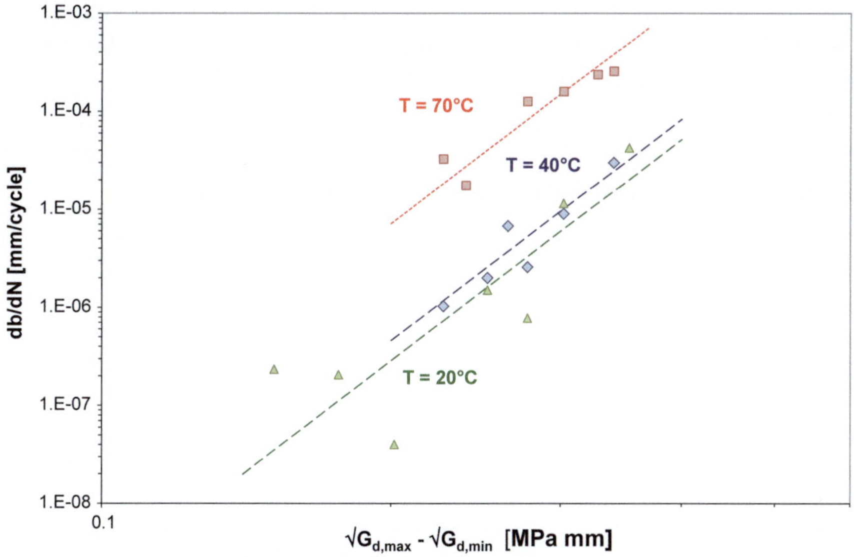

Fig. 11.8 Paris-type relationship plotting the delamination growth rates against the strain energy release rate, illustration of the effect of temperature [19, 23, 24, 29]

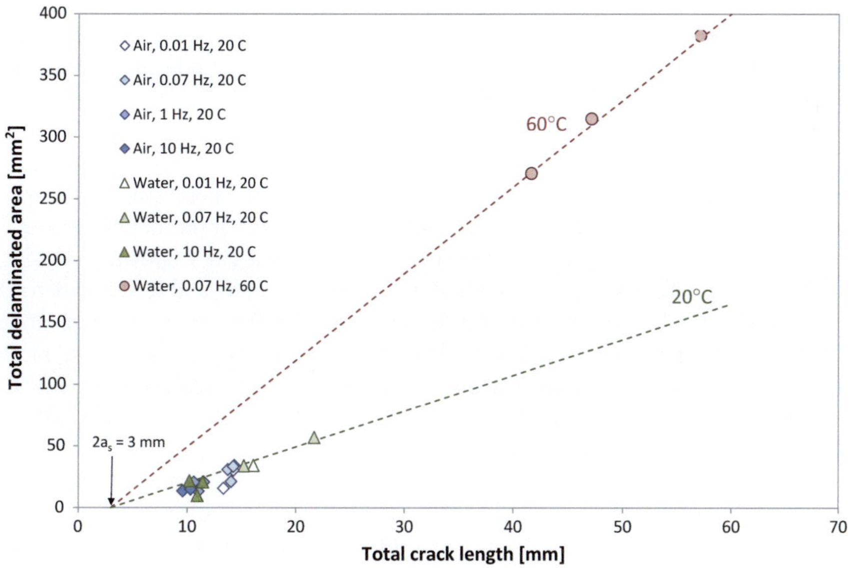

Fig. 11.9 Difference between delamination growth at room temperature and at 60 °C based on the correlation between total delaminated area and total crack length, data from [28]

Fig. 11.9. The environment, i.e. air or water, seems to have negligible influence on the development of the delamination shapes. However, increasing the ambient temperature to 60 °C almost doubled the recorded delamination areas.

11.5.3 Influence of Temperature on Damage Growth in FMLs

The two previous sections focussed on the effect of temperature on the individual constituents, i.e. metal layers, and fibre/metal interfaces. Combining these two aspects allows the evaluation of the influence of the temperature on the fatigue damage growth characteristics of FMLs.

In theory, the methodology explained in Chap. 9, with input from the studies discussed in the previous two sections, should be able to explain the observations on centre crack tension fatigue crack specimens.

Various studies have been performed into the influence of temperature on the fatigue damage growth in FMLs. The earlier studies, for example by van der Hoeven [30] and by Huijzer [31], mostly reported the experimental observations, because a proper theory to compare the results with was not available.

The strength of the theory explained in Chap. 9, in combination with the constituent input characteristics discussed in two previous sections, has been demonstrated by Rans et al. [19]. The comparison between the predicted and experimentally measured fatigue crack growth for both standard GLARE laminates, and FMLs based on titanium sheets and carbon fibre/epoxy layers, has proven the validity of the methodology, and with that highlighted the understanding of the fatigue damage mechanisms in FMLs.

In particular, the case of titanium–carbon FMLs in [19] has underlined the influence of the history of fatigue loading in combination with the temperature; in the experiments to which the predictions are compared, there was a stepwise increase of both temperature and load amplitudes. In these tests, both increase in load amplitude and an increase in temperature accelerated crack growth.

This implies that FML panels with identical crack lengths and tested under the same loading conditions may reveal different crack growth rates if the temperature history of both specimens is different. This observation is illustrated in Fig. 11.10 where a test initially performed at room temperature was continued at −40 °C. Comparison of the observed crack growth rates with the test that was made entirely at −40 °C clearly shows this influence.

In this context it is noteworthy that Beumler [13] performed fatigue crack opening measurements on specimen at particular temperature levels, for fatigue cracks that had initiated and propagated at different temperature levels. Hence, the work illustrates the influence of different delamination sizes on the crack opening, which further explains the observation in Fig. 11.10. Obviously, both studies lead

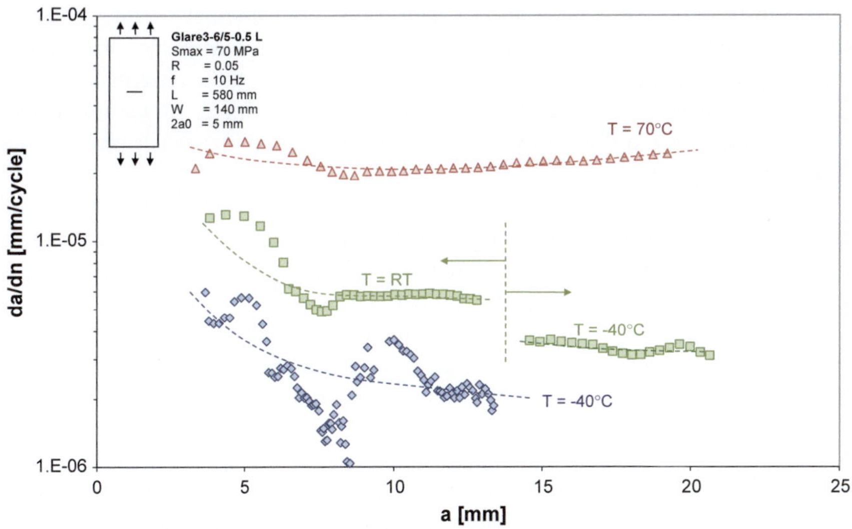

Fig. 11.10 Illustration of fatigue crack growth rates measured in GLARE3-6/5-0.5 tested in the L direction for three different temperatures; continuation of a RT test at −40 °C resulted in a crack growth rate higher than the crack growth rate measured in the −40 °C test [19]

to the question how to approach the complexity of crack propagation (and delamination growth) at variable temperatures.

This influence of delamination size on subsequent crack growth in fact seems similar to the case presented by van der Hoeven and Schra [9], who interrupted constant amplitude fatigue crack growth tests on GLARE3-5/4-0.4 with 1000 h of accelerated exposure. The influence of the exposure cycles on the delamination shapes, and the influence these shapes have on the crack growth, seems somewhat similar [32], see also Sect. 12.3.4.

11.6 Thermal Fatigue

Considering the effect of temperature on the residual stresses within each layer of the laminate, discussed in the previous section, the effect of thermal cycling might be an aspect to be addressed even when no additional mechanical loads are applied. This aspect may become particularly important when heating and cooling are applied to FMLs, as is the case, for example, with anti-icing and de-icing systems [33].

For such applications, the method of heating and cooling should be considered. This is explained by Müller et al. [33–35] who investigated the performance of GLARE under the application of thermal cycling. One could, for example, heat and cool the FML externally, by changing the ambient temperature. The effect of such a thermal load is smaller than for the case where the heating is induced with a mesh positioned internally within the laminate.

To illustrate this, Müller et al. [33] performed both thermal cycling methods. They reported that external heating and cooling for 12,000 cycles has limited influence on the interlaminar shear properties. In fact, they observed that the interlaminar shear strength tends to increase with the number of thermal cycles applied, which they attributed to the microstructure, based upon their observations during heat capacity tests.

This increasing trend seems supported by recent work of Li et al. [36] who applied up to 1000 thermal cycles to a novel FML based on an aluminium–lithium alloy. They could not observe any degradation in either the tensile strength, flexural strength or the fatigue crack growth properties. Instead, they also report a slight increase in tensile and flexural strength with the number of applied thermal cycles. Interestingly enough, the authors attributed the increase in strength values to positive age hardening behaviour of the aluminium–lithium rather than the ageing of the matrix material.

The slight increase in strength values was also observed by Da Costa et al. [37] who applied up to 2000 thermal cycles to FMLs with a 2/1 lay-up containing 1.6-mm aluminium outer layers. They concluded based on their experiments that thermal cycling does not have a noticeable effect on the static strength and the interlaminar shear strength. However, when discussing the results, their first explanation is related to potential thickness variations. Only as a second possibility do they identify the possible influence of precipitation hardening.

In the thermal cycling tests of Müller et al. [33], in which the heating is applied internally using a heater mesh while cooling from the outside, a reduction in strength of about 5–8% was observed. Partly, the reduction is induced by the presence of the mesh, as the authors explain. They report a 1.6% reduction in interlaminar shear strength due to the mesh.

The internal heating with a mesh imposes high stresses locally, in particular in the beginning of the heating cycle. The local high heating rates increase the temperature of the material in the direct vicinity of the heater mesh faster than that of the surrounding material. The differences in coefficients of thermal expansion then cause significant stresses. These heating rate differences are smaller, when heating is applied externally.

References

1. Laliberté J, Mahendran M, Djokic D, Li C, Kratz J (2007) Effect of process-induced residual stresses on mechanical properties and fatigue initiation in fibre metal laminates. In: Proceedings of the 24th ICAF symposium, Napels, Italy
2. Mensink JLM (1994) Thermal properties and possibilities for space applications for fibre metal laminates. Master's thesis, Delft University of Technology, Delft
3. Graafmans GMH (1995) Thermal behaviour of fibre metal laminates. MSc thesis, Delft University of Technology, Delft
4. Verolme JL (1996) Thermal behavior of fibre metal laminates, report TD-R-95-024 (issue 2). Structural Laminates Company, Delft

5. Sjöström J (2012) Thermal properties of GLARE: heat capacity, anisotropic thermal conductivity, and thermogravimetric analysis—Benchmark material WP1. Report NMP3-LA-2010-246037 Project FIRE-RESIST
6. Horst P (1995) Methods fibre metal laminates (subtask 5.1), brite/euram-2040. Technical report MBB TN-TK536-5/95, DBAA
7. Boertien MFHC (1996) Strength of GLARE after exposure to moisture. Master's thesis, Delft University of Technology, Delft
8. van der Hoeven W, Schra L (1999) Interim report, qualification testing for the Glare panel in the A310p/F MSN 484 aircraft. Results of tests on non-exposed specimens. Technical report NLR-CR-99269, National Aerospace Laboratory NLR
9. van der Hoeven W, Schra L (2000) Final report Glare durability program, Results of tests carried out at NLR. Technical report NLR-CR-2000-237, National Aerospace Laboratory NLR
10. Hagenbeek M (2005) Characterisation of fibre metal laminates under thermo-mechanical loadings. PhD dissertation, Delft University of Technology, Delft
11. MMPDS (2003) Metallic materials properties development and standardization (MMPDS-01). U.S. Department of Transportation
12. Kieboom O (2000) Fatigue crack initiation and early crack growth in glare at different temperatures. MSc thesis, Delft University of Technology, Delft
13. Beumler T (2004) Flying GLARE®, a contribution to aircraft certification issues on strengths properties in non-damaged and fatigue damaged GLARE® structures. PhD dissertation, Delft University of Technology
14. Homan JJ (2006) Fatigue initiation in fibre metal laminates. Int J Fatigue 28:366–374
15. Alderliesten RC (2007) Analytical prediction model for fatigue crack propagation and delamination growth in Glare. Int J Fatigue 29(4):628–646
16. Schijve J (2001) Fatigue of structures and materials. Kluwer Academic Publishers, Dordrecht
17. Bär H (1992) Verifikation und Ergänzung von Berechnungsmethoden für die statische und dynamische Auslegung von Glare-Strukturen. Diplomarbeit, Institut für Flugzeugbau der Universität Stuttgart
18. Beumler T (2010) Influence of frequency and variable temperature on the fatigue crack initiation and crack propagation behaviour of Standard-GLARE riveted joints. Airbus report L53D05006967
19. Rans CD, Alderliesten RC, Benedictus R (2011) Predicting the influence of temperature on fatigue crack propagation in Fibre Metal Laminates. Eng Fract Mech 78:2193–2201
20. Vogelesang LB (1979) The effect of environment on the transition from tensile mode to shear mode during fatigue crack growth in aluminium alloys—a model for environmentally assisted crack growth, report LR-286. Delft University of Technology, Delft
21. Homan JJ (2001) Crack growth properties of thin aluminium sheets at various temperatures, report B2V-02-39. Delft University of Technology, Delft
22. Homan JJ (2001) Crack growth properties of thin aluminium sheets, report B2V-01-16 (issue 2). Delft University of Technology, Delft
23. van Kesteren R (2005) Elevated temperature and moisture absorption effects on delamination growth inn GLARE. Graduation report, Hogeschool INHOLLAND, Delft University of Technology, Delft
24. Schut JE (2006) Glare: delamination growth at low temperatures. Internship report, Delft University of Technology, Delft
25. Doevendans LP (2008) Delamination behaviour of adhesive films, Report B2v-07-03. Delft University of Technology, Delft, The Netherlands
26. Pascoe JA, Rans CD, Benedictus R (2013) Characterizing fatigue delamination growth behaviour using specimens with multiple delaminations: The effect of unequal delamination lengths. Eng Fract Mech 109:150–160
27. Alderliesten RC, Schijve J, Van der Zwaag S (2006) Application of the energy release rate approach for delamination growth in Glare. Eng Fract Mech 73:697–709
28. Deutekom MJ (1994) The effect of frequency, moisture and temperature on the constant amplitude fatigue behaviour of Glare 3. Master's thesis, Delft University of Technology, Delft

29. Schut JE, Alderliesten RC (2006) Delamination growth rate at low and elevated temperatures in GLARE. In: Proceedings of the 25th international congress of the aeronautical sciences, ICAS2006
30. van der Hoeven W (1984) Temperature effects on the fatigue crack growth behaviour of ARALL. NLR report TR-84048, NLR
31. Huijzer EL (1992) Durability of GLARE—The influence of water and temperature on the low-frequency behaviour of GLARE. MSc thesis, Delft University of Technology, Delft
32. Alderliesten RC (2001) Fatigue. In: Vlot A, Gunnink JW (eds) Fibre metal laminates—an introduction. Kluwer Academic Publishers, Dordrecht, the Netherlands
33. Müller B, Hagenbeek M, Sinke J (2016) Thermal cycling of (heated) fibre metal laminates. Compos Struct 152:106–116
34. Müller B, Anisimov AG, Sinke J, Groves RM (2015) Analysis of thermal strains and stresses in heated fibre metal laminates. In: Proceedings of the 6th international conference on emerging technologies in non-destructive testing (ETNDT), pp 1–6
35. Müller B, Teixeira De Freitas S, Sinke J (2015) Thermal cycling fiber metal laminates: Considerations, test setup and results. In: Proceedings of the 20th international conference on composite materials (ICCM), Copenhagen, Denmark, 2015, paper no. 4212-3,1-11
36. Li H, Hu Y, Liu C, Zheng X, Liu H, Tao J (2016) The effect of thermal fatigue on the mechanical properties of the novel fiber metal laminates based on aluminum–lithium alloy. Compos A 84:36–42
37. Da Costa AA, Da Silva DFNR, Travessa DN, Botelho EC (2012) The effect of thermal cycles on the mechanical properties of fiber–metal laminates. Mater Des 42:434–440

Chapter 12
Effect of Environment

Abstract The advantageous properties of FMLs with respect to environmental conditions are explained based on the different diffusion characteristics compared to composites. The chapter illustrates how various environments influence the mechanical, fatigue and damage tolerance properties. In this context, the effect of test frequency on the fatigue performance is illustrated.

12.1 Introduction

The assessment of durability aspects of FMLs comprises not only the influence of temperature, as discussed in the previous chapter, but also the assessment of strength reduction over time due to environmental effects.

For metallic materials, this generally incorporates the assessment of corrosion and corrosion fatigue as a result of humid environments. Advocates of composites, or fibre-reinforced plastics, often emphasize the advantage of these materials in this context, stressing that composites do not suffer from corrosion. However, despite the abundant use of this argument, the truth is that composites suffer from environmental circumstances as well. In particular, where hot and humid conditions are considered. The difference in response to this environment between metals and composites is that composites absorb moisture from the environment, which depending on the chemical and/or physical stability of the material may cause serious degradation of properties.

The degradation of properties is thus related to the amount of moisture absorbed and its distribution in the material. Hence, durability assessments often evaluate the moisture absorption characteristics of composites and their effect on the mechanical and physical properties. The results are then translated into knock-down factors on the pristine material mechanical properties.

Because FMLs comprise both metallic layers and also composite plies, this aspect must also be considered in the durability assessment of FMLs. This chapter discusses studies that have been performed in the past to evaluate the influence of environment on the performance of FMLs, in particular the FMLs ARALL and GLARE.

R. Alderliesten, *Fatigue and Fracture of Fibre Metal Laminates*, Solid Mechanics and Its Applications 236, DOI 10.1007/978-3-319-56227-8_12

12.2 Moisture Absorption

With respect to moisture absorption, an important difference between fibre-reinforced plastics and FMLs must be considered. Moisture absorption is a process that is not present in metal. Hence, the metallic layers protect the composite plies against moisture absorption, reducing the problem from a three-dimensional diffusion process (fibre-reinforced plastics) to a two-dimensional diffusion process (FMLs). This aspect is illustrated in Fig. 12.1.

12.2.1 Planar Diffusion of Moisture

Where exposure to wet environments easily results in moisture ingress and saturation of composite materials, FMLs significantly delay the diffusion, because the moisture can only access the composite constituent from the free edges. As illustrated in Fig. 12.1, these edges are only present at panel edges and at any location where a hole or cut-out has been created.

In this case, it becomes important not only to quantify the amount of moisture absorbed by the material, but to consider the structural consequence of the diffusion. The distribution of moisture in FMLs is entirely different to that in composites.

To capture the difference at a material level, Verbruggen [1, 2] performed various studies addressing the durability aspects of FMLs, in particular ARALL. He performed exposure tests where FML samples were immersed in distilled water while monitoring the moisture absorption over the time of exposure. The

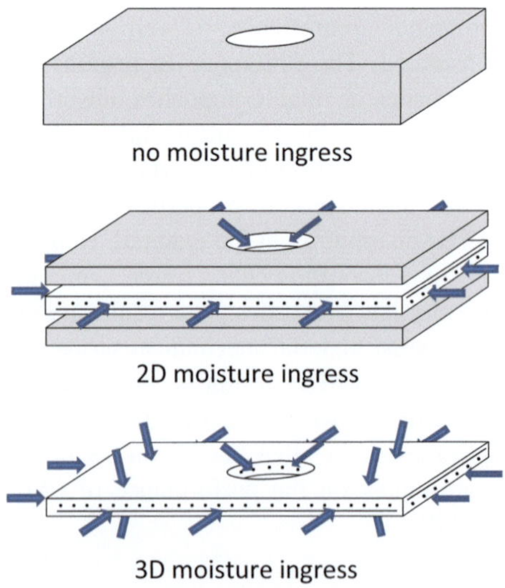

Fig. 12.1 Comparison between the moisture ingress characteristics of metals, fibre metal laminates and composites

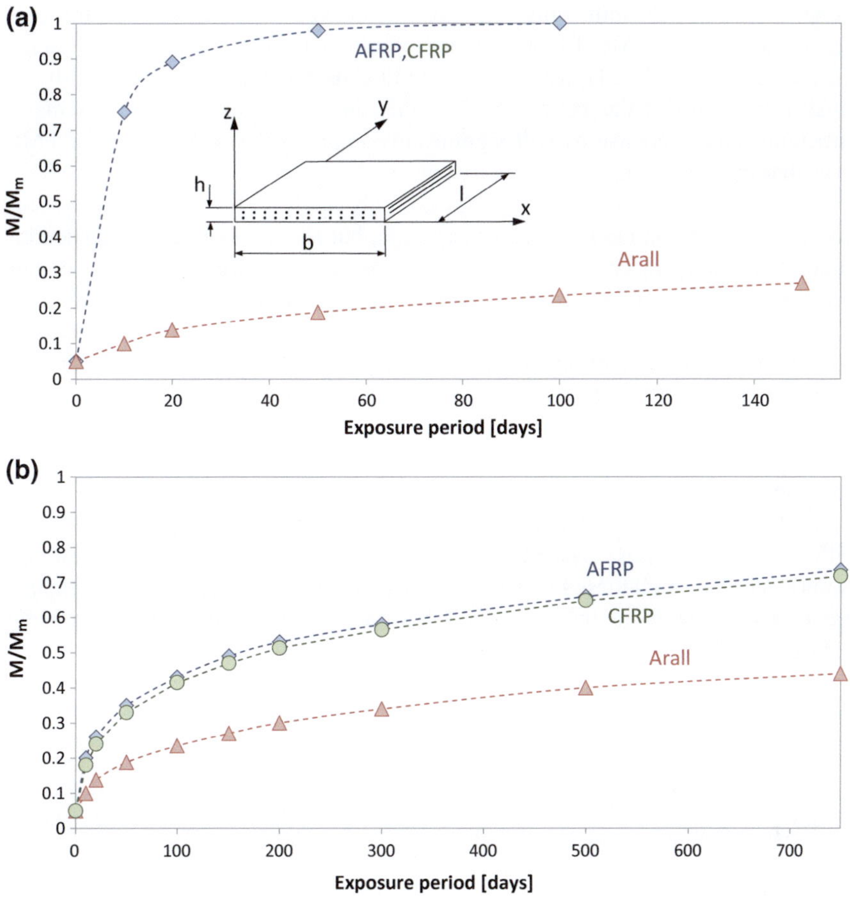

Fig. 12.2 Comparison between the moisture absorption of ARALL (expressed as moisture content M normalized by the maximum moisture content M_m), AFRP (aramid fibre-reinforced plastic) and CFRP (carbon fibre-reinforced plastic) for a 60 °C distilled water environment; specimen dimensions: b × l × h = 50 × 50 × 1 mm (**a**) and b × l × h = 50 × 50 × 10 mm (**b**), data from [1]

comparison with the moisture absorption characteristics of aramid- and carbon fibre-reinforced plastics is given in Fig. 12.2.

The tests were performed with samples of different thicknesses to illustrate the influence of effective distance to free surfaces if the volume changes, while the diffusion rates are specific for the material. Figure 12.2 clearly shows that when the samples are thin, i.e. 1 mm thickness, the moisture quickly ingresses into the composite samples. In the ARALL samples however, the moisture uptake is substantially lower, because diffusion only takes place from the edges of the sample.

The difference becomes relatively smaller when the sample thickness increases, which is discussed by Verbruggen in [2]. The exposure tests reported in [1] use samples of 50 × 50 mm while varying the thickness. Increasing the thickness brings the curves for ARALL and the fibre-reinforced plastic counterparts closer, as illustrated in Fig. 12.2. However, one should consider how this is affected by the small dimensions of the sample. Scaling the sample dimensions to representative structural panel dimensions will significantly improve the performance of FMLs over that of composites.

This aspect of test dimensions does not only apply to the amount of moisture taken up over the amount of time of exposure, but also apply to the corresponding effect it has on the mechanical or physical properties. Take for example the strength tests performed by Boertien [3] on various unidirectional and cross-ply laminates.

The tensile tests were slender specimens that had a width of only 12.5 mm in the section of interest. Exposure of these samples to distilled water or salt-spray environments significantly reduces the advantage of protection by the outer metal layers as observed in the tests of Verbruggen, see Fig. 12.2. Hence, the moisture uptake from the environment is more significant in Boertien's tests, with a clear reduction in strength of the material as result.

Boertien [3] exposed GLARE2-3/2-0.3 and GLARE3-3/2-0.3 laminates with different fibre and epoxy systems to salt-spraying. The result on the tensile strength is fairly similar for all these laminates as illustrated in Fig. 12.3. The longer samples are exposed, the more moisture ingresses and the more mechanical properties reduce.

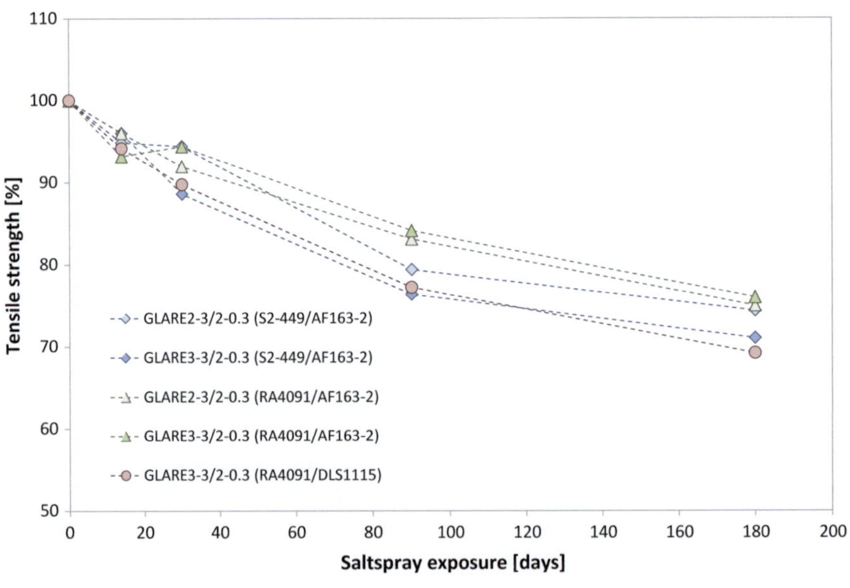

Fig. 12.3 Effect of salt-spray exposure on the tensile strength of GLARE2-3/2-0.3 and GLARE3-3/2-0.3 [3]

However, one should keep in mind that the reduction of 20–30% in strength after 180 days of exposure applies to the specific geometry of this test; the results are at most representative in these areas of FML structures that are in proximity of free edges.

12.2.2 Relevance of Exposure Type

Another aspect that should be considered is the nature or type of the exposure. Boertien [3] compared the exposure to salt-spray with exposure to distilled water by immersion. An example is illustrated in Fig. 12.4.

However, both types of exposure constitute each a different form of accelerated exposure, which is an overestimation of what happens during outdoor exposure over the years of structural operation.

To indicate the relevance of exposure to salt-spray or distilled water, illustrated in Fig. 12.4, Boertien also performed outdoor exposure tests. Samples were put on the roof top of the Faculty of Aerospace Engineering, TU Delft, representing a typically Dutch moderate sea climate. The comparison between the outdoor-exposed samples and the ones exposed to distilled water immersion is given in Fig. 12.5. Comparing this figure with Fig. 12.4 illustrates the range in properties one may obtain depending on the nature or type of exposure applied.

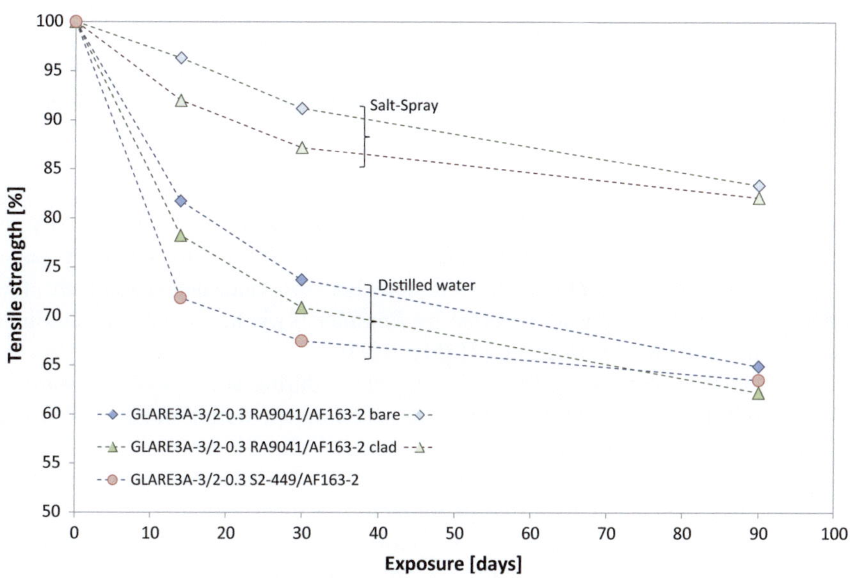

Fig. 12.4 Comparison of the strength reduction of GLARE3-3/2-0.3 as result of salt-spray exposure and exposure to distilled water, data from [3]

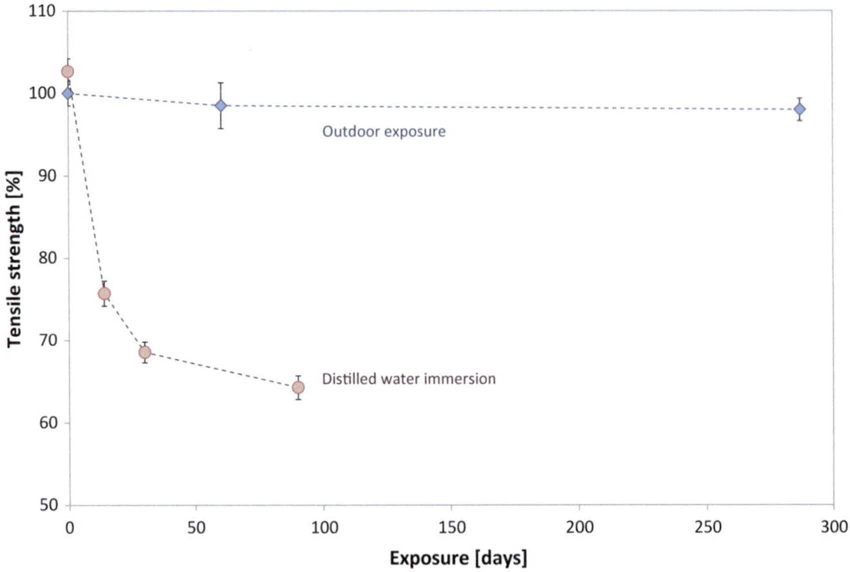

Fig. 12.5 Comparison of the strength reduction of GLARE3-3/2-0.3 as result of outdoor exposure (moderate sea climate TU Delft) and distilled water immersion, data from [3]

Most studies on environmental aspects of FMLs for the development of the aircraft fuselage structure of the A380 considered accelerated exposure at 70 °C/85%RH for about 3000 h. However, as discussed in more detail by Beumler [4] this still is a considerable exaggeration of the influence of environment in operational conditions.

For example, Beumler discusses the correlation between accelerated exposure to 70 °C/85%RH with outdoor exposure results from a programme where GLARE samples were exposed at the Aeronautical & Maritime Research Laboratory in Queensland, Australia (site located at latitude 17.4° south and 146° longitude east).

Considering the different climate conditions between Queensland and Delft, the influence of outdoor exposure reported by Beumler [4] is likely more severe than the influence reported by Boertien [3] (Fig. 12.5).

But even then, comparing the moisture uptake during accelerated exposure at 70 °C/85%RH for about 3000 h with the moisture uptake during outdoor exposure Queensland, Australia [5], Beumler [4] concluded that 30 years of operation of short-range aircraft could be simulated with 500 h of accelerated exposure, while for long-range aircraft the exposure period could be even less.

Here, it should be noted that the moisture reference panels exposed in Queensland clearly revealed variations in weight depending on the season. Hence, the diffusion of moisture into the specimens is not constant and in fact is even reversible [5]. This is entirely different from the accelerated exposures where moisture continuously diffuses into the samples.

Beumler [4] further illustrates that the weight gain as result of accelerated exposure at 30 °C/70%RH is significantly less severe compared to that at 70 °C/85%RH, but still more severe than outdoor exposure. This seems in agreement with the influence of the level of relative humidity discussed Verbruggen [1], who compared levels between 30 and 80%.

The other aspect also discussed by Beumler [4] concerns the influence of dry-out time. Comparison of strength values obtained after exposure with those obtained after subsequent dry-out time reveals that dry-out improves the strength. This is in agreement with the improvement reported by Boertien [3], illustrated in Fig. 12.13.

Hence, one may conclude that most studies looking at the environmental aspects of GLARE for the application to the Airbus A380 using accelerated exposure at 70 °C/85%RH are significantly conservative in comparison with true operational conditions. However, one has to keep in mind that he correlation between accelerated exposure and outdoor exposure includes diffusion into and out of the specimens depending on the seasons. That being said, even aircraft operated in tropical environments spends a significant amount of time during cruise in ambient environments at low temperatures where diffusion does not occur [5]. This indicates that for each FML application one has to define the appropriate correlation with accelerated exposure tests, based on anticipated operational conditions.

12.3 Effects of Moisture Ingress

12.3.1 Static Strength

The static strength of FMLs decays over time when exposed to various environmental conditions, as observed and reported in various studies. Figures 12.3 and 12.4 clearly illustrate this observation. However, the reduction in strength observed when exposing slender tensile test specimens to various environments is exaggerated compared to realistic structural applications in operational conditions. This can be illustrated with the comparison between tensile strength reduction and blunt notch strength reduction, reported by Boertien [3]. The relative difference between the reduction in tensile strength and blunt notch strength is illustrated in Fig. 12.6.

This difference can easily be explained considering the different moisture diffusion that occurs in both specimen geometries. The slender tensile test specimens rapidly pick up the moisture and are more quickly saturated in the area of interest, i.e. failure, than is the case for blunt notch.

Hence, one could wonder what the relevance is of exposing tensile test specimens to various environments, if one considers that the geometry represents a case that is not representative at all for the structural applications of FMLs.

In reference to Fig. 12.1, one has to consider that this lack of relevance is characteristic for elementary FML specimens, while for composites it may be considered more representative. With moisture ingression through the thickness (in

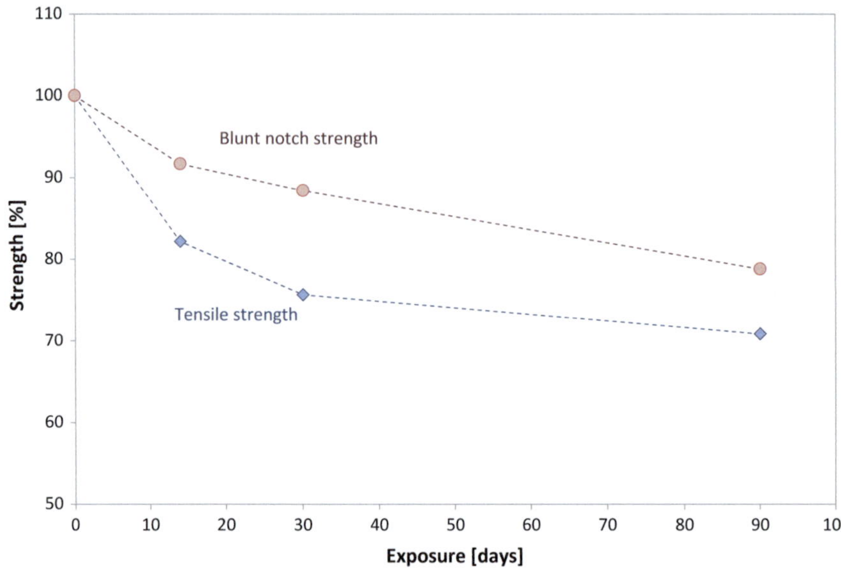

Fig. 12.6 Comparison between the reduction in tensile strength and blunt notch strength of GLARE3-3/2-0.3 after distilled water immersion, data from [3]

FMLs prevented by the outer metal layers), entire composite structures may exhibit the observed strength reduction. In FMLs, this at most can be concluded for the areas in proximity of free edges.

12.3.2 Blunt Notch Strength

The influence of environment on the blunt notch strength could be considered more representative than its effect on tensile strength. The many fasteners in FML structures represent structural locations where environment may have access to the inner layers of the FML and thus may affect the properties.

Various studies have reported the effect of exposure on blunt notch strength. The diffusion rates are different in the direction of the continuous fibres and perpendicular to that direction. Hence, moisture ingress into the specimen through the access of an open circular hole yields different degrees of ingress in the various directions. For cross-ply laminates, the influence may be small because the equal number of fibres is oriented in the two perpendicular directions. However, for unidirectional laminates, like GLARE2, this effect may be more significant.

In any case, considering the assumed relevance for fastener holes in real structures, the open hole configuration should always be accompanied by a specimen that contains fasteners installed in the holes. Filling the hole with a fastener

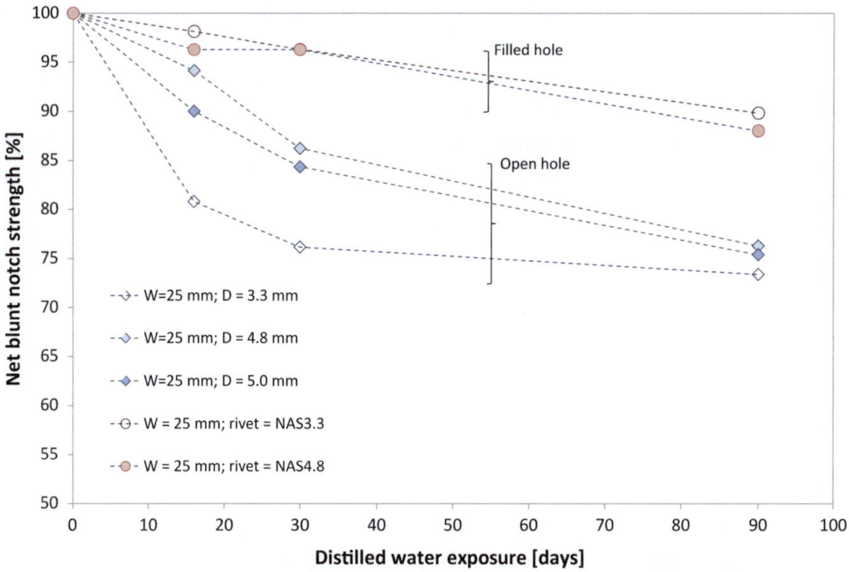

Fig. 12.7 Comparison between open hole and filled hole blunt notch strength reduction of GLARE3-3/2-0.3 after distilled water exposure, data from [3]

significantly affects the access environment has on the inner layers of the FML. This is illustrated by Fig. 12.7.

Beumler reports the observation that the blunt notch strength reduction seems more affected by the diffusion rate in the direction of loading, rather than the diffusion in the net section of the blunt notch specimens.

This can be understood, because the blunt notch strength is mostly dominated by the properties of the continuous fires in loading direction, see also Sect. 5.2.3. Moisture mostly affects the epoxy, rather than the glass fibres. Hence, the influence moisture absorption has in the net section seems mostly limited to the effective load introduction into the fibres in the direct notch vicinity. Here, it appears that the diffusion along the fibres in loading direction influences the shear properties of the epoxy resulting in more splitting and earlier failure of the specimens.

The investigation of the effect of moisture ingress in open hole or filled hole specimens aims to establish the influence of the access the hole provides to the inner layers. This means that the edges of the specimens should be protected from the environment. Otherwise, moisture will also diffuse from the edges into the specimen, which exaggerates the influence of environment. The investigation by Beumler [4], for example, illustrated that artificial exposure to 70 °C/85%RH of open hole blunt notch specimens will link up diffusion fronts in the sample after 500 h.

This aspect of testing was also discussed by Borgonje and Beumler [6] who observed that isolation of free edges of samples never completely prevents moisture diffusion. Hence, they also tested a specimen configuration containing a row of open holes, of which they milled away one hole at both sides directly after exposure. The challenge obviously is to keep the time window between the exposure and testing very small in order to reduce the amount of moisture diffusing out of the specimens, which improves the properties again, see for example Fig. 12.13.

The investigation of the influence of exposure on the blunt notch strength revealed that the effect is nil for the tested cross-ply GLARE4A laminate both for artificial exposure to 70 °C/85%RH and for the outdoor exposure, see Fig. 12.8. For the unidirectional GLARE2B tested in the fibre direction, however, a slight strength reduction can be observed for artificial and 2-year outdoor exposure, but a considerable reduction for the 6-year outdoor exposure. This reduction is not well understood nor explained [6].

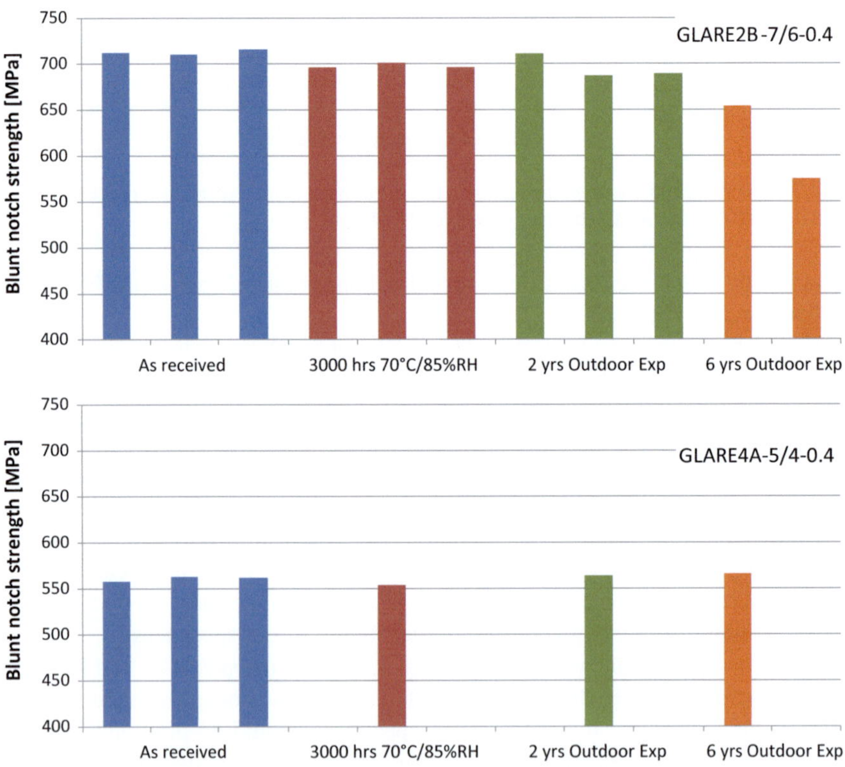

Fig. 12.8 Illustration of the influence of artificial and outdoor exposure on the blunt notch strength of unidirectional and cross-ply GLARE [6]

12.3.3 Delamination Resistance

The observed effect moisture has on the reduction of shear strength of the adhesive, resulting in earlier splitting in blunt notch specimens, could also indicate an effect on the delamination resistance at the metal fibre ply interfaces.

To investigate the effect for mode II fatigue delamination growth, ply interrupt specimens were tested to obtain delamination resistance curves similar to Fig. 8.6. The specimens were straight samples of 30 mm width, which have been exposed for 1400 h at 70 °C/85%RH [7].

Figure 12.9 illustrates that the exposure of the delamination samples of 30 mm wide for 14 h did not yield an apparent influence on the delamination resistance. The few data obtained with these tests coincide with the data obtained for un-exposed specimens.

Because the expectation was that exposure to 70 °C/85%RH would reduce the delamination resistance, the specimens were etched post-mortem to reveal the delaminated areas. An illustration of such an etched surface is given in Fig. 12.10.

What can be seen in this figure is that the delamination front is not straight and perpendicular to the loading direction. This was observed several times with this ply interrupt specimen configuration. Because the growth rate, representing the derivative of the crack growth rate, is correlated to the theoretical strain energy release rate, the absolute difference between the delamination lengths at both sides of the specimen was deemed to be of lesser importance.

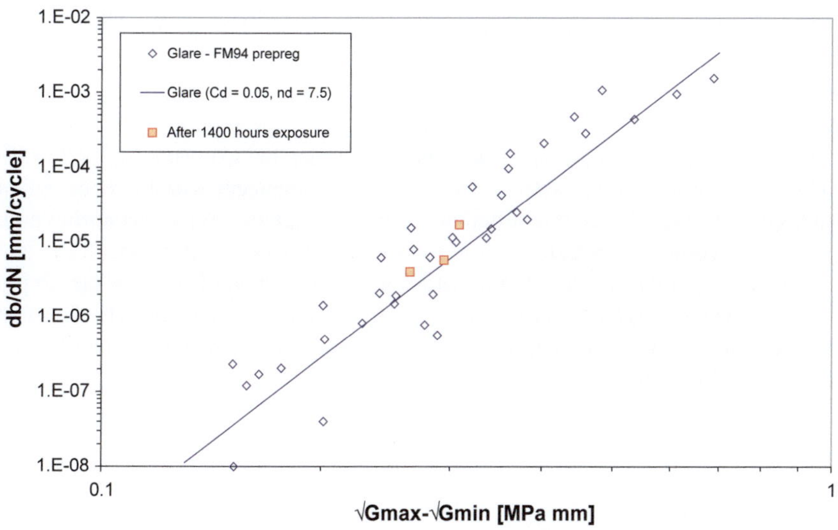

Fig. 12.9 Correlation of the delamination resistance of standard GLARE after 1400-h exposure at 70 °C/85%RH [7]

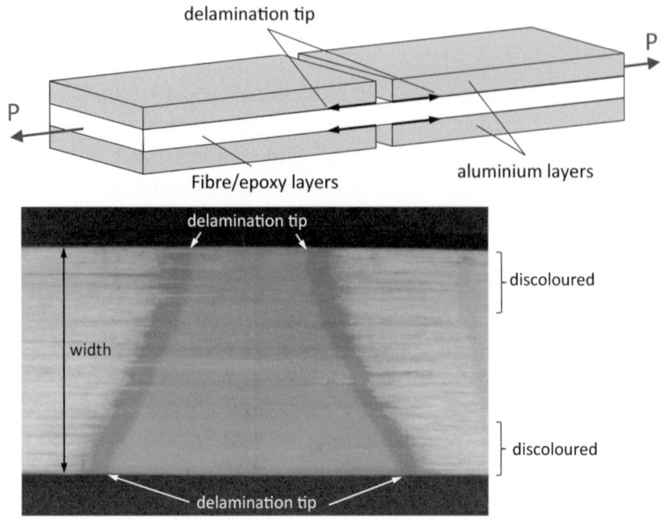

Fig. 12.10 Post-mortem appearance (obtained after removing the aluminium layers) of the discoloured edges of the illustrated ply interrupt delamination specimens after 1400-h exposure at 70 °C/85%RH and subsequent fatigue delamination testing [7]

Figure 12.10 further illustrates that the discolouration observed at the interface, attributed to the moisture ingress during exposure, had not yet reached the entire width of the sample. Due to the low diffusion rates perpendicular to the fibre direction, 1400 h of exposure was insufficient to diffuse the moisture completely through the sample's width.

This could explain why no effect of exposure is visible in Fig. 12.9 in case exposure would have an influence. To reveal this influence however, diffusion through the entire width should be obtained, which could be achieved by testing smaller samples after 300 h of exposure.

On the other hand, the limited influence of accelerated exposure to 70 °C/85% RH on the delamination resistance seems to be in agreement with the observation reported by Bosker [8], who reported no growth of edge delaminations during both tension and compression fatigue after 1000 h of exposure to 70 °C/85%RH.

The case of compression fatigue yields effectively a mode I opening due to ply buckling, which usually is more critical than the mode II opening. The fact that even in that case no delamination growth was observed seems to indicate that the effect could be limited.

12.3.4 Fatigue Crack Growth

An alternative method to verify this observation is to perform fatigue crack growth tests in combination with accelerated exposure. Van der Hoeven and Schra [9] report the results of fatigue crack growth tests on both centre crack tension specimens and double-edge-notched specimens.

What seems particularly interesting in their results is that accelerated exposure of specimens for 3000 h prior to fatigue testing hardly reveals any influence on the growth of the fatigue cracks. If the fatigue crack growth tests are interrupted a few times to expose the fatigue-cracked panel further for another 1000 h, then an influence can be observed.

This is illustrated in Fig. 12.11, where the crack growth curves for non-exposed and exposure of 3000 h prior to the test almost coincide. However, the growth rate of the test that is interrupted for 1000 h of accelerated exposure seems to incrementally increase with every exposure cycle.

Thus, the fatigue crack with corresponding delamination between metal and fibre layers provides access to the environment for further moisture absorption. This is illustrated in Fig. 12.12. After initial exposure, the cracks propagate out of the region of moisture diffusion, resulting in crack growth behaviour similar to the un-exposed case. The interruption with another 1000 h of accelerated exposure diffuses more moisture into the specimen, in particular in the region of the fatigue crack and delamination.

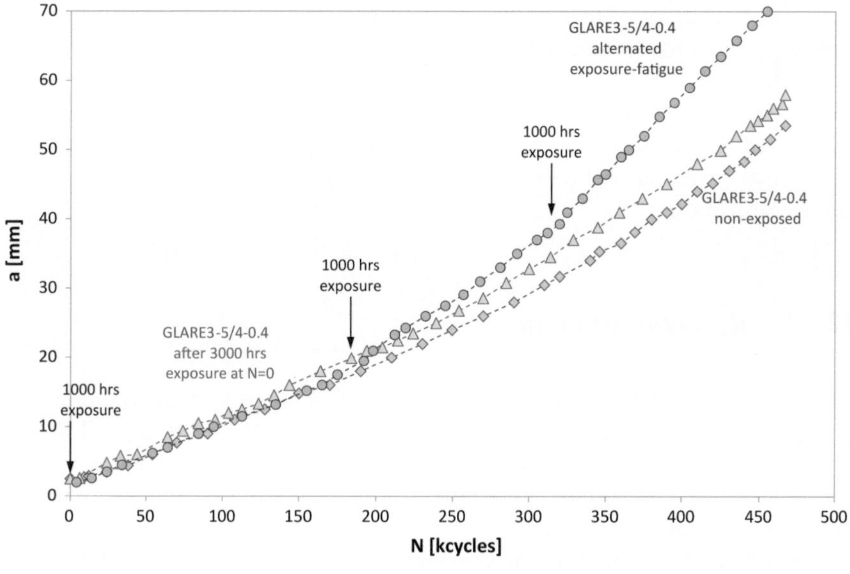

Fig. 12.11 Effect of exposure to 70 °C/85%RH on the fatigue crack growth behaviour of GLARE3-5/4-0.4 [9]

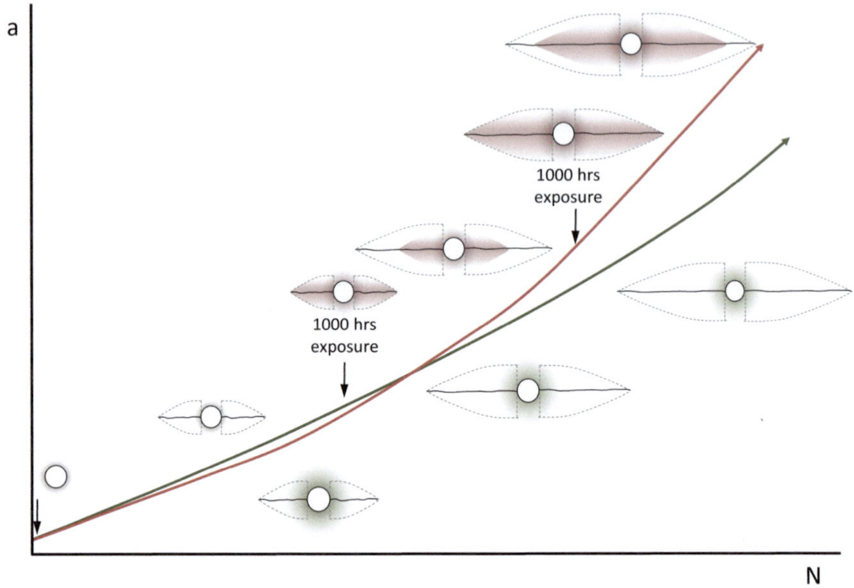

Fig. 12.12 Illustration of the different influence of exposure in the tests reported in Fig. 12.11

In addition, Beumler [4] reports that crack opening in fatigued FML specimen increases after artificial exposure in a 70 °C/85%RH environment. The explanation for this increased opening is that the shear strength and stiffness of the prepreg decrease. Nonetheless, Beumler concludes that this strength reduction is not considered relevant for an aircraft structure during a life time of 30 years.

Figure 12.11 illustrates that the exposure cycles do increase the growth, but that the increase is not substantial. Considering that the influence is induced with accelerated exposure which, as discussed before, exaggerates the influence of operational environmental conditions, one can understand the limited relevance for aircraft structures.

12.3.5 Residual Strength

Boertien [3] reported the results of three residual strength tests on 400-mm-wide centre crack tension specimens with an initial saw-cut length of $2a_0 = 100$ mm. Two of the three samples were exposed to salt-spraying, after which residual strength tests were performed. Very little to no influence of exposure was observed on the residual strength results.

This can be understood, if one considers the failure mechanisms that dominate the residual strength sequence, discussed in Chap. 10. The initial blunting of the crack tip and the development of plasticity dominate the early phase of stable crack extension, after which the crack has propagated out of the region where moisture has diffused into the laminate. Hence, the general conclusion is that environmental conditions hardly influence the residual strength of FMLs.

What remains of interest is the strength of mechanically fastened joints after exposure. The difference between for example lap-joints and the centre crack tension configuration is that the row of fastener holes forms multiple sites where moisture can penetrate the laminate and diffuse into the laminate.

Illustrative here are the results of Boertien [3] on riveted lap-joints of GLARE3-3/2-0.3 which were exposed to distilled water by immersion. The results comprise lap-joint strength reduction up to 90 days of exposure, but also present the influence of subsequent dry-out time. The dry-out time results in fairly substantial recovering of strength, which seems to underline again the exaggeration of accelerated exposures. The results are given in Fig. 12.13.

The residual strength of these riveted joints can be predicted based on the damage ratio originally proposed by Müller [10], as discussed in Chap. 10. To investigate the influence of exposure to the environment, various studies have been performed looking either at accelerated or artificial exposure, or at outdoor exposure.

A very interesting study is the earlier-mentioned outdoor exposure programme, where specimens were exposed outside in a representative warm and humid environment in Queensland, Australia. Specimens were tested to various degrees of fatigue damage accumulation to be exposed subsequently for 2 or 6 years. After the

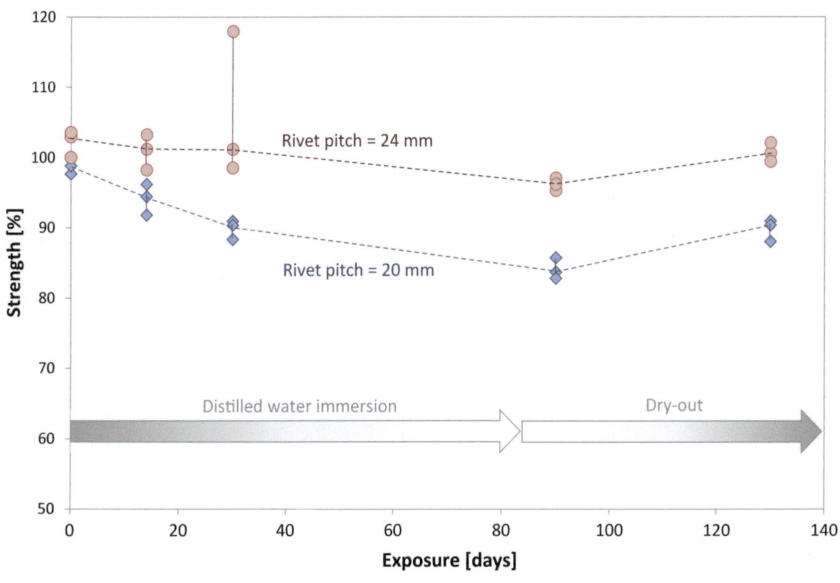

Fig. 12.13 Influence of distilled water immersion and subsequent dry-out on GLARE3-3/2-0.3 lap-joint strength; two different rivet pitches [3]

exposure, the samples were tested in Delft for their residual strength. The results were reported and discussed extensively by Borgonje and Beumler [4, 6, 11]. In particular here, it is important to understand whether the damage ratio discussed in Chap. 10 applies also to cases where environmental exposure took place.

To this aim, the results presented by Borgonje and Beumler in [6] on unidirectional GLARE2B-7/6-0.4 and cross-play GLARE4A-5/4-0.4 are illustrated in, respectively, Fig. 12.14a, b. In these figures, the strength for non-exposed

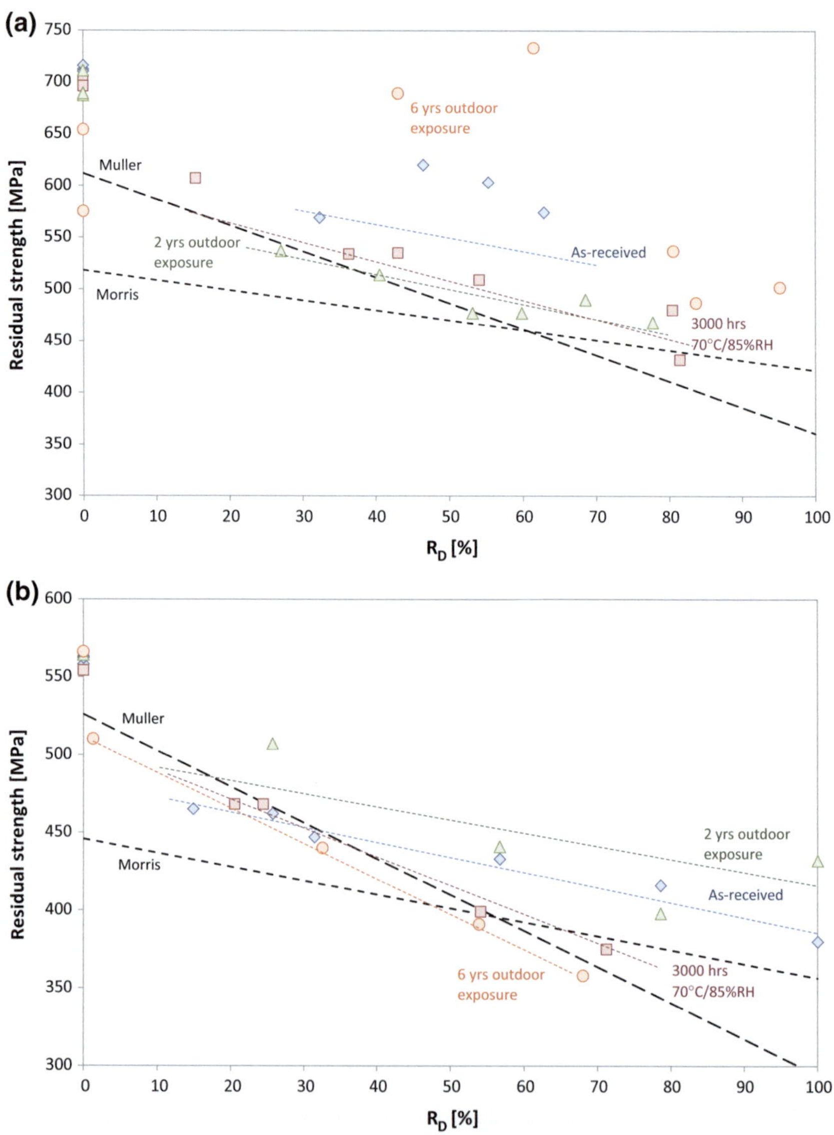

Fig. 12.14 Influence of artificial and outdoor exposure on the residual strength of a row of open holes for GLARE2B-7/6-0.4 (**a**) and GLARE4B-5/4-0.4 (**b**) [6]; $R_D = 0$ corresponds to Fig. 12.8

as-received GLARE is compared to that to of 3000-h exposure to 70 °C/85%RH, and 2 years or 6 years of outdoor exposure in Queensland.

In addition, minimum values are plotted estimated with either the Müller or Morris method, discussed in Chap. 10. The minimum curve labelled Müller in Fig. 12.14 is obtained with

$$S_{res} = S_{BN,FML,min} - MVF \cdot R_D \cdot S_{BN,m} \quad (12.1)$$

where

$$S_{BN,FML,min} \frac{S_{BN,FML}}{1.1} \quad (12.2)$$

The curve labelled Morris is obtained with [11]

$$S_{res} = 0.77\left(S_{BN,FML} - \frac{1}{2} MVF \cdot R_D \cdot S_{BN,m}\right) \quad (12.3)$$

In both cases, the damage ratio R_D has been determined based on the measurement of fatigue cracks in the outer surface [6], rather than being based on fractographic investigation of all individual cracks, including the ones in inner layers, as reported in [10, 12].

Evident in both figures is that the Müller curve seems to follow the trend of the exposed specimens better than the Morris curve, which is in particular conservative up to a damage ratio of R_D = 50%. The Morris relationship was originally developed based on non-exposed samples, in particular addressing the difference between pristine blunt notch strength values (R_D = 0%) and the strength of blunt notch samples containing small fatigue cracks at the notch root. The influence of sharp fatigue cracks instead of the blunt notch root induced a significant reduction with respect to the blunt notch strength, which was not well predicted by the Müller relation.

With the evidence from accelerated and outdoor exposure however, the conclusion may be revisited and the Müller method may be more practical to not only include fatigue damage, but also include the influence of environmental exposure.

12.4 Effect of Frequency

In the past, various studies have reported addressing the influence of frequency and wave shape of the fatigue load cycles applied. Schijve [13] provides an excellent overview, explaining that the influence of frequency and wave shape on the corresponding fatigue properties should be considered an effect of environment. Low frequencies or wave shapes that slowly open fatigue cracks enable the environment to access the crack tip vicinity, resulting in a different crack increment despite the same mechanical loads applied.

In particular, when the relevance of aircraft fuselage pressurization cycles is considered in the fatigue tests, the influence of frequency and wave shape should be considered. Although most fatigue tests in laboratory utilize sinusoidal wave shapes with often higher frequencies, the wave shape of loads induced by pressurization of the fuselage may be better resembled by trapezoidal wave shapes (Fig. 12.16).

To investigate the effect of environment in combination with the test frequency, and the wave shapes of the load cycles, Deutekom [14, 15] performed fatigue test at various test frequencies in both laboratory air and in distilled water (Fig. 12.15).

The frequencies corresponding to the two different wave shapes, i.e. trapezium and sinusoidal, are compared according to the illustration in Fig. 12.6. The trapezium wave shapes had a fixed ramp rate in loading and unloading. Hence, the test frequency was decreased by applying longer dwell times. The dwell time here represents the cruise phase of the flight, where the aircraft fuselage is pressurized with a constant level of pressure.

The results of Deutekom [15] are in agreement with the results reported by Chen [16] and Van Dijk [17] for ARALL laminates, in that at the same test frequency, the trapezium wave shape yields faster crack growth than the sinusoidal wave shape. The influence of the wave shape seems more pronounced than the influence of the test frequency over the frequency range tested.

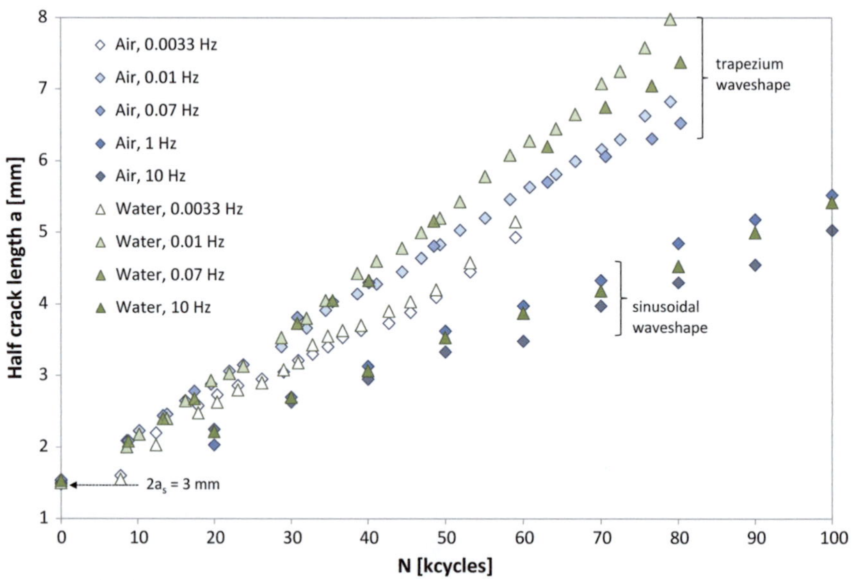

Fig. 12.15 Crack growth curves for GLARE3-3/2-0.3clad (2024-T3 with S2-449/AF163-2) tested at different frequencies in distilled water and air [15]

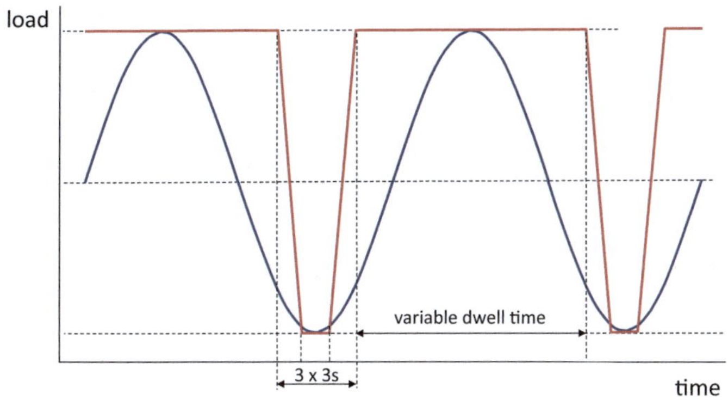

Fig. 12.16 Correlation between sinusoidal and trapezium wave shape at the same frequency as adopted in the tests from [15]

References

1. Verbruggen MLCE (1984a) Moisture absorption of Arall. Report LR-474, Delft University of Technology, Delft
2. Verbruggen MLCE (1984b) The moisture absorption of Arall compared to carbon and aramid reinforced composites. Memorandum M-506, Delft University of Technology, Delft
3. Boertien MFHC (1996) Strength of GLARE after exposure to moisture. Master's thesis, Delft University of Technology, Delft
4. Beumler T (2004) Flying GLARE®, a contribution to aircraft certification issues on strengths properties in non-damaged and fatigue damaged GLARE® structures. PhD dissertation, Delft University of Technology
5. Beumler T, Borgonje B (2008) GLARE outdoor exposure program: moisture reference specimens. Report B2V-02-06, Faculty of Aerospace Engineering, Delft University of Technology
6. Borgonje B, Beumler T (2009) GLARE outdoor exposure program: Blunt notch strength. Report B2V-02-40, Delft University of Technology, Delft
7. Alderliesten RC (2006) Effect of elevated temperature and moisture on fatigue delamination in the fibre metal laminate Glare. In: Proceedings of the 9th joint FAA/DoD/NASA conference on aging aircraft, Atlanta, USA
8. Bosker OJ (2000) Growth of debonds in glare, part II (with moisture absorption). Report B2V-00-08, Delft University of Technology, Delft
9. Hoeven W, van der Schra L (2000) Final report Glare durability program, results of tests carried out at NLR. Technical report NLRCR-2000-237, National Aerospace Laboratory NLR
10. Müller RPG (1995) An experimental and analytical investigation on the fatigue behaviour of fuselage riveted lap joints. PhD dissertation, Faculty of Aerospace Engineering, Delft University of Technology
11. Beumler T, Borgonje B, Sinke J (2009) A contribution of environmental investigations for GLARE riveted joint sizing. In: Proceedings of the 25th ICAF symposium, Rotterdam, The Netherlands
12. de Rijck JJM (2003) Residual strength prediction of XJ-advanced specimens. Report B2V-03-07, Faculty of Aerospace Engineering, Delft University of Technology
13. Schijve J (2001) Fatigue of structures and materials. Kluwer Academic Publishers
14. Deutekom MJ (1994) Environmental effects on the constant amplitude fatigue behaviour of Aluminium, Arall and Glare. Preliminary thesis, Delft University of Technology, Delft

15. Deutekom MJ (1994) The effect of frequency, moisture and temperature on the constant amplitude fatigue behaviour of Glare 3. Master's thesis, Delft University of Technology, Delft
16. Chen D (1987) Some aspects of test frequency influence on the fatigue behaviour of ARALL. Report LR-549, Delft University of Technology, Delft
17. Van Dijk GJ (1986) Frequency influence on the fatigue behaviour of Arall (in dutch). Internship report, Delft University of Technology, Delft

Chapter 13
Acoustic Fatigue

Abstract This chapter briefly addresses the typical damping characteristics of FMLs in context of the acoustic fatigue properties. Results of high frequency bending experiments are discussed to illustrate the excellent performance of FMLs with respect to this mode of excitation.

13.1 Introduction

Acoustic fatigue or high bending fatigue is an aspect related to thin-walled stiffened structures subjected to variable high-frequency loads due to random noise and vibrations. These vibrations may be caused by engines (like turbofans) or operation of equipment. The requirements with respect to increased engine performance for turbojets and minimization of airframe weight result in structural components subjected to high stresses at high-intensity pressure fluctuations.

Most structural parts sensitive to acoustic fatigue are located in regions of separated flow behind protuberances such as air brakes, in surface areas near the plane of a propeller rotation and in bomb bays, where pressure fluctuations occur during bay door opening at high speed.

Compared to the subjects discussed in previous chapters, the topic of acoustic fatigue in FMLs has received only limited attention. The topic mostly received attention in the early days when developing ARALL for wing applications or studying the application of ARALL to the empennage of turboprop aircraft like the Fokker F27. This chapter briefly summarizes the available research and highlights the main results and observations.

R. Alderliesten, *Fatigue and Fracture of Fibre Metal Laminates*, Solid Mechanics and Its Applications 236, DOI 10.1007/978-3-319-56227-8_13

13.2 Damping Characteristics

To understand the phenomenon of acoustic fatigue, two factors affecting the behaviour should be distinguished: the mechanical response and the fatigue characteristics of a structure. The complete description of the mechanical structural response to high-frequency excitation requires knowledge of all vibration modes of that particular structure, the resonant frequencies and the damping characteristics. The damping characteristics of various FMLs (ARALL and GLARE) have been investigated by several researchers [1, 2]. These studies have indicated that adhesively bonded laminates exhibit improved vibrational damping properties compared to monolithic counterparts. In general, FMLs show two to three times better damping behaviour than aluminium [3]. This benefit of FMLs can be attributed to the internal damping capability of the visco-elastic layers, which reduces the stress by a factor 2 compared to monolithic aluminium 2024-T3 [1].

Nonetheless, the work by Waleson [1, 4, 5] has illustrated that one should take care in correlating results like in Fig. 13.1. For example, in [5], Waleson evaluated the vibrating beam, the impedance or direct method and the resonance method. To extract the loss factor from the vibrating beam, he evaluated the half-energy bandwidth method, the Kennedy-Pancu method and the resonance dwell method. In correlating the results, he observed that some methods resulted in values that could be up to 2.5 times higher than the results of other methods.

In addition, he could clearly identify the influence of air as environment on the damping characteristics, the influence of specimen clamping and the influence of normal stresses in the visco-elastic layers. Here, he observed that the internal damping characteristics of ARALL2-2/1 were fairly small, similar to monolithic aluminium, opposite to the findings in Fig. 13.1 [3].

In case of ARALL1 [3], the internal residual stress system after curing has been modified with post-stretching (explained in Sect. 3.4.1). This means that comparison between the acoustic fatigue properties of ARALL1 and monolithic aluminium

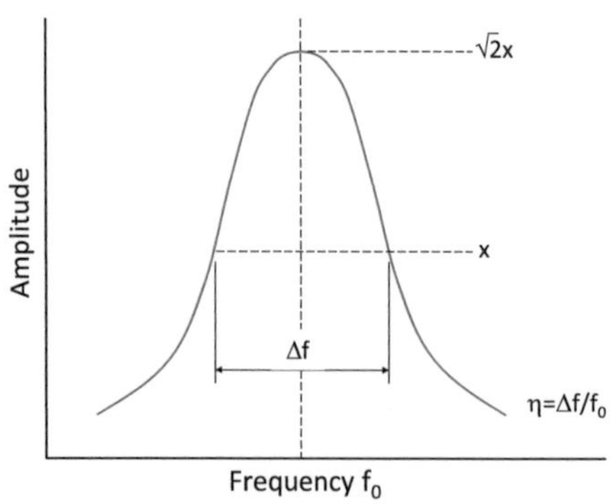

Material	Loss factor η
Arall 1-3/2 (longitudinal)	$21.0 \cdot 10^{-4}$
Arall 1-3/2 (transverse)	$16.0 \cdot 10^{-4}$
Aluminium 2024	$7.2 \cdot 10^{-4}$
Aluminium 6061	$6.7 \cdot 10^{-4}$

Fig. 13.1 Correlation of damping characteristics of ARALL1-3/2 laminates in both the longitudinal and transverse direction with those of aluminium 2024 and 6061 [3]

mostly demonstrated the influence of favourable compressive residual stresses in the aluminium layers of ARALL1.

For Waleson [1], this was the reason to perform similar damping experiments on ARALL2-2/1 (without post-stretching), which was in the as-cured condition with tensile residual stresses in the aluminium layers. In that correlation, the benefit of ARALL2 over monolithic aluminium is still there, but evidently much smaller. Waleson reports that the transmission loss of ARALL2-2/1 is slightly lower than that of 1-mm-thick aluminium 2024-T3.

For the studied case, the bending stiffness of the ARALL laminates is about 45% of that of 1-mm-thick aluminium, with a surface density of about 30%.

Exploring the influence of different prepregs illustrated that a similar lay-up but with a thermoplastic resin improved both the loss factor and bending stiffness by at most 20%. Based on these findings, Waleson [1] recommended to focus on the acoustic fatigue response and damage tolerance, rather than the damping characteristics.

13.3 Acoustic Fatigue

There are various ways to experimentally assess the acoustic fatigue behaviour of a structure or material. Structural parts may be subjected to in-service acoustic loading induced by high power engines, while they are assembled in the complete airframe, or structures may be placed in a wind tunnel set-up to induce the aerodynamic loading. This type of tests is very expensive and can only be performed when the structure is assembled.

Without assembling into a structure, parts can also be subjected to acoustic loading induced by reverberant chambers or sirens. A simple and most often applied experimental method is high-frequency bending testing with shaker equipment. This type of test can only be performed on a material level, enabling generic comparison of different materials. The experimental data presented here are based on this shaker type of experiments, as discussed in more detail in the next section.

13.4 High-Frequency Bending Fatigue Experiments

13.4.1 Specimen Configuration and Test Set-up

High-frequency bending fatigue tests were performed by Syamaun [2] on shaker test specimens, which are illustrated in Fig. 13.2 [2]. The specimens were clamped at the mid-span stiffener to a shaker, with which a sinusoidal waveform was generated using a waveform generator. All specimen configurations were tested at a

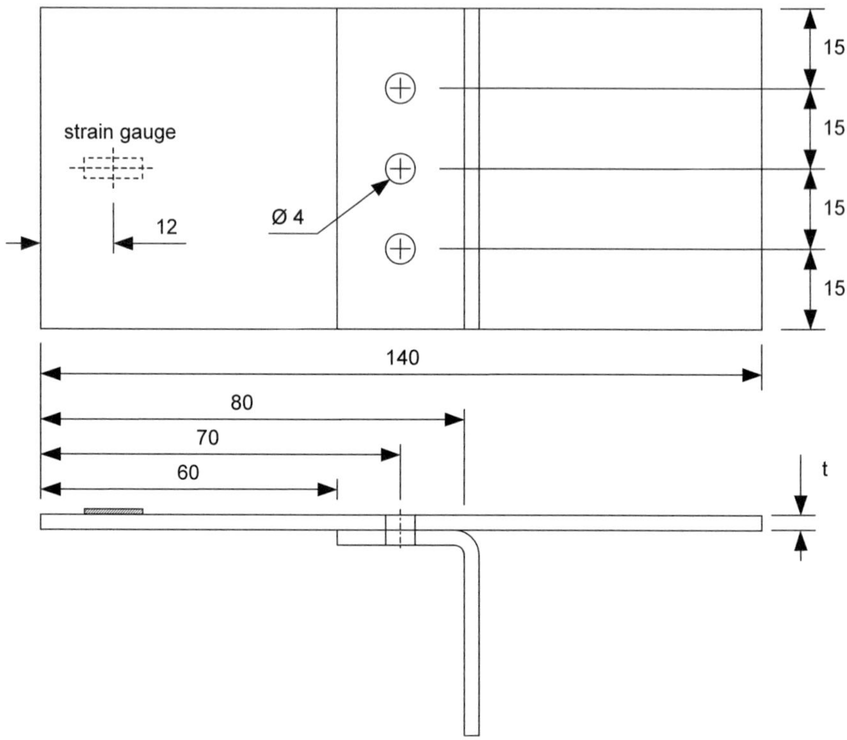

Fig. 13.2 High-frequency bending test specimen; high-frequency excitation induced through the mid-span stiffener using a shaker (all dimensions in mm) [2]

constant strain level of 200 micro-strain at the strain gauge location indicated in Fig. 13.2.

13.4.2 Test Procedure

The specimens were tested in a shaker test set-up with the objective to generate S-N data for different joint configurations and materials. The specimens were vibrated at a resonance frequency to generate alternating bending stresses in the beam.

As the fatigue damage increases, the frequency decreases due to the change in beam stiffness. When the frequency drop was observed, the generator frequency was adjusted in such way that the strain amplitude reached again 200 micro-strain. This procedure to maintain the strain amplitude appeared to be somewhat difficult because conventional strain gauges do not have sufficient fatigue life themselves to last the entire fatigue life of the specimens at the given strain amplitude levels.

The time history of the specimen frequency is a measure for the fatigue damage developed in the specimen. The high-frequency bending tests were terminated at the moment that either the crack reached the edges of the specimen, the resonant frequency dropped below 100 Hz, or when the total testing time exceeded 900 min.

13.4.3 Performed Tests

Syamaun [2] tested several FML configurations based on either aramid or glass fibres, with in addition an FML based on carbon fibres. Because the tested ARALL and GLARE laminates contain either 2024-T3 or 7075-T6 as aluminium constituent, similar tests were performed on the monolithic alloys, with a plate thickness of 1 mm, both with and without clad layers. The entire set of tested materials is listed in Table 13.1.

Table 13.1 Overview of the high-frequency bending experiments reported in [2]

Group	Material	Lay-up	Post-stretch	Thickness (mm)	Weight (g)	Primed	Number of specimen
A B C D	Al 2024 bare Al 2024 clad Al 7075 bare Al 7075 clad			1.01 1.02 1.00 1.02	40.0 40.1 39.5 40.8		3 3 3 3
E F	GLARE1 (7R32) ARALL1 (7H32)	2/1 2/1	✓ ✓	0.92 0.94 0.82 0.84	36.1 37.1 33.2 33.5	✓ ✓ ✓ ✓	3 2 3 2
G H	GLARE2 (2R32) ARALL2 (2H32)	2/1 2/1		0.69 0.76	30.9 32.5		3 3
I J	GLARE1 (7R33) ARALL1 (7H32)	3/2 3/2	✓ ✓	1.55 1.55 1.36 1.40	47.1 47.1 42.1 43.1	✓ ✓ ✓ ✓	3 2 3 2
K L	GLARE2 (2R33) ARALL2 (2H33)	3/2 3/20		1.10 1.25	41.2 41.1		3 3
M	Carbon ARALL	2/1		0.82	34.5		2
N	GLARE3	3/2		1.35	46.6		2

13.5 Results and Observations

An illustration of the test results on the high-frequency bending tests on monolithic aluminium and two GLARE2 laminates is given in Fig. 13.3. Here, one should note that a direct comparison between the high-frequency bending fatigue behaviour of aluminium and GLARE is in fact not possible, since parameters such as thickness and, more importantly, weight are not equal for these tests, see Table 13.1.

However, from Fig. 13.3, it can be seen that the monolithic aluminium specimens have resonant frequencies at the start of the test that are comparable to GLARE2-3/2-0.3. The difference in thickness between the monolithic aluminium specimen and the 3/2 lay-up is fairly small. However, the 2/1 lay-up has significantly lower bending stiffness and as a consequence also a substantially lower resonant frequency at the start of the test.

The aluminium 2024-T3 bare specimen failed after 36 min, slightly longer than the clad specimen, which dropped below the 100-Hz level after 28 min. The reduction in the resonant frequency of both GLARE laminates seems similar in trend and illustrates the lower rate with which fatigue damage develops in the materials in comparison with monolithic aluminium. Because GLARE2-2/1-0.3 started at a lower frequency, it obviously required less time to drop below the 100-Hz level compared to the 3/2 lay-up

A comparison between the frequency bending behaviour of the various ARALL and GLARE laminates is illustrated in Fig. 13.4. From these results, it can be

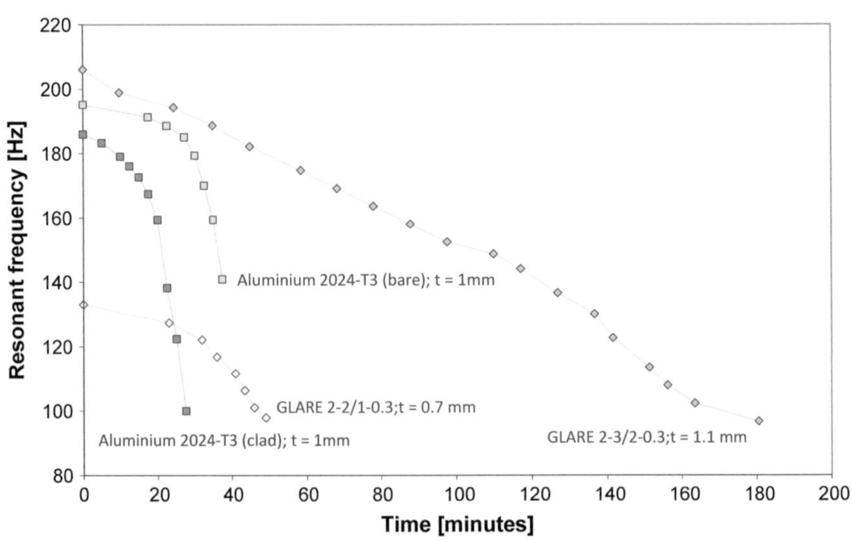

Fig. 13.3 Frequency versus time comparison between GLARE laminates and monolithic aluminium 2024-T3 using the shaker test on specimens illustrated in Fig. 13.2 with round-head rivets [2]

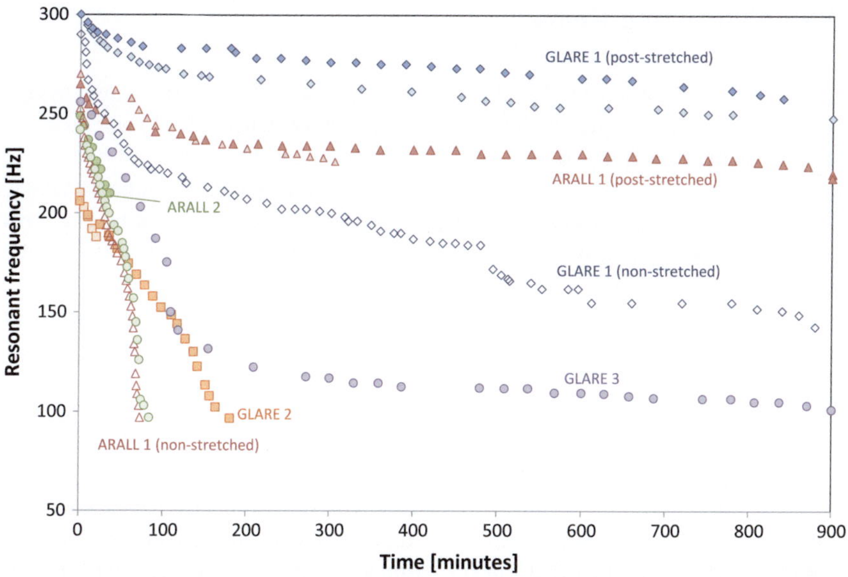

Fig. 13.4 Frequency versus time comparison between various ARALL and GLARE laminates with and without post-stretching using the shaker test on specimens illustrated in Fig. 13.2 with round-head rivets [2]

concluded that the unidirectional GLARE laminates performed better than the unidirectional ARALL laminates. This could be attributed to the greater strain to failure of the glass fibres in comparison with the aramid fibres.

Another observation one can make is that the post-stretched laminates perform significantly better than the laminates in as-cured conditions. At the start of the test, the as-cured laminates have the same resonant frequency as their post-stretched counterparts. However, the post-stretched laminates easily sustained 900 min, while maintaining high frequencies. This is obviously related to the favourable residual stress system in the post-stretched laminates that delays the initiation and progression of fatigue damage in comparison with the as-cured laminates.

The only cross-ply laminate in Fig. 13.4, GLARE3-3/2-0.3, initially quickly dropped in frequency, almost similar to the as-cured ARALL laminates, but after about 200 min the degradation reduced substantially. In the end, the resonance frequency of the GLARE3 specimen still exceeded the 100-Hz level after 900 min.

13.6 Concluding Remarks

The studies by Waleson [1, 5] and Syamaun [2] have illustrated various aspects related to the high-frequency bending fatigue performance of FMLs. First, the increase of acoustic fatigue life of FMLs over monolithic aluminium becomes larger for thicker laminates. The thin 2/1 lay-up hardly yields an improvement in either damping characteristics or acoustic fatigue properties. This can be explained with the position of the fibre layers in thickness direction. In a 2/1 lay-up, all fibre layers are located near the neutral line, which makes them fairly ineffective in restraining crack opening by bridging. Only for 3/2 lay-ups or thicker laminates, do the fibres become effective and evidently contribute to the acoustic fatigue performance.

Furthermore, the studies made clear by comparing non-stretched and post-stretched FMLs that post-stretching substantially increases the fatigue life due to the favourable residual stress system. In addition, the failure strain of the fibres likely has a dominant contribution, because the GLARE laminates performed better than both the ARALL laminates and the laminate based on the carbon fibres.

With respect to the fact that thicker laminates perform better than their thinner counterparts, one may assume that for structural applications where the laminate lay-up is at least 3/2, damping characteristics and acoustic fatigue likely are not a subject of concern when compared to monolithic aluminium. Only when very thin laminates are considered, like the 2/1 lay-up, do the properties demand assessment of the acoustic fatigue characteristics.

References

1. Waleson JEA (1987) The acoustic properties of ARALL-material, Part A: The damping of ARALL-material in a structure (in dutch). Master Thesis, Delft University of Technology, Delft
2. Syamaun M (1991) Bending fatigue of metal laminates (ARALL and GLARE). Thesis, Delft University of Technology, Delft
3. Bucci RJ, Mueller LN, Schultz RW, Prohaska JL (1987), ARALL laminates-results from a cooperative test program. Technical Report, Alcoa Laboratories, Alcoa Center, PA 15069
4. Waleson JEA (1987) The acoustic properties of ARALL-material, Part A: The damping of ARALL-material in a structure (in dutch). Preliminary Thesis, Delft University of Technology, Delft
5. Waleson JEA (1987), The acoustic properties of ARALL-material, Part B: Establishing damping of plater material (in dutch). Master Thesis, Delft University of Technology, Delft

Index

R. Alderliesten, *Fatigue and Fracture of Fibre Metal Laminates*, Solid Mechanics and Its Applications 236, DOI 10.1007/978-3-319-56227-8

MIX
Papier aus verantwortungsvollen Quellen
Paper from responsible sources
FSC® C105338
FSC
www.fsc.org

If you have any concerns about our products,
you can contact us on
ProductSafety@springernature.com

In case Publisher is established outside the EU,
the EU authorized representative is:
Springer Nature Customer Service Center GmbH
Europaplatz 3, 69115 Heidelberg, Germany

Printed by Libri Plureos GmbH
in Hamburg, Germany